AF358822

Wearable Sensing Devices and Technology

Wearable Sensing Devices and Technology

Editors

Lei Jing
Jiehan Zhou
Zhan Zhang

Basel • Beijing • Wuhan • Barcelona • Belgrade • Novi Sad • Cluj • Manchester

Editors

Lei Jing
Graduate School of Computer
Science and Engineering
The University of Aizu
Aizuwakamatsu
Japan

Jiehan Zhou
School of Computer
Science and Engineering
Shandong University of
Science and Technology
Qingdao
China

Zhan Zhang
School of Computer Science
and Engineering
Harbin Institute of
Technology
Harbin
China

Editorial Office
MDPI AG
Grosspeteranlage 5
4052 Basel, Switzerland

This is a reprint of articles from the Special Issue published online in the open access journal *Electronics* (ISSN 2079-9292) (available at: www.mdpi.com/journal/electronics/special_issues/ Wearable_Sensing_Devices_Technology).

For citation purposes, cite each article independently as indicated on the article page online and as indicated below:

Lastname, A.A.; Lastname, B.B. Article Title. *Journal Name* **Year**, *Volume Number*, Page Range.

ISBN 978-3-7258-1984-3 (Hbk)
ISBN 978-3-7258-1983-6 (PDF)
doi.org/10.3390/books978-3-7258-1983-6

Contents

About the Editors

Lei Jing

Lei Jing (Member, IEEE) received his Ph.D. degree in computer science and engineering from the University of Aizu in Japan in 2008. He is currently a senior associate professor at the UoA, where he leads the Human Intelligent Sensing Group and the team dedicated to human motion tracking. His particular interest lies in hand motion detection, employing multimodal sensor fusion methods. The applications of his work include activity abnormality detection, sign language recognition, and skill heritage. He has published over 100 papers and holds 6 patents in related areas. Lei Jing has been a member of IEEE and ACM since 2010, and a member of IPSJ since 2012.

Jiehan Zhou

Professor Jiehan Zhou is currently a professor at the School of Computer Science, Shandong University of Science and Technology, and a tenured professor at the Department of Electronics and Information Engineering, University of Oulu, Finland. He has a rich academic background and extensive professional experience, having earned his Ph.D. in Mechanical Automation from Huazhong University of Science and Technology in 2000 (under the supervision of academician Shuzi Yang) and a Ph.D. in Computer Engineering from the University of Oulu, Finland, in 2011. Professor Zhou has worked at several internationally renowned universities and research institutes, including Tsinghua University, the VTT Technical Research Centre of Finland, INRIA in France, LIST in Luxembourg, the University of Oulu, and the University of Toronto. With over 24 years of international research experience, he has held positions such as Senior Scientist, Team Leader, Laboratory Director, Tenured Professor at the University of Oulu, General Manager of Aerospace Cloud Network Germany for the China Aerospace Science and Industry Corporation, and Engineer at Huawei Research Institute in Canada. He also serves as a visiting professor at several Chinese universities, including Huazhong University of Science and Technology. Professor Zhou has conducted in-depth research in intelligent manufacturing, system modeling and simulation, industrial internet, and industrial big data analysis, achieving notable success in both academic and industrial fields. He has led and participated in numerous European Union projects, including the ERCIM Fellowship Program of the European Academy, the EU FP6-Amigo project, and the EU ITEA4-CAM4Home project.

Zhan Zhang

Zhang Zhan received his Ph.D. in computer science and technology from the Harbin Institute of Technology (HIT), China. He is now a professor at the Research Center of Fault-Tolerance and Mobile Computing, School of Computer Science and Technology, HIT. His research interests include wearable computers, fault-tolerant computers, and system evaluation theory and technology.

 *electronics*

Editorial

Wearable Sensing Devices and Technology

Lei Jing [1,*], **Jiehan Zhou** [2] **and Zhan Zhang** [3]

[1] School of Computer Science and Engineering, The University of Aizu, Aizu-Wakamatsu 965-8580, Fukushima, Japan
[2] Computer Science and Engineering, Shandong University of Science and Technology, Qingdao 266590, China; jiehan.zhou@ieee.org
[3] School of Computer Science and Technology, Harbin Institute of Technology, Harbin 150001, China; zhangzhan@hit.edu.cn
* Correspondence: leijing@u-aizu.ac.jp

Citation: Jing, L.; Zhou, J.; Zhang, Z. Wearable Sensing Devices and Technology. *Electronics* **2024**, *13*, 2951. https://doi.org/10.3390/electronics13152951

Received: 22 July 2024
Accepted: 23 July 2024
Published: 26 July 2024

The field of wearable sensing devices and technology has witnessed remarkable advancements, driven by the need for more integrated, efficient, and user-friendly systems. This Special Issue has compiled a collection of significant contributions that address various aspects of wearable technology, ranging from healthcare applications to novel sensing mechanisms and data processing techniques.

The papers in this Special Issue have collectively addressed several critical gaps in the current body of knowledge:

1. **Authentication and security:**

   - **MetaEar**: This paper introduces an innovative approach to continuous authentication using imperceptible acoustic side channels, enhancing security without compromising user experience. This method leverages the Enhanced Reverberation Transfer Function (ERTF) for imperceptible authentication, filling the gap in secure yet user-friendly authentication methods for wearables (Contributor 1).

2. **Localization in wireless sensor networks:**

   - **A novel anchor-free localization method**: This study presents a cross-technology communication-based method for anchor-free localization in wireless sensor networks. It bridges the gap in efficient and accurate localization techniques necessary for robust wearable applications in dynamic environments (Contributor 2).

3. **Sign language recognition:**

   - **Data glove with bending sensor and inertial sensor**: the introduction of a weighted DTW fusion technique for sign language recognition demonstrates significant improvements in accuracy and usability, addressing the need for more effective communication aids for the hearing-impaired (Contributor 3).

4. **Healthcare systems:**

   - **MDP-based MAC protocol for WBANs**: this paper discusses a new MAC protocol designed for wireless body area networks (WBANs) in edge-enabled eHealth systems, enhancing the reliability and efficiency of health monitoring (Contributor 4).
   - **Lidom**: a predictive model for disease risk in nursing homes using LightGBM, providing a valuable tool for proactive healthcare management (Contributor 5).
   - **Usability of inexpensive optical pulse sensors**: this study evaluates the integration of optical pulse sensors into textiles, highlighting advancements in affordable and non-invasive heart rate monitoring (Contributor 6).
   - **Dynamic fall detection**: using graph-based spatial–temporal convolution and attention networks, this research improves fall detection systems' accuracy and response times, crucial for elderly care (Contributor 7).

- ○ **Gait for Parkinson's Disease**: This study investigates the reduced upper-limb swing capabilities during gait in Parkinson's Disease (PD) patients. It employs a viscoelastic model to assess forearm swing, comparing PD patients to healthy individuals (Contributor 8).
- ○ **IoT-based home telehealth system**: this preliminary study explores smart healthcare by monitoring sleep and water usage, offering insights into home-based health monitoring innovations (Contributor 9).

5. **Enhancing user experience:**
- ○ **Enhanced wearable force-feedback mechanism**: the development of a free-range haptic experience extended by mixed reality showcases advancements in user interaction and immersion in virtual environments (Contributor 10).
- ○ **VR drumming pedagogy**: this paper explores action observation and virtual co-embodiment in VR drumming, providing innovative approaches to music education and training (Contributor 11).

6. **Positioning and movement analysis:**
- ○ **Smarter positioning through GNSS and multi-sensor data**: this research enhances positioning accuracy using raw GNSS and multi-sensor data, crucial for navigation and tracking applications in wearables (Contributor 12).
- ○ **PIFall**: a novel fall detection system using pressure insoles and ResNet3D, addressing the need for reliable fall detection in elderly care (Contributor 13).

Future Research Directions

The advancements presented in this Special Issue open several avenues for future research:

- **Interdisciplinary integration**: combining insights from materials science, data science, and human–computer interaction to develop more sophisticated wearable devices.
- **Scalability and miniaturization**: further miniaturizing sensors and improving battery life to enhance wearability and convenience.
- **Enhanced security**: continuing to explore secure, non-intrusive authentication methods to protect user data in wearable devices.
- Personalized healthcare: advancing predictive models and real-time monitoring to provide personalized healthcare solutions tailored to individual needs.
- User experience: improving user interfaces and feedback mechanisms to make wearable technology more intuitive and engaging.

This Special Issue has provided a comprehensive overview of the current state of wearable sensing devices and technology, addressing key challenges and setting the stage for future innovations. We look forward to witnessing the continued evolution of this dynamic field, driven by the collaborative efforts of researchers and practitioners worldwide.

Conflicts of Interest: The authors declare no conflicts of interest.

List of Contributions

1. Chang, Z.; Wang, L.; Li, B.; Liu, W. MetaEar: Imperceptible Acoustic Side Channel Continuous Authentication Based on ERTF. *Electronics* **2022**, *11*, 3401.
2. Jing, N.; Zhang, B.; Wang, L. A Novel Anchor-Free Localization Method Using Cross-Technology Communication for Wireless Sensor Network. *Electronics* **2022**, *11*, 4025.
3. Lu, C.; Amino, S.; Jing, L. Data Glove with Bending Sensor and Inertial Sensor Based on Weighted DTW Fusion for Sign Language Recognition. *Electronics* **2023**, *12*, 613.
4. Su, H.; Pan, M.-S.; Chen, H.; Liu, X. MDP-Based MAC Protocol for WBANs in Edge-Enabled eHealth Systems. *Electronics* **2023**, *12*, 947.
5. Zhou, F.; Hu, S.; Du, X.; Wan, X.; Lu, Z.; Wu, J. Lidom: A Disease Risk Prediction Model Based on LightGBM Applied to Nursing Homes. *Electronics* **2023**, *12*, 1009.

6. Richter, N.; Tuvshinbayar, K.; Ehrmann, G.; Ehrmann, A. Usability of Inexpensive Optical Pulse Sensors for Textile Integration and Heartbeat Detection Code Development. *Electronics* **2023**, *12*, 1521.

7. Egawa, R.; Musa Miah, A.S.; Hirooka, K.; Tomioka, Y.; Shin, J. Dynamic Fall Detection Using Graph-Based Spatial Temporal Convolution and Attention Network. *Electronics* **2023**, *12*, 3234.

8. Pietrosanti, L.; Verrelli, C.M.; Giannini, F.; Suppa, A.; Fattapposta, F.; Zampogna, A.; Patera, M.; Rosati, V.; Saggio, G. A Viscoelastic Model to Evidence Reduced Upper-Limb-Swing Capabilities during Gait for Parkinson's Disease-Affected Subjects. *Electronics* **2023**, *12*, 3347.

9. Tang, Z.; Jiang, L.; Zhu, X.; Huang, M. An Internet of Things-Based Home Telehealth System for Smart Healthcare by Monitoring Sleep and Water Usage: A Preliminary Study. *Electronics* **2023**, *12*, 3652.

10. Kudry, P.; Cohen, M. Enhanced Wearable Force-Feedback Mechanism for Free-Range Haptic Experience Extended by Pass-Through Mixed Reality. *Electronics* **2023**, *12*, 3659.

11. Pinkl, J.; Cohen, M. VR Drumming Pedagogy: Action Observation, Virtual Co-Embodiment, and Development of Drumming "Halvatar". *Electronics* **2023**, *12*, 3708.

12. Grenier, A.; Lohan, E.S.; Ometov, A.; Nurmi, J. Towards Smarter Positioning through Analyzing Raw GNSS and Multi-Sensor Data from Android Devices: A Dataset and an Open-Source Application. *Electronics* **2023**, *12*, 4781.

13. Guo, W.; Liu, X.; Lu, C.; Jing, L. PIFall: A Pressure Insole-Based Fall Detection System for the Elderly Using ResNet3D. *Electronics* **2024**, *13*, 1066.

Article

PIFall: A Pressure Insole-Based Fall Detection System for the Elderly Using ResNet3D

Wei Guo [1,†], Xiaoyang Liu [1,†], Chenghong Lu [1] and Lei Jing [2,*]

1   Graduate School of Computer Science and Engineering, The University of Aizu, Aizuwakamatsu 965-8580, Fukushima, Japan; d8232103@u-aizu.ac.jp (W.G.)
2   School of Computer Science and Engineering, The University of Aizu, Aizuwakamatsu 965-8580, Fukushima, Japan
*   Correspondence: leijing@u-aizu.ac.jp
†   These authors contributed equally to this work.

Abstract: Falls among the elderly are a significant public health issue, resulting in about 684,000 deaths annually. Such incidents often lead to severe consequences including fractures, contusions, and cranial injuries, immensely affecting the quality of life and independence of the elderly. Existing fall detection methods using cameras and wearable sensors face challenges such as privacy concerns, blind spots in vision and being troublesome to wear. In this paper, we propose PIFall, a Pressure Insole-Based Fall Detection System for the Elderly, utilizing the ResNet3D algorithm. Initially, we design and fabricate a pair of insoles equipped with low-cost resistive films to measure plantar pressure, arranging 5×9 pressure sensors on each insole. Furthermore, we present a fall detection method that combines ResNet(2+1)D with an insole-based sensor matrix, utilizing time-series 'stress videos' derived from pressure map data as input. Lastly, we collect data on 12 different actions from five subjects, including fall risk activities specifically designed to be easily confused with actual falls. The system achieves an overall accuracy of 91% in detecting falls and 94% in identifying specific fall actions. Additionally, feedback is gathered from eight elderly individuals using a structured questionnaire to assess user experience and satisfaction with the pressure insoles.

Keywords: fall detection; pressure sensor; e-textile sensor; ResNet; HAR; insole sensor; elder healthcare; abnormal detection; wearable computing; motion capture

Citation: Guo, W.; Liu, X.; Lu, C.; Jing, L. PIFall: A Pressure Insole-Based Fall Detection System for the Elderly Using ResNet3D. *Electronics* **2024**, *13*, 1066. https://doi.org/10.3390/electronics13061066

Academic Editor: Daniele Riboni

Received: 29 January 2024
Revised: 11 March 2024
Accepted: 12 March 2024
Published: 13 March 2024

1. Introduction

Falls are a major public health problem, accounting for an estimated 684,000 deaths worldwide each year [1]. They are the second leading cause of unintentional injury deaths and are especially common among adults over 60 years of age. Even non-fatal falls can lead to serious health issues, such as fractures, contusions, and head injuries, often necessitating medical care and potentially resulting in long-term health and quality-of-life impacts.

Given the severity and prevalence of falls in senior citizens, it is important to have effective methods to detect falls and provide timely assistance. Our goal is to develop effective fall detection strategies for the elderly that respect their independence and privacy, while also improving their health and quality of life. Fall detection is essential for providing timely assistance in the event of a fall and to help reduce the potential consequences of falls in the elderly. It can also help identify the root causes of falls and provide valuable information for developing fall prevention strategies.

Current traditional detection methods include image-based cameras and wearable sensors [2], and a large number of publicly available datasets at this stage. However, these established detection methods still have limitations, such as not guaranteeing the independence and privacy of the elderly, and the burden that additional wearable devices impose on the elderly. Therefore, several key factors must be considered when building fall detection solutions for the elderly:

1. Protecting the privacy of the elderly: Protecting the independence and privacy of the elderly is crucial to avoid solutions that may infringe on their habits and sense of privacy. According to previous study [3], even the smallest infrared cameras can cause discomfort to the elderly and invade their privacy.
2. Continuous detection: A fall detection system must be able to continuously detect falls without being obstructed or obscured by other objects in the environment. This is critical to ensure that the system is continuously active and ready to respond in the event of a fall.
3. Portability: A fall detection system should be independent of its surroundings so that the individual does not need something else to utilize it. It is necessary to ensure that the system is available for user action detection in all situations.
4. Cost-effective and low-maintenance: Traditional fall detection systems can be expensive and require a lot of maintenance, which can be a barrier for the elderly. Low-cost, low-maintenance fall detection solutions that respect the independence and privacy of senior citizens are needed.

To address the four problems mentioned above, this study proposes a solution. We design a pair of insoles with embedded pressure sensors to collect pressure data from the user, addressing the four problems outlined earlier, which are as follows:

1. Protecting the privacy of the elderly: We develop a pair of soft insoles with embedded E-Textile pressure sensors, designed to be easily incorporated into regular footwear, which negates the necessity for extra wearable sensor devices. These insoles only collect plantar pressure data, thus preserving the user's privacy without capturing any personal identifiers. This approach effectively addresses the discomfort and privacy issues associated with camera-based systems or wearable sensors technologies.
2. Continuous detection: The insoles can be conveniently placed within everyday footwear to continuously monitor plantar pressure in real-time, with the pressure data being transmitted promptly via Wi-Fi. This method overcomes the challenges of discontinuous monitoring due to blind spots or inadequate lighting that are common with camera-based systems.
3. Portability: When monitoring environmental changes, RF-based methods like Wi-Fi and infrared require data re-collection and model re-training. In contrast, the insole-based fall detection system exhibits environmental independence, ensuring stable operation despite changes in the usage environment.
4. Cost-effective and low-maintenance: Compared to existing fall monitoring systems, the developed insoles significantly reduce manufacturing costs, requiring only a chip, fabric insole, and the circuitry integrated on the insole. The system has also been optimized for reduced maintenance complexity. A sleep feature is introduced where the insole enters a low-power state when no pressure is detected and reactivates upon sensing pressure, thereby conserving battery life and extending operational longevity. An MD5 checksum has been implemented to ensure the integrity of data transmission. To further ease maintenance efforts, the insoles are equipped with an OTA (Over-The-Air) update capability, allowing for automatic firmware upgrades without manual intervention, simplifying the user experience for the elderly.

In existing studies [4–9], off-the-shelf Force Sensitive Resistors (FSR), accelerometers, and gyroscopes are utilized to fabricate pressure insoles designed to protect user privacy and enable continuous monitoring capabilities. However, due to the size and shape constraints of FSRs, sensors can only be placed at critical positions on the sole of the foot, allowing for the detection of falls and gait but limiting the ability to capture a richer set of motion information from the insole. To acquire comprehensive gait information, study [10] expanded the number of pressure sensors in each insole to 96. However, customizing such a pair of insoles comes at a higher cost. To reduce the cost of pressure insoles, study [11] proposes a deep learning model capable of predicting the entire foot's pressure distribution using data from a limited number of insole pressure sensors, offering potential applications in diagnosing foot deformities, pathological gait, falls, and pressure sores, especially in diabetic patients. It is evident that embedding a greater number of sensors in pressure insoles at a lower cost is crucial, as a dense array of sensors provides richer plantar pressure data for improved analysis of falls and activity. In this paper, we introduce a fall detection system based on pressure insoles, utilizing resistive film to manufacture pressure sensors densely distributed across the insole, thereby reducing production costs and simplifying fabrication. Our key contributions are as follows:

First, we propose PIFall, a Pressure Insole-Based Fall Detection System for the Elderly, utilizing the ResNet3D algorithm. We design and fabricate a pair of insoles equipped with resistive films to measure plantar pressure, arranging 5×9 pressure sensors on each insole. Compared to capacitive pressure sensors, resistive films offer a cost-effective alternative, enabling the dense arrangement of sensors at a lower expense.

Second, we present a fall detection method by combining a unique insole-based sensor matrix with the ResNet(2+1)D architecture. Our system utilizes 'stress video', a time-series dataset from pressure maps, as network input. To address the challenges of low-resolution data, we employ upsampling. Interpolation makes the pressure map closer to real-world smoothness and enhances data granularity, thereby improving recognition accuracy.

Lastly, we collect data on 12 different actions from five subjects, including fall risk activities specifically designed to be easily confused with actual falls. These actions are efficiently classified using a ResNet(2+1)D neural network model, achieving an overall accuracy of 85.61% and a fall detection accuracy of 94%.

2. Related Works

2.1. Human Action Recognition

Human action recognition is an area of computer science and artificial intelligence that involves the development of algorithms and systems to recognize and classify human actions in digital videos and other media.

According to a recent review [12], mainstream data collection methods are visual-based, such as using video frames or images [13], as well as sensor-based methods that collect data on various modalities, such as acceleration [14], body metrics and pressure. Data can also be collected using wireless bands and infrared signals. For image-skeleton-based fall recognition [15], RGB data from the internet and multiple sources and formats are used to train recognition models. From a skeleton-based perspective, Ref. [16] adds classification based on inter-joint connection relationships and joint trajectory information to the ST-GCN algorithm for enhanced recognition. In the domain of non-image recognition, Ref. [17] employs accelerated data and compares the performance of three deep learning model architectures on a dataset for senior citizens fall detection, achieving improved results by using gender and age as auxiliary outputs. Furthermore, Ref. [18] utilizes ultra-wideband radar to detect falls by capturing wireless channel fluctuations and applying the ConvLSTM recognition algorithm for device-free fall detection. Recently, multimodal fusion methods have been proposed, where multiple features such as video features and acceleration features are combined for enhanced learning [19].

However, in the context of fall detection for senior citizens, ensuring sufficient independence and privacy remains a significant concern. Visual modalities not only present technical challenges but also raise ethical privacy issues. Wearable devices that rely on accelerometer sensors can cause stress and discomfort for seniors, who may then reject the device or forget to carry it consistently. Wireless channel-based signals may offer a promising approach, yet their effectiveness is limited by the structural layout of the room, and a single model may not perform well in different environments. These challenges must be addressed to improve fall detection for senior citizens.

2.2. Fall Detection for Senior Citizens

The primary target audience for fall detection is senior citizens, as falls are the second leading cause of accidental or unintentional injury deaths within this group. Effective fall detection can facilitate timely medical care and potentially save lives. In this case, falls are considered a subset of human actions, which means that methods used in motion detection, such as acceleration sensors and camera/depth cameras, can also be applied to fall detection.

There are currently several challenges in the field of senior citizen fall detection, such as privacy, user adoption and challenges in maintaining privacy. Wearable devices that collect personal motion and activity data may raise privacy concerns, especially when the data is shared with third parties. Moreover, seniors may exhibit reluctance towards employing wearable devices or technologies that are unfamiliar to them, potentially undermining the efficacy of fall detection systems.

To tackle these challenges, a novel direction distinct from normal human motion detection has been proposed in [20], which entails the use of environmental devices. This method involves installing a series of sensors near the person of interest, such as walls, floors, beds, etc. Data gathered from these sensors are analyzed by an algorithm to ascertain the occurrence of a fall. Subsequently, the event is reported to the caregiver. Since there is no need to wear any sensors, this becomes a solution to both of the challenges. For instance, Ref. [18] introduces a technique utilizing ultra-wideband (UWB) single-station radar for data acquisition and convolutional long short-term memory (LSTM) networks for detecting falls within a room. Additionally, Desai Kimaya's study [21] presents a novel machine learning approach for fall detection using a wearable belt. Another study [22] details an improved threshold-based fall detection method applied to smartphones, which uses collected acceleration data to identify falls during everyday activities in four different directions.

2.3. Pressure-Sensing-Based HAR

In action detection of environmental devices, pressure data can also be used as a data that can be collected by environmental devices. Pressure-based action detection in the field of action recognition, ref. [23] proposed by Sundaram, presents an electronic textile glove as an electronic textile glove using pressure variable resistance film. Likewise, Ref. [24] develops a large-scale pressure-sensing carpet that utilizes the same mechanism to reconstruct the human skeletal structure from pressure distributions, then applies established skeleton-based recognition techniques for action classification. Concurrently, studies such as [25] involve extensive data gathering from a significant number of senior citizens via plantar pressure sensors, which assists in identifying falls and frailty through classification methods.

The effectiveness of pressure-based data for action recognition has been demonstrated, and this paper further distinguishes between fall actions and normal actions using plantar pressure data. This study identifies 12 actions, including 5 daily safety actions, 4 actions that pose a risk of falling, and 3 different directional fall actions.

In contrast to approaches that place pressure sensors directly on the sole of the foot, we present a uniquely crafted insole. It features a three-layer electrical circuit structure—comprising wire, piezoresistive film, and wire—arranged in both horizontal and vertical orientations to form a 5×9 pressure sensing matrix. This matrix enables pressure detection at 45 distinct locations on the sole of the foot, aiding the model in learning the dynamics of the subject's center of gravity shifts. Our design incorporates a greater number of sensors and a higher pressure data sampling frequency than similar studies [25,26] that classify actions based on plantar pressure data. Capable of detecting pressure changes at 50–125 Hz sampling rate across 45 points on the foot's underside, our insole offers an enhanced range of detection capabilities.

3. Application Model

The application model is shown in Figure 1. It is a high-resolution and -sampling-rate pressure-detecting insole worn by senior citizen living alone to detect in real-time whether fall-related actions have occurred. When the elderly person falls while wearing the shoe with the pressure-detecting insole, the data during this time is transmitted to a computer/server via a wireless chip mounted on the insole. A convolutional neural network learning algorithm is used in the computer/server to classify the input data to determine if the action is a fall. In addition, the system detects and records daily actions, creating a life log of the elderly' actions. These data can provide caregivers or medical professionals with valuable information to monitor the health and well-being of the senior citizen. The computer/server then analyzes these data with deep learning algorithms and calls for emergency medical assistance and notifies the guardian when the features of a fall are detected.

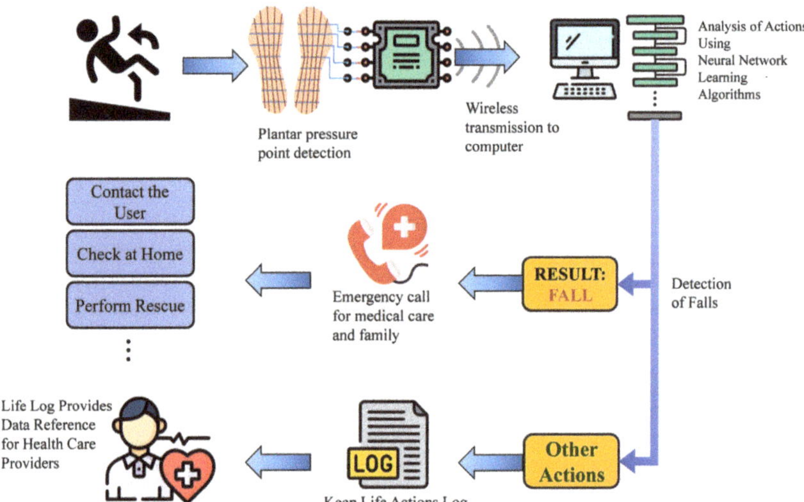

Figure 1. Application model. The insole is worn by the senior citizen for real-time fall detection. The insole transmits data to a computer/server via the wireless chip to detect falls. The computer/server uses a neural network learning algorithm to classify the movements. A guardian is notified when a fall is detected and emergency medical assistance is called. And creates a life log of daily senior behaviors that can provide valuable information to caregivers or medical professionals to improve the health and well-being of senior citizen.

4. System Design

4.1. System Outline

To solve the falls problem, the fall detection system developed in this research uses pressure sensors installed in shoe insoles to detect falls for senior citizens living alone. The system process is shown in Figure 2, and the key processes are as follows:

1. The user produces different voltages according to different pressures when wearing the shoes. This allows the system to indirectly obtain the pressure distribution of the user over time. The voltage is transmitted to the chip through a wire, connected to the pressure sensors in the insoles.

2. The pressure distribution data are transmitted to a computer via wireless communication, using a Wi-Fi access point device. This allows the data to be analyzed in real-time, enabling the system to detect a fall as soon as it occurs.

3. The computer analyzes the data to determine whether the user is likely to fall. The analysis is based on machine learning algorithms that have been trained on a dataset of falls and non-falls. If a fall is detected, the system provides timely assistance to the user, such as alerting a caregiver or emergency contact.

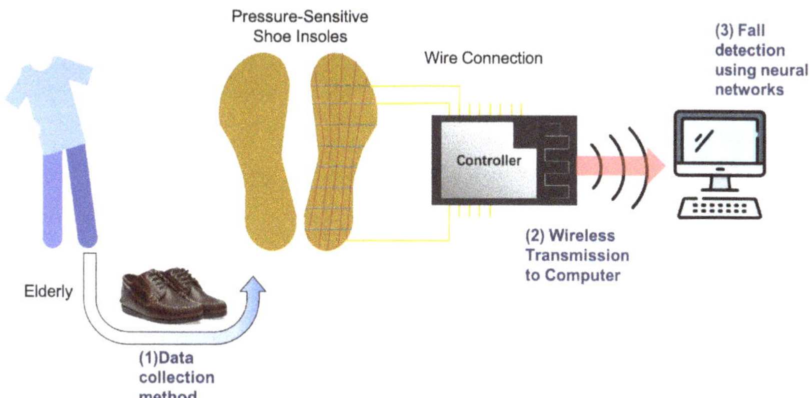

Figure 2. The overall flow of the system. The system detects the user's pressure distribution over time using a pressure sensor in the shoe, which is then transmitted to a computer for analysis. To determine whether the user is likely to fall, the computer employs a machine learning algorithm trained on the dataset collected by the insole. If a fall is detected, the system responds quickly by alerting caregivers or emergency contacts.

4.2. Plantar Pressure Collection Method

To measure the pressure, our insole has a three-layer structure, which measures the pressure through a pressure variable resistor and a path formed by horizontal and vertical lines, as shown in Figure 3. A GPIO pin that outputs 3.3 V is connected to the horizontal line of the first layer. Current flows through a pressure variable resistor layer in the second layer. A vertical line connected to an ADC pin forms the third layer.

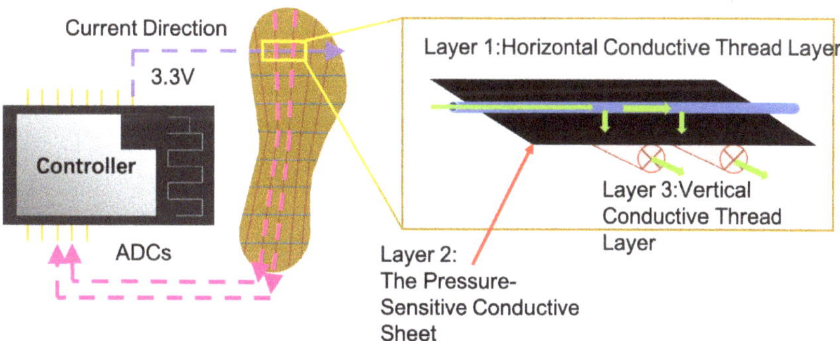

Figure 3. Design of data collection.

Current flows horizontally through the first layer and vertically through the variable resistor layer being introduced into the third layer in order to measure pressure. The current flows into the ADC pin to measure the resistance of the variable resistor and determine the corresponding pressure value. The ADC pin has a 900 Ω sampling resistor.

Figure 4 shows the circuit design diagram, and the physical circuit board is shown in Figure 5. The circuit includes four modules, the description of each module is as follows:

(a)   5 ADCs and 9 GPIOs, with 900 Ω pull-down sampling resistors under each ADC.
(b)   ESP32C-12F chip: A microprocessor that supports high-speed computing and low power consumption with wireless and Bluetooth functions, can be written in code using Arduino IDE 2.2.1. ESP is described in detail in the next paragraph.

(c) Charging and low power consumption circuit design: It can automatically switch between high/low power consumption mode through the code (high power consumption mode when downloading data and detecting the presence of pressure; low power consumption mode when no pressure is detected for a certain period of time, reducing the sampling frequency to save power).

(d) USB to serial input: Before the implementation of OTA, there is a need to rely on this part of the input compiled binary file.

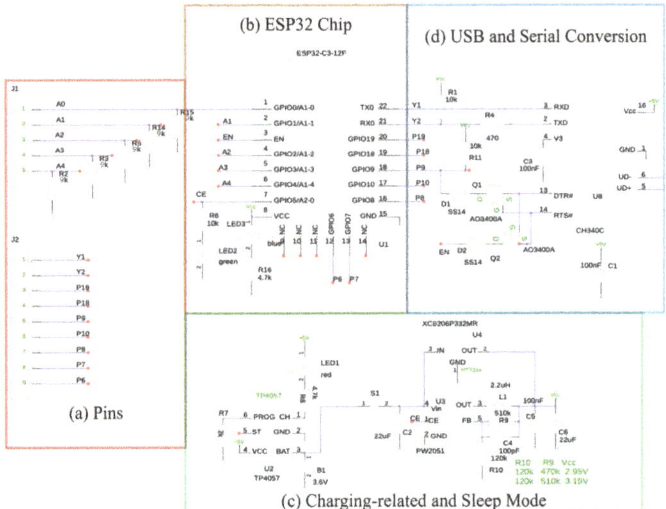

Figure 4. Circuit design diagram.

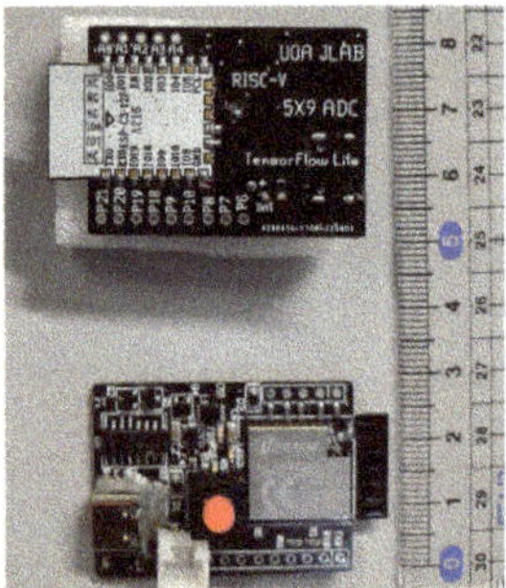

Figure 5. Physical circuit diagram.

We use the ESP32 as the board's processor, using five ADCs and seven GPIOs. Compared to other circuit devices that use serial communication, we have the following two advantages:

1. The traditional circuit relies on serial communication for data transmission, as the maximum bandwidth of serial communication is 115,200 bps stable, resulting in the sampling frequency receives a bandwidth limit, while the bandwidth of wireless communication is far greater, so a higher sampling frequency can be achieved. The device can reach 45 pressure points up to 125 Hz sampling frequency.
2. The device has a function called OTA. It can realize the function of version upgrade through Wi-Fi and compile upload without connecting any transmission line. Since wireless communication replaces serial communication, it means the pins of RX and TX can be multiplexed as GPIO to realize the pressure point detection of 5×9 matrix. Also, OTA can solve the challenge of the unmanned maintenance of intelligent textiles.

4.3. Fall Recognition System

Figure 6 shows the flow chart of the fall detection process, including data collection, data preprocessing and implementation of machine learning model.

1. Data collection: Incoming data from the insoles are collected and sent to a computer/server for analysis.
2. Data pre-processing: The collected data is pre-processed to ensure that it is ready for analysis. This includes normalization, adding default values and removing outliers.
3. Machine learning model: A ResNet(2+1)D neural network is used to predict the probability of various actions based on the pre-processed data and is used to classify the stress action data.

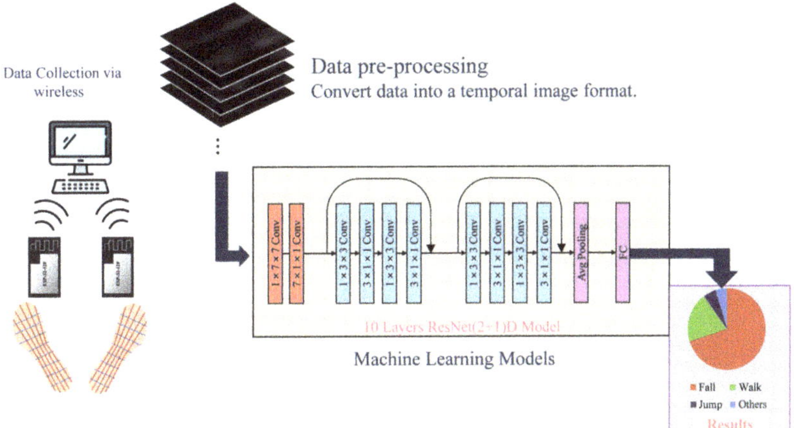

Figure 6. Recognition process of the fall detect system.

4.3.1. Data Collection

The ESP32 can either establish a wireless network by itself as an access point or connect to a wireless network as a client. In this research, we use a router to open a wireless network and let the ESP32 and the computer connect to this wireless network, with each of the two ESP32s connected to a different port of the computer. Since high-frequency network transmission wastes a lot of chip computing time, we let the ESP32 cache the measured data and send it as a batch when it reaches 1500 bytes in size.

4.3.2. Data Pre-Processing

We utilize a sewing machine to fix conductive sewing thread and resistive film with non-conductive cotton thread to make 45 pressure sensors. However, due to human factors, the 45 sensors are unevenly stressed, resulting in different pressure responses when no load is applied. To solve this problem, we calculate the average no-load pressure for each pressure sensor and then subtract this average to zero out the sensor reading.

For the arrangement of the pressure sensors, each insole contains sensors laid out in five columns and nine rows. When two insoles are combined, this layout doubles to 10 columns and 9 rows. To standardize the data format, we add an extra row of zeros, resulting in a final distribution of 10 columns by 10 rows, which gives us a uniform pressure image data matrix. We initially sample the pressure data at a frequency of 100 Hz within a 5-s timeframe for each action. To optimize data processing, we resample this data to 50 Hz. From this resampled data, we then generate a 250-frame grayscale video to visually represent the pressure changes over time.

4.3.3. Machine Learning Models

To take into account the possibility of more action types and more data volumes in the future, the choice of a suitable neural network model is important. Focusing on the purpose of preventing overfitting, we chose the ResNet, which utilizes residual connectivity to make it easier for the model to learn complex functions and reduce the risk of overfitting.

In our study, we analyze stress video data, which necessitates a model capable of interpreting both spatial and temporal dimensions. Therefore, we adopted ResNet(2+1)D [27], a variant of ResNet inspired by the work of Du Tran. Tran's research demonstrated the effectiveness of replacing ResNet's 2D convolutional layers with 3D convolutional layers. They decomposed the 3D convolution into two independent consecutive operations, a 2D spatial convolution and a 1D temporal convolution. This change brings an additional nonlinear correction (additional nonlinear correction for spatial convolution followed by spatial convolution), which is easier to optimize and thus reduce losses compared to full 3D convolution. So in this research, we are using ResNet(2+1)D as the neural network model for the classifier.

Given the small size of the data collected in this research and the lower resolution than traditional video data (112×112 for traditional video data, while we use 10×10), we chose to use only a 10-layer ResNet(2+1)D model for training. If in the future we obtain more data and action types, it may be more effective to use a model with more complex layers. In addition, we use 75% of the original dataset for training and 25% for testing to ensure the robustness and accuracy of the model.

5. Implementation

We implement a pressure-based fall detection system in three key steps: design of the tactile insoles, data collection, and analysis.

5.1. Pressure Sensors Distribution in Insoles

We used the Japanese size standard for insoles, which is measured in cm with 0.5 cm intervals between each size. When designing the insoles for the fall detection system, the distribution of pressure points was a key consideration, because an uneven distribution of pressure points could affect the accuracy and reliability of the fall detection system. The irregular shape of the insole presented a challenge in this regard, as it is difficult to arrange the pressure points in a way that will ensure an even distribution.

To address this issue, two solutions were considered. The first solution involves giving up pressure points in the narrower part of the width and focusing on ensuring an even distribution in the wider part of the insole. This approach has the advantage of being relatively easy to implement and requiring no special sewing operations. However, it may result in some areas of the plantar being left uncovered by pressure points.

The second solution involves carefully controlling the direction of the stitches to ensure that each pressure point is evenly distributed on the horizontal line. This approach provides better coverage of the plantar, but requires more time and effort in the sewing process.

Ultimately, we chose the second solution, depicted in Figure 7, as it provides the best balance between coverage and accuracy. However, both solutions have their advantages and limitations, and further research may be needed to determine the optimal approach for distributing pressure points in insoles for fall detection systems.

5.2. Stitching Details

The stitching of the pins is also a key factor to consider when sewing the wires of the insoles. As shown in Figure 8, stitches are used to hold wires in place and prevent them from moving or shifting during use. However, if the stitches are placed too close together, a number of problems can arise.

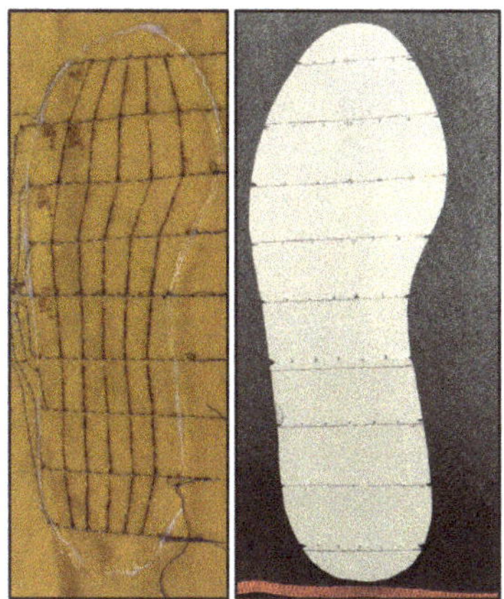

Figure 7. Pressure point distribution.

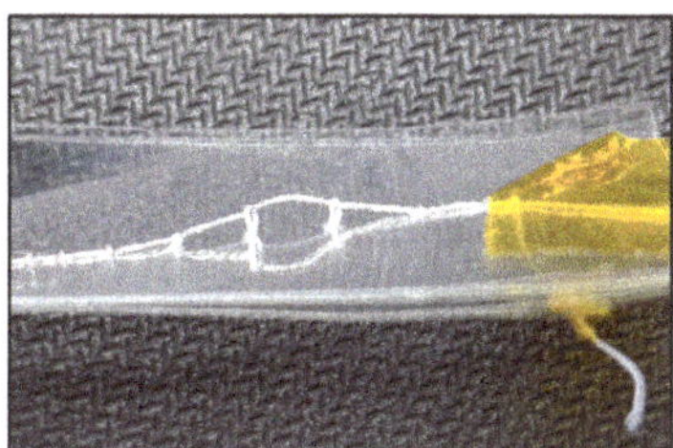

Figure 8. Stitching details. The horizontal and vertical conductive threads may not make stable contact with the conductive threads (poor contact due to blocking by the cotton threads).

First, deformation of the fabric can cause the wire to be pulled by the cotton thread to the other side of the fabric, possibly making direct contact with another crossed wire and causing a short circuit. A short will cause all pressure points on this crossed wire to bypass the resistor going through here, which affects the accuracy and reliability of the device. Second, the wire at the pin may be tied up in cotton thread, which may prevent the horizontal and vertical wires from crossing, resulting in poor contact between the wire and the pin. This can affect the system's ability to accurately detect pressure changes and interpret them as drops.

To avoid these problems, we carefully considered the location of the pins at the crossover of the horizontal and vertical wires to ensure that they are spaced far enough apart at the crossover time to prevent short circuits and maintain good contact between the wires and pins. In our designs, we tried to avoid pins at the intersection of horizontal and vertical wires (as shown in the Figure 9, the pin spacing at the intersection is manually stretched to prevent problems). Ensuring that good contact is maintained between the wires results in more accurate and reliable data collection.

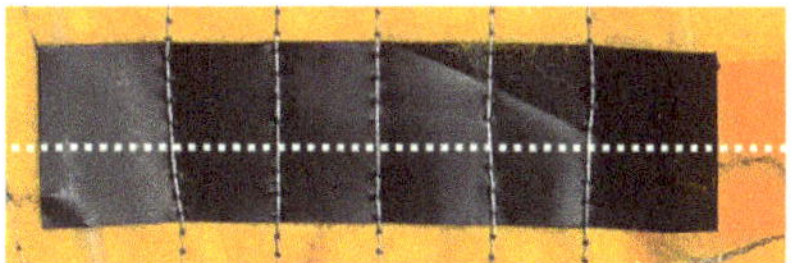

Figure 9. The stitch spacing of the vertical line is intentionally large at the location of the horizontal cross (dotted line).

5.3. Conductive Wiring

However, the link between the circuit board and the insole became a challenge. The main problem is the flexibility of the conductive sewing wire, which made it impossible to use soldering on the conductive wire, as well as making it not easy to insert the wire into the terminals for connection. The wire itself is also very fragile and prone to breakage.

To solve this problem, we stripped the outer shell of the DuPont wire and twisted the hard wire of the DuPont wire to the conductive sewing wire, creating a strong, stable connection. The details are shown in Figure 10. To further ensure the integrity of the connection, we sealed and secured it with hot melt adhesive and heat shrink tubing. The other end of the wire was attached to the circuit board and secured with tape. The final design also takes into account that collecting data during use impacts the wiring connections, so we needed to ensure that the battery and wires do not come off during strenuous action.

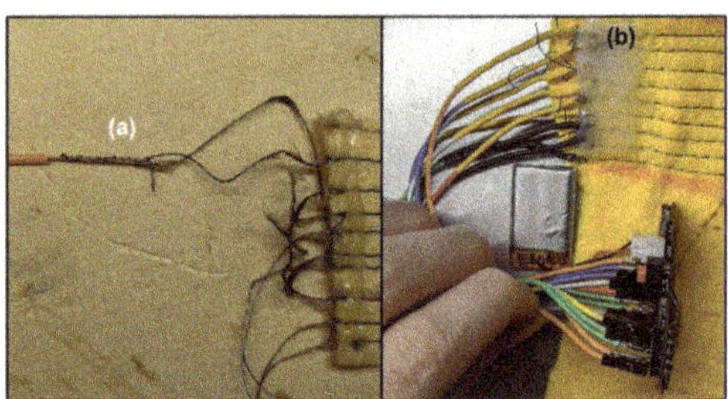

Figure 10. (a) Set out the copper wire of the DuPont wire and the wire stranded together. (b) Use hot melt adhesive to fix the copper wire and wire contact part.

5.4. Controller

The controller part of the code was written using C++ code for the Arduino platform. The workflow of controller is shown in Figure 11. The Arduino code provides two main functions, `setup()` and `loop()`. The setup function runs once after the controller is powered up, and then the code loops through in `loop()`. The controller's setup handles some initialization functions, such as setting the state of each pin and connecting to the wireless network. In the loop function, two tasks are performed: running a check for code updates and executing a sleep task.

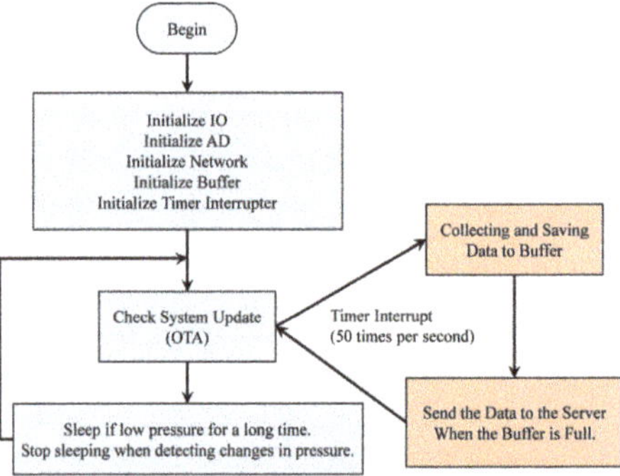

Figure 11. Flow chart of controller. The process that interrupts the loop with a fixed frequency inserts the code that executes to the right. The global buffer avoids avoiding delays due to high-frequency network transmissions.

To improve the accuracy and stability of the sampling frequency, a special function called interrupter was implemented in the controller, which hangs the current process for a fixed period of time and then inserts a piece of code to execute it. The right-hand part of the controller performs the data collection and sending functions after the interrupt. Furthermore, in order to optimize the performance of the system, a global buffer was added in front of the saved data in order to transfer several rounds of data at once and to avoid the network response time affecting the computing time due to multiple high-frequency TCP transfers.

5.5. Classification Algorithm

We utilize the ResNet(2+1)D algorithm as the classifier for fall detection, as shown in Figure 12. These sensors capture a 10×10 resolution pressure map for each frame, resulting in a unique form of time-series data, termed "stress video".

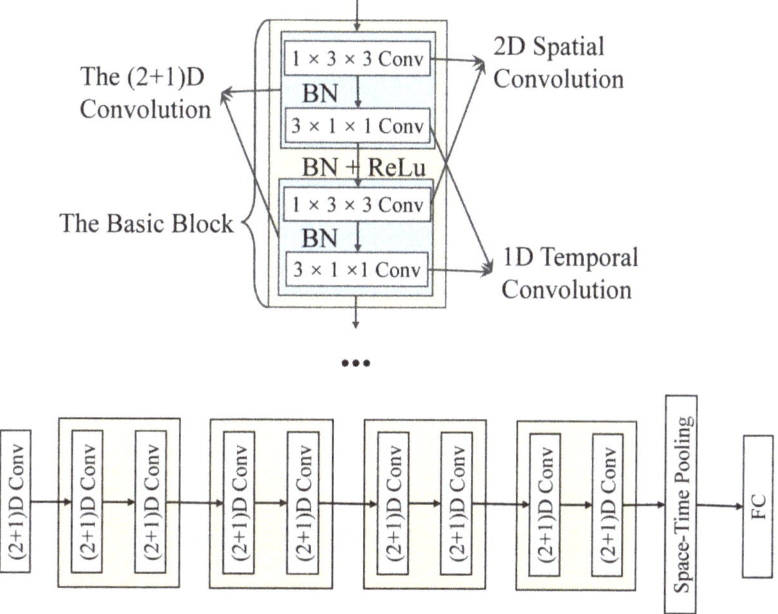

Figure 12. The structure of the ResNet(2+1)D algorithm. The top figure displays the BasicBlock structure, while the bottom image illustrates the system's neural network structure.

We identify ResNet3D as a potential solution due to its proficiency in learning time-varying relationships in video data. Nevertheless, the standard 3D convolution operations in ResNet3D are computationally intensive and less effective at disentangling spatial and temporal features. Hence, we adopted the ResNet(2+1)D architecture, which introduces a novel (2+1)D convolution block. This block separates the 3D convolution into two distinct operations: a 2D spatial convolution to process each frame independently, followed by a 1D temporal convolution to model the temporal dynamics across frames. This decomposition not only simplifies the learning of spatial and temporal features but also reduces the model's complexity.

Initially, our approach involved directly feeding the 10×10 resolution data from the pressure maps into the network. However, preliminary training results plateaued at a modest 50% accuracy. We hypothesized that the application of the $1 \times 3 \times 3$ convolution kernel on the low-resolution data ($250 \times 10 \times 10$) might result in substantial information loss, hindering the model's learning capability. To counter this, we employed an upsampling strategy, resizing the data to $250 \times 112 \times 112$. This transformation significantly enhanced the model's performance, as the increased resolution provided a richer representation of the pressure patterns, allowing the convolutional networks to capture more nuanced spatial and temporal features. We know that the pressure of a foot stepping on an insole in the real world is gradual and smooth, so upsampling can actually restore the pressure distribution in the real world through an appropriate difference method, providing data closer to reality in the network.

6. Experiment and Evaluation

In this section, we first introduce the experimental scenario setup, followed by a description of the types of actions and the dataset utilized in the experiment. Subsequently, we present the analysis results of pressure values for two specific actions: falling and walking. Finally, an analysis of the system's performance is conducted.

6.1. Experimental Setup

Figure 13 illustrates our indoor experimental setup. We invited five volunteers who wore the pressure insoles and performed specified actions within a 2 × 3 m action area, as depicted in the diagram. Considering the interior floor was hard, for safety purposes, soft mats were utilized during the fall experiments to provide cushioning and ensure participant safety. The insoles collect plantar pressure data at a frequency of 100 Hz, which was wirelessly transmitted to a laptop via Wi-Fi device. A camera was used to record the actions as ground truth. The collected data was processed using Python 3.7.4.

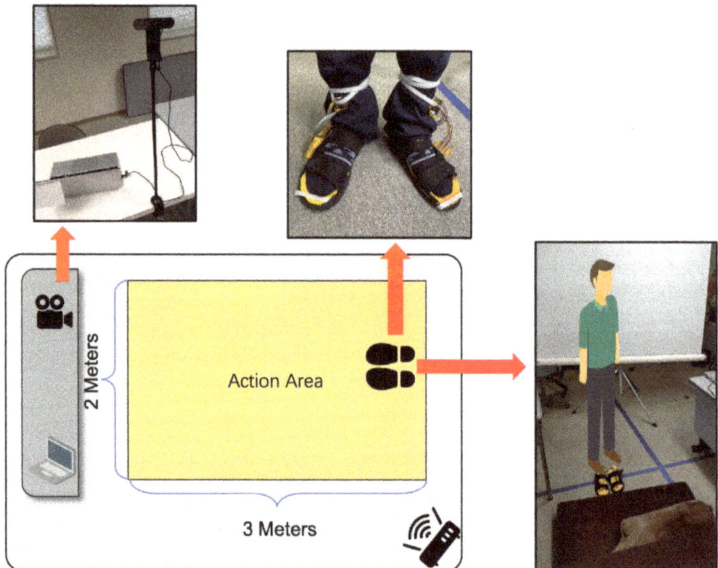

Figure 13. Experiment environment.

6.2. Dataset

To assess the system performance, we designed twelve actions categorized into three types, as detailed in Table 1. The first type includes five daily activities: walking, sweeping, seating, standing, and walking with a cane. The second type involves four actions associated with a risk of falling, including body leaning forward, backward, left, and right. The third type encompasses three falling actions, specifically falling forward, backward, and to the left. Each action consists of 65 samples.

Participants performed each action for a duration of 5 s. To ensure data consistency and integrity, they were instructed to complete a full set of actions from a stationary position within 5 s and then return to a stationary position. Following this, a pause of 5–10 s was observed to prepare for the next action. To ensure participant safety during fall data collection, we used a mat and allowed participants to fall in any safe manner towards a designated direction, without restricting their landing posture or position.

Table 1. Details of actions.

Actions	Number of Samples	Description
Walking	65	Each participant starts from a standing position, takes two steps forward, and then returns to a stationary position at the end of the two steps, waiting for the test time to end.
Sweeping	65	Participants hold a stick to simulate a sweeping motion. They begin by leaning slightly forward and shifting their weight slightly, then perform a sweeping motion to simulate sweeping.
Seating	65	A chair is placed behind the participant. The participant starts standing still and then slowly and naturally sits in the chair until the test period ends, simulating a real-life sitting down situation.
Standing	65	A chair is used in this action. Participants initially sit in the chair, then slowly and naturally stand up from the chair and remain standing until the end of the trial period, simulating a real-life standing up situation.
Walking with a cane	65	The participant holds a cane and simulates an older person walking with a cane. The participant, holding the cane with the right hand, starts from a stationary position, walks with the cane ensuring that part of the pressure on the right foot is taken by the cane. The rest of the requirements are the same as the walking action.
Body leaning	65 × 4	The participant is instructed to perform leaning in one direction, exhibiting a posture with an unstable center of gravity. This action is divided into four categories corresponding to different directions of tilt: front, back, left, and right, to simulate real-life body leaning directions.
Falls	65 × 3	The participants perform falls in three different directions: forward, backward, and leftward. They have the freedom to perform the falls in any manner, as long as the direction of the fall aligns with the specified direction.

6.3. Pressure Sensor Data Analysis

To illustrate the differences in plantar pressure distribution across various actions, we analyze the data using two actions: forward falling and walking. Figure 14 depicts the relationship between different phases of these actions and plantar pressure distribution. Figure 14a illustrates the action of forward falling, while Figure 14b depicts walking.

In Figure 14a, the forward fall is divided into four stages: (a) stationary standing, (b) forward-leaning with both forefoot and heel on the ground and center of gravity shifting towards the forefoot, (c) further leaning with only the forefoot touching the ground, and (d) falling, where both forefoot and heel are off the ground, resulting in zero plantar pressure. Each insole is equipped with 45 sensors, and we selected 3 sensors from both the forefoot and heel areas of each insole to plot the real-time pressure value change curves. The x-axis represents time, while the y-axis corresponds to voltage values. The data collected from the pressure sensors are outputted as a distribution of voltage values ranging from 0–4095, with a measurement range of 0–2.450 V. Upper curves in the figure indicate forefoot pressure, while lower curves represent heel pressure, with dashed lines for the left insole and solid lines for the right. The pressure curves reveal distinct trends across different stages. During stage (a), both the forefoot and heel exhibit stable, minor fluctuations in pressure. In stage (b), there is an increase in forefoot pressure accompanied by a decrease in heel pressure. Moving to stage (c), the forefoot pressure continues to rise, reaching a peak, while heel pressure drops to zero. By stage (d), pressure in both the forefoot and heel registers as zero.

In Figure 14b, the walking action is divided into five stages: (a) stationary standing, (b) right foot lifted with left foot on the ground, (c) right foot touching the ground while left foot is lifted, (d) left foot on the ground with right foot lifted, and (e) similar to stage (a). During each walking cycle, the center of gravity on the foot in contact with the ground shifts from the heel to the forefoot. Similarly, we select three sensors from the forefoot and heel of each insole to plot the curves depicting the change in pressure over time. From the figure, we can see the obvious changes in pressure at each stage. In stage (a), participants stand still with evenly distributed pressure on both feet. In stage (b), lifting the right foot reduces its pressure to zero. Stage (c) sees a similar pattern with the left foot, shifting weight to the right. Finally, stages (d) and (e) rebalance the pressure across both feet as walking continues. This sequence of alternating high and zero pressure illustrates a typical walking gait.

Each action is composed of multiple stages, each stage exhibiting distinct characteristics in terms of pressure distribution and variations. Data from 90 pressure sensors, distributed with 45 sensors per insole, are fed into the ResNet3D model. The ResNet3D model is trained to recognize distinct pressure patterns, improving its ability to differentiate between falling and other types of actions.

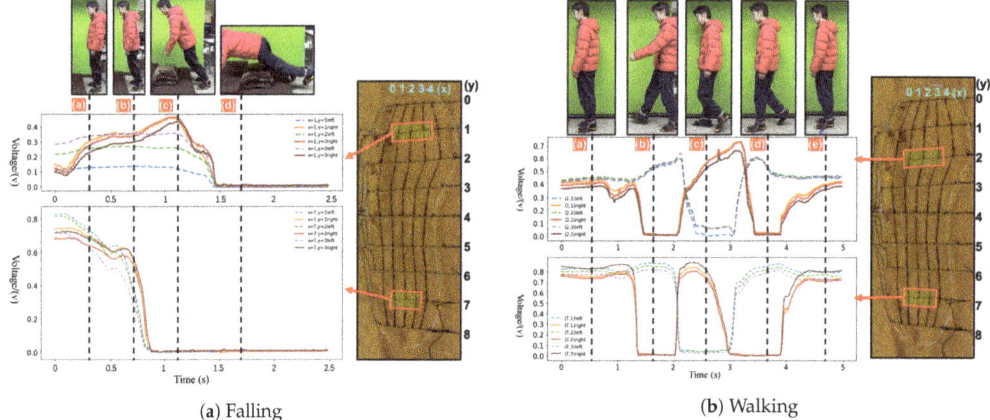

(**a**) Falling

(**b**) Walking

Figure 14. Plantar pressure distribution during falling and walking.

6.4. Evaluation

The experimental data from this study were used to train a neural network classifier and evaluate its performance. Figure 15 shows the curves of loss and accuracy for the classification of 10 classes of actions. In Figure 16, the classifier achieves an overall accuracy of 91% on the dataset when the three falls are grouped into one fall action. For the detection of falls, the accuracy is 94%, which is a very effective result.

We also trained the classification model for 12 actions, splitting falls into 3 categories (forward, backward, and leftward) for classification. Figure 17 shows the curves of loss and accuracy for the classification of 12 classes of actions. In Figure 18, the classification results achieved a recognition rate of 85.61%. This demonstrates that our system can accurately classify most of the actions in the dataset. However, it is also observed that the major incorrect classifications occur between the three fall categories.

It is notable that there are only 65 samples per action in the current experiment, which is a very small amount of data. This could affect the learning of the neural network and also have a significant impact on the classification of the actions. Despite this limitation, the results of this research are promising and suggest the potential for further development and improvement of this method. Future research should aim to increase the amount of data and test classifiers on larger and more diverse groups of people.

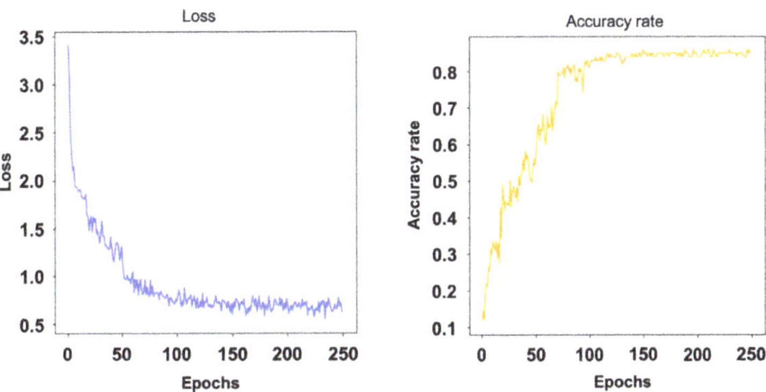

Figure 15. Curvesof loss and accuracy for the classification of 10 classes of actions.

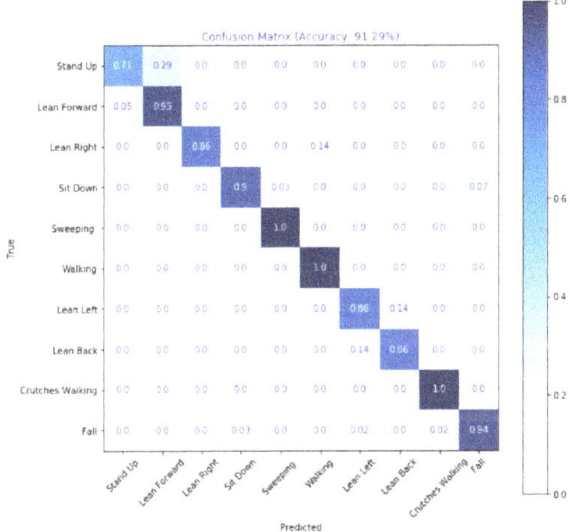

Figure 16. Confusion matrix for the classification of 10 classes of actions.

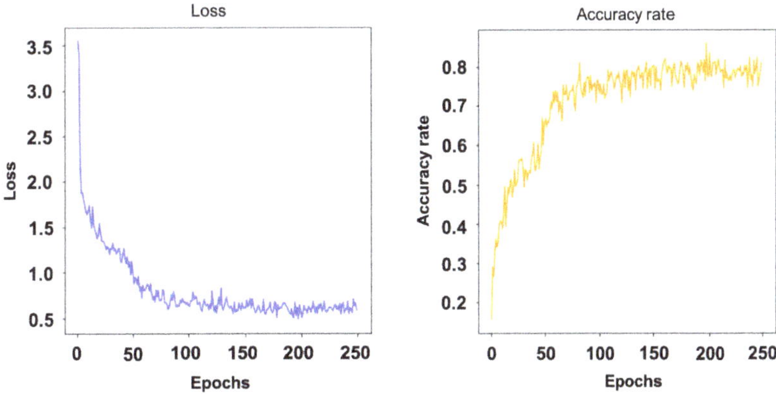

Figure 17. Curvesof loss and accuracy for the classification of 12 classes of actions.

The results of the experimental data in this research show that the proposed approach to fall detection using pressure sensors in insoles and ResNet(2+1)D is promising. It provides a non-invasive, privacy-respecting and efficient solution for fall detection in the senior citizens. The results suggest that this approach has the potential to be further developed and improved to make it more accurate and reliable in detecting falls.

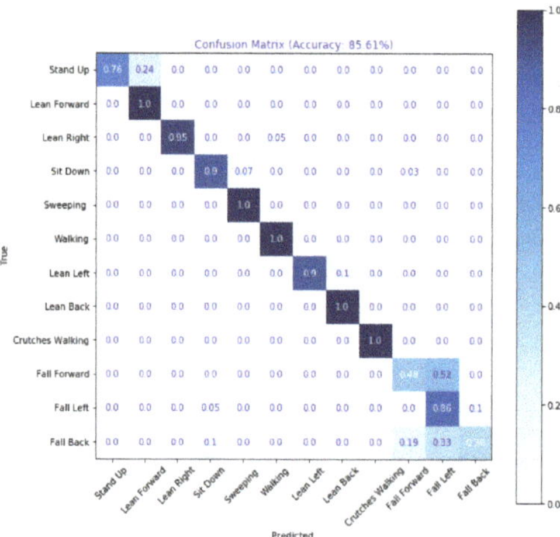

Figure 18. Confusion matrix for the classification of 12 classes of actions.

7. Discussion

To assess the system's performance in real-world scenarios, we engaged eight volunteers (five females, three males) aged between 37 and 85, with an average age of 61, for testing. Participants were instructed to perform actions listed in Table 1 wearing shoes equipped with the smart insole. Foot pressure data were collected in real-time and visualized in MATLAB R2022a to depict pressure distribution. To gather authentic user feedback, a questionnaire was designed, with questions as presented in Table 2. We framed questions around five system characteristics: privacy, utility, portability, usability, and comprehensiveness. Responses were scaled from 1 to 5, with 1 indicating strong disagreement and 5 indicating strong agreement. In addition, in the actual questionnaire, the order of the questions is different from that in Table 2.

We designed this questionnaire to comprehensively assess the multifaceted impact of insole systems on the end user. Each question was carefully designed to gather insights into a specific dimension of user experience, as detailed below:

Privacy (Q1-1 and Q1-2): These two questions sought to understand users' perceptions of protecting their privacy and how comfortable they are with third parties, such as medical personnel or family members, accessing their data. These questions are critical to assessing the level of trust in the system's data management and identifying any privacy issues that may need to be addressed.

Utility (Q2-1 to Q2-4): These questions are integral to evaluating the functional benefits of a system. They explored the system's effectiveness in monitoring foot pressure, helping to prevent falls, enhance understanding of gait, and help improve walking habits, allowing the practical application and value of the system to be evaluated.

Portability (Q3-1 to Q3-4): These questions assessed ease of carrying, ease of wearing, ease of integration of the insole into a variety of footwear, and user willingness to wear it for extended periods of time. This assessed whether product design meets user needs for mobility and convenience.

Usability (Q4-1 to Q4-4): Since user-friendliness is critical for product adoption, the questionnaire included inquiries about ease of use, comparison to other insoles, and intuitiveness of the user interface. These questions help us understand the user's adaptation and comfort level with the insole.

Comprehensive (Q5-1 to Q5-3): Evaluated the overall fit and perceived benefit of the system to the user's lifestyle with respect to questions such as likelihood of daily use, comfort during use, and willingness to recommend the product to others. Helped us measure overall user satisfaction and acceptance.

Table 2. List of questions in the questionnaire.

Category	Question
Privacy	Q1-1 It protects your privacy when collecting and processing your foot pressure data. Q1-2 You would feel uncomfortable if your data were accessed by third parties (e.g., doctors or family members).
Utility	Q2-1 It effectively helps you monitor your foot pressure. Q2-2 It is helpful in preventing falls. Q2-3 You have a better understanding of your own gait and walking style after using it. Q2-4 It is helpful in improving your walking habits and preventing potential health issues.
Portability	Q3-1 It is easy to carry and wear. Q3-2 It meets your daily shoe needs in terms of weight and size. Q3-3 It is convenient for you to switch between multiple shoes. Q3-4 You would be willing to wear it for extended periods of time.
Usability	Q4-1 You are able to use it without any trouble. Q4-2 The feeling of using this insole is almost the same as using other insoles. Q4-3 Its user interface is intuitive and clear. Q4-4 It is easy to use.
Comprehensive	Q5-1 You would like to use such an insole in your daily life. Q5-2 It causes discomfort to the soles of your feet. Q5-3 You would like to recommend this insole to others who may be at risk of falling.

7.1. Questionnaire Results

We collected questionnaire responses from eight volunteers and calculated scores based on their answers to each question. Responses ranged from 1–5, with each option representing its respective score. The average score for each category was computed, and the survey results are depicted in Figures 19 and 20.

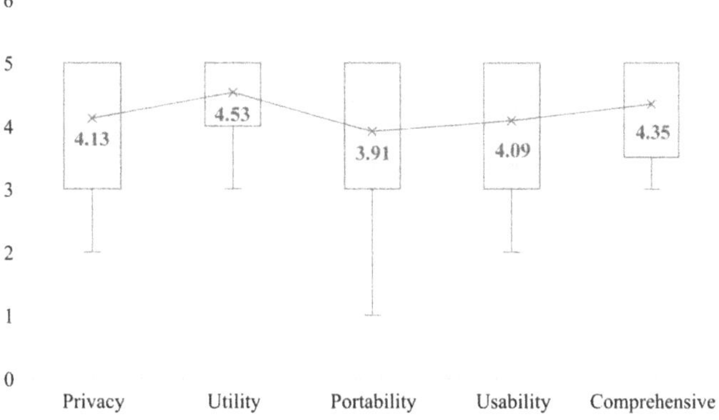

Figure 19. Boxplot of system characteristic's scores.

In Figure 19, the boxplot illustrates the distribution of scores for the system characteristics. The x-axis denotes these characteristics, while the y-axis represents scores. The line within the boxplot indicates the mean score for each characteristic. With a maximum score of 5, the results are as follows: privacy scored 4.13, utility 4.53, portability 3.91, usability 4.09

and comprehensive 4.35. Utility and privacy receive high user ratings, highlighting their positive reception. Conversely, portability shows potential for improvement. The system's usability and comprehensive characteristics score 4.09 and 4.35, respectively, underscoring its ease of use and users' perception of it as a well-rounded and beneficial tool.

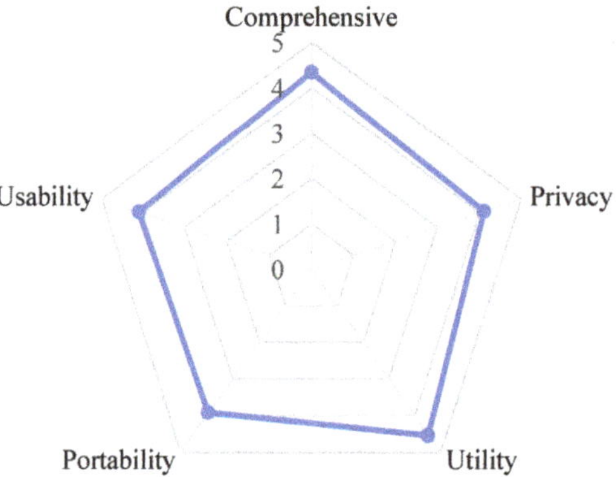

Figure 20. Radar chart of system characteristics.

In Figure 20, the radar chart demonstrates that the system receives favorable feedback across all characteristics. Utility and privacy emerge as significant strengths, while portability represents a primary area for enhancement.

7.2. Privacy Analysis

In privacy assessment of the system, we queried eight participants with Q1-1 and Q1-2. Q1-1 evaluates system privacy perceptions, and Q1-2 probes attitudes on third-party data access. Figure 21 displays user ratings for system privacy, with the x-axis representing user IDs and the y-axis representing scores. Blue bars correspond to Q1-1 ratings, and orange bars to Q1-2. The chart reveals that six users strongly oppose third-party data access, whereas two find it acceptable. Six users give the system's privacy a score of 4.17, and two rate it 4, underscoring user approval of the system's privacy.

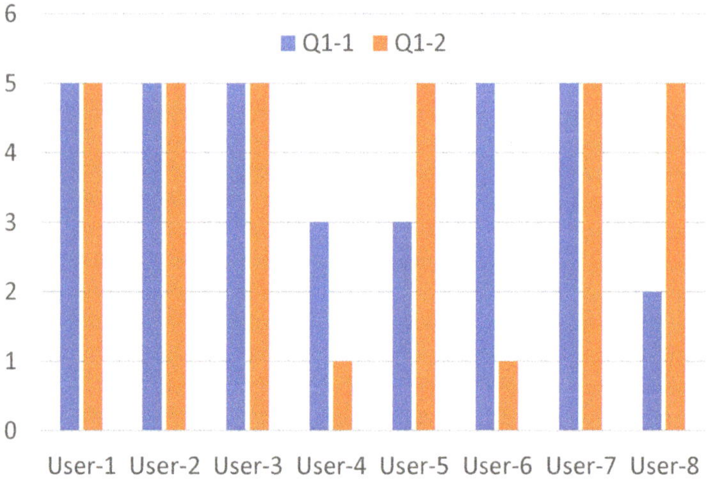

Figure 21. System privacy protection analysis diagram.

7.3. Utility Analysis

For system utility evaluation, we implement four questions (Q2-1 to Q2-4) that address foot pressure monitoring, fall prevention, gait understanding and walking habit improvement. Figure 22 shows that participants consistently rate the system's utility highly, especially for foot pressure monitoring and walking habit improvement, underscoring the system's effectiveness in fall prevention.

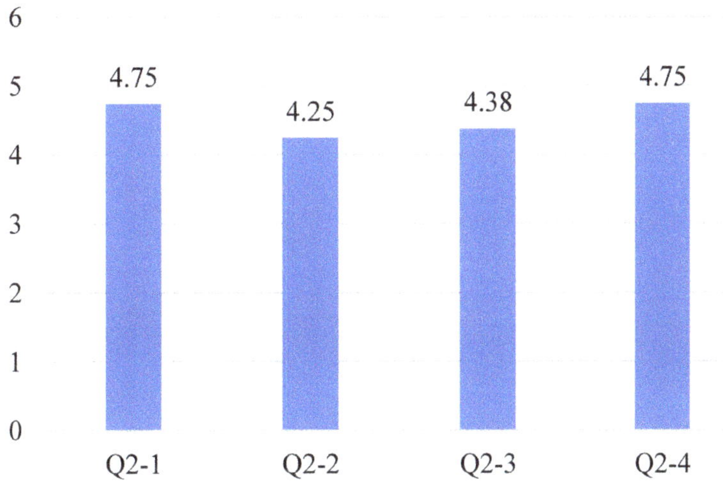

Figure 22. System utility analysis diagram.

Additionally, of the eight participants surveyed, five have experienced falls while three have not. Categorizing participants based on fall experience, we analyze their ratings on system utility, as depicted in Figure 23. In the figure, orange bars represent ratings from users with fall experience, while blue bars denote those without. Participants with fall experience have an average rating of 4.5, whereas those without score an average of 4.58, further highlighting the system's effectiveness in fall prevention.

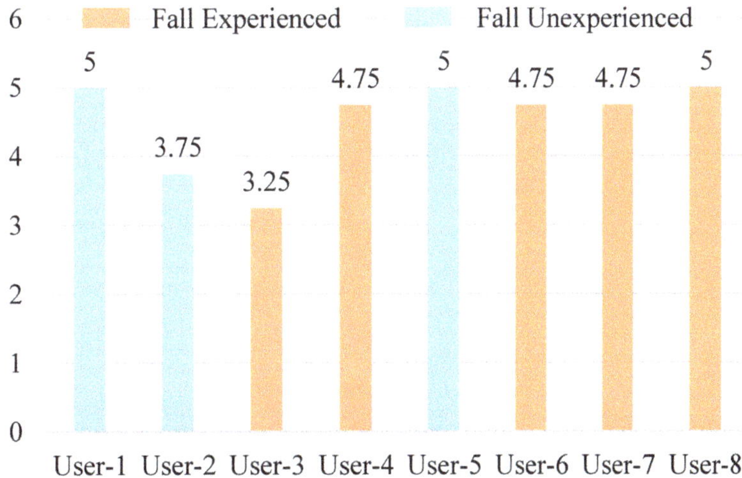

Figure 23. Comparison between users with and without fall experience.

7.4. Portability Analysis

For the system portability assessment, we introduced questions Q3-1 to Q3-4, asking participants to rate the convenience of carrying and wearing the insole, its weight and size, ease of swapping it between different pairs of shoes, and willingness for prolonged use. Figure 24 presents the results. Q3-2 scores the highest at 4.5, followed by Q3-1 and Q3-4, both at 3.75, with Q3-3 being the lowest at 3.43. The data suggest users appreciate the insole's size and quality, yet its usability requires enhancement, pointing towards further miniaturization in hardware design.

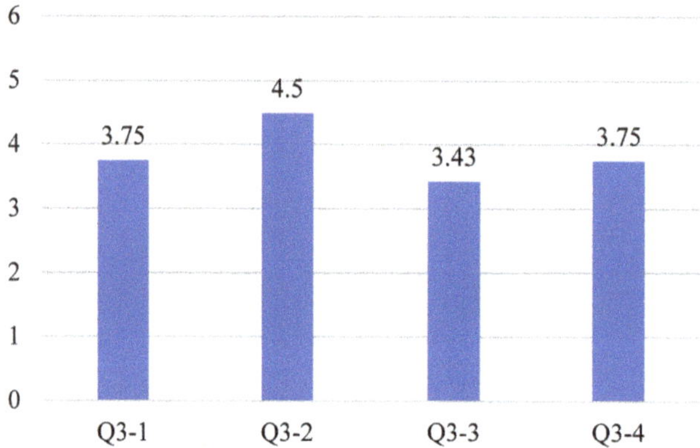

Figure 24. System portability analysis diagram.

7.5. Usability Analysis

We set up four questions to count the usability of the system, and users scored the system in terms of whether they encountered difficulties in using the system, whether the insole was the same as normal insoles, whether the user interface of the system was intuitive and clear, and whether it was easy to use. Figure 25 shows the scoring results, with Q4-4 scoring the highest at 4.63, followed by Q4-3 at 4.13, Q4-1 at 3.43, and Q4-2 at the lowest at 2.33. It is worth noting that the users' ratings of Q4-2 are polarized, with some users scoring the insoles in terms of shape and size, believing that the shape is almost the same as that of ordinary insoles, and others scoring the insoles in terms of their functionality and considered the functionality to be completely different from that of a regular insole, thus leading to lower ratings.The ratings in Q4-4 illustrate that the users gave high ratings to the ease of use of the system.

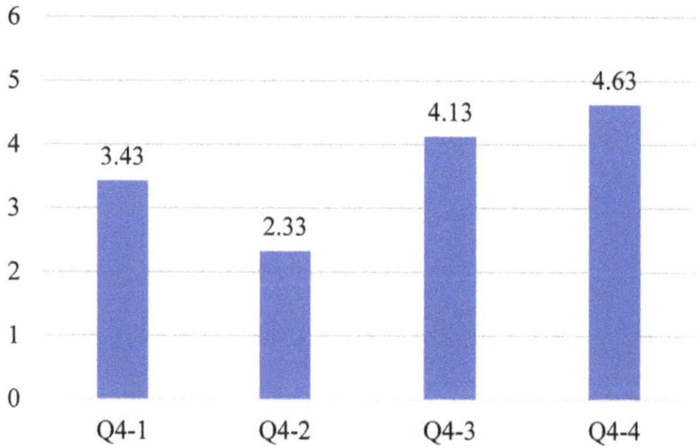

Figure 25. System usability analysis diagram.

7.6. Comprehensive Analysis

For the comprehensive evaluation of the system, we designed a total of three questions from Q5-1 to Q5-3, inviting users to rate the insole in terms of whether they would like to use this insole in their daily lives, whether the insole caused discomfort, and whether they would recommend this insole to a friend who is at risk of falling. Please note that question Q5-2 was a reverse question and we converted its rating. The results of the research are shown in Figure 26, where users gave a high score of 4.71 for the comfort of the insole, 4.13 for the willingness to use this insole in daily life, and 4.13 for the willingness to recommend this insole to a friend with a risk of falling, which further illustrates the user's recognition of the system's comprehensiveness.

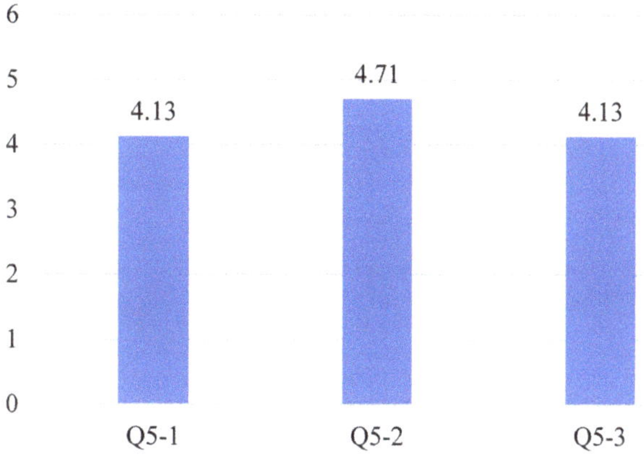

Figure 26. System comprehensive analysis diagram.

8. Conclusions

In conclusion, this research proposes a fall detection system using a shoe insole with a pressure sensor. The experimental results show that our proposed system can accurately classify 12 different actions and detect falls with 94% accuracy. The overall recognition rate of the 12 actions is 85.61%. These results demonstrate the potential of our proposed system for fall detection in the senior citizens.

However, the major false identifications occurred between the three fall categories, suggesting that there is still room for improvement in fine-grained fall detection. In addition, the experiment used a small sample size, which may have an impact on the results.

In future work, we aim to enhance the performance of our system by utilizing larger sample sizes and applying more advanced deep learning algorithms. Additionally, we plan to explore the reduction of sensor quantity and optimization of sensor distribution in the insoles to lower the model complexity and computational time. For the production of insoles, we will use an embroidery machine to improve the accuracy of the distribution of sensors and the efficiency of the production, and to reduce the influence of human factors on the variability of each pair of insoles. At the same time, we will consider the effect of plantar temperature on the conductivity of the sensors, and add heat-insulating materials to the insoles. Finally, we intend to integrate the developed insoles with Wi-Fi devices to enable human activity recognition and location tracking within the context of smart home applications.

Author Contributions: W.G. was responsible for the experimental setup, insole fabrication, data acquisition, and analysis, as well as the design and analysis of the questionnaire. X.L. was responsible for the system design, experimental setup, insole fabrication, data acquisition, data analysis, as well as the drafting of the manuscript. C.L. provided valuable assistance in the collection and organization of experimental data and also played a significant role in the design of the algorithms. L.J. provided guidance throughout the research, helped shape the study's design, offered critical feedback on the manuscript, and contributed essential financial support. All authors have read and agreed to the published version of the manuscript.

Funding: This work was supported in part by JSPS KAKENHI under Grant 22K12114, in part by JKA Foundation, and in part by NEDO Younger Research Support Project under Grant JPNP20004.

Data Availability Statement: The data presented in this study are available on request from the corresponding author.

Conflicts of Interest: The authors declare no conflicts of interest.

References

1. World Health Organization. "Falls", Who.int (World Health Organization: WHO, April 26). 2021. Available online: https://www.who.int/news-room/fact-sheets/detail/falls (accessed on 13 January 2024).
2. Subramaniam, S.; Faisal, A.I.; Deen, M.J. Wearable sensor systems for fall risk assessment: A review. *Front. Digit. Health* **2022**, *4*, 921506. [CrossRef] [PubMed]
3. Alwan, M.; Rajendran, P.; Kell, S.; Mack, D.; Dalal, S.; Wolfe, M.; Felder, R. A Smart and Passive Floor-Vibration Based Fall Detector for Elderly. In Proceedings of the 2nd International Conference on Information & Communication Technologies, Damascus, Syria, 24–28 April 2006; Volume 1, pp. 1003–1007. [CrossRef]
4. Das, R.; Kumar, N. Investigations on postural stability and spatiotemporal parameters of human gait using developed wearable smart insole. *J. Med. Eng. Technol.* **2015**, *39*, 75–78. [CrossRef] [PubMed]
5. Di Rosa, M.; Hausdorff, J.M.; Stara, V.; Rossi, L.; Glynn, L.; Casey, M.; Burkard, S.; Cherubini, A. Concurrent validation of an index to estimate fall risk in community dwelling seniors through a wireless sensor insole system: A pilot study. *Gait Posture* **2017**, *55*, 6–11. [CrossRef] [PubMed]
6. Hu, X.; Zhao, J.; Peng, D.; Sun, Z.; Qu, X. Estimation of foot plantar center of pressure trajectories with low-cost instrumented insoles using an individual-specific nonlinear model. *Sensors* **2018**, *18*, 421. [CrossRef] [PubMed]
7. Ayena, J.C.; Chioukh, L.; Otis, M.J.D.; Deslandes, D. Risk of falling in a timed up and go test using an UWB radar and an instrumented insole. *Sensors* **2021**, *21*, 722. [CrossRef] [PubMed]
8. Cates, B.; Sim, T.; Heo, H.M.; Kim, B.; Kim, H.; Mun, J.H. A novel detection model and its optimal features to classify falls from low-and high-acceleration activities of daily life using an insole sensor system. *Sensors* **2018**, *18*, 1227. [CrossRef] [PubMed]
9. Saidani, S.; Haddad, R.; Bouallegue, R.; Shubair, R. Smart Insole Monitoring System for Fall Detection and Bad Plantar Pressure. In Proceedings of the International Conference on Advanced Information Networking and Applications, Sydney, Australia, 13–15 April 2022; Springer: Berlin/Heidelberg, Germany, 2022; pp. 199–208.
10. Chen, D.; Asaeikheybari, G.; Chen, H.; Xu, W.; Huang, M.C. Ubiquitous fall hazard identification with smart insole. *IEEE J. Biomed. Health Informat.* **2020**, *25*, 2768–2776. [CrossRef] [PubMed]
11. Mun, F.; Choi, A. Deep learning approach to estimate foot pressure distribution in walking with application for a cost-effective insole system. *J. Neuroeng. Rehabil.* **2022**, *19*, 4. [CrossRef] [PubMed]
12. Sun, Z.; Ke, Q.; Rahmani, H.; Bennamoun, M.; Wang, G.; Liu, J. Human Action Recognition From Various Data Modalities: A Review. *IEEE Trans. Pattern Anal. Mach. Intell.* **2022**, *45*, 3200–3225. [CrossRef] [PubMed]
13. Pareek, P.; Thakkar, A. A survey on video-based human action recognition: Recent updates, datasets, challenges, and applications. *Artif. Intell. Rev.* **2021**, *54*, 2259–2322. [CrossRef]
14. Guo, W.; Yamagishi, S.; Jing, L. Human Activity Recognition via Wi-Fi and Inertial Sensors with Machine Learning. *IEEE Access* **2024**, *12*, 18821–18836. [CrossRef]
15. Duan, H.; Zhao, Y.; Xiong, Y.; Liu, W.; Lin, D. Omni-Sourced Webly-Supervised Learning for Video Recognition. In Proceedings of the Computer Vision—ECCV 2020, Glasgow, UK, 23–28 August 2020; Vedaldi, A., Bischof, H., Brox, T., Frahm, J.M., Eds.; Springer: Cham, Switzerland, 2020; pp. 670–688.
16. Li, B.; Li, X.; Zhang, Z.; Wu, F. Spatio-temporal graph routing for skeleton-based action recognition. In Proceedings of the AAAI Conference on Artificial Intelligence, Honolulu, HI, USA, 27 January–1 February 2019; Volume 33, pp. 8561–8568.
17. Nait Aicha, A.; Englebienne, G.; Van Schooten, K.S.; Pijnappels, M.; Kröse, B. Deep learning to predict falls in older adults based on daily-life trunk accelerometry. *Sensors* **2018**, *18*, 1654. [CrossRef] [PubMed]
18. Ma, L.; Liu, M.; Wang, N.; Wang, L.; Yang, Y.; Wang, H. Room-level fall detection based on ultra-wideband (UWB) monostatic radar and convolutional long short-term memory (LSTM). *Sensors* **2020**, *20*, 1105. [CrossRef] [PubMed]
19. Yadav, S.K.; Tiwari, K.; Pandey, H.M.; Akbar, S.A. A review of multimodal human activity recognition with special emphasis on classification, applications, challenges and future directions. *Knowl.-Based Syst.* **2021**, *223*, 106970. [CrossRef]
20. Tanwar, R.; Nandal, N.; Zamani, M.; Manaf, A.A. Pathway of trends and technologies in fall detection: A systematic review. *Healthcare* **2022**, *10*, 172. [CrossRef] [PubMed]
21. Desai, K.; Mane, P.; Dsilva, M.; Zare, A.; Shingala, P.; Ambawade, D. A novel machine learning based wearable belt for fall detection. In Proceedings of the 2020 IEEE International Conference on Computing, Power and Communication Technologies (GUCON), Greater Noida, India, 2–4 October 2020; IEEE: Piscataway, NJ, USA, 2020; pp. 502–505.

22. Lee, J.S.; Tseng, H.H. Development of an enhanced threshold-based fall detection system using smartphones with built-in accelerometers. *IEEE Sens. J.* **2019**, *19*, 8293–8302. [CrossRef]
23. Sundaram, S.; Kellnhofer, P.; Li, Y.; Zhu, J.Y.; Torralba, A.; Matusik, W. Learning the signatures of the human grasp using a scalable tactile glove. *Nature* **2019**, *569*, 698–702. [CrossRef] [PubMed]
24. Luo, Y.; Li, Y.; Foshey, M.; Shou, W.; Sharma, P.; Palacios, T.; Torralba, A.; Matusik, W. Intelligent Carpet: Inferring 3D Human Pose From Tactile Signals. In Proceedings of the IEEE/CVF Conference on Computer Vision and Pattern Recognition (CVPR), Nashville, TN, USA, 20–25 June 2021; pp. 11255–11265.
25. Anzai, E.; Ren, D.; Cazenille, L.; Aubert-Kato, N.; Tripette, J.; Ohta, Y. Random forest algorithms to classify frailty and falling history in seniors using plantar pressure measurement insoles: A large-scale feasibility study. *BMC Geriatr.* **2022**, *22*, 746.
26. Zhang, M.; Liu, D.; Wang, Q.; Zhao, B.; Bai, O.; Sun, J. Gait Pattern Recognition Based on Plantar Pressure Signals and Acceleration Signals. *IEEE Trans. Instrum. Meas.* **2022**, *71*, 4008415. [CrossRef]
27. Tran, D.; Wang, H.; Torresani, L.; Ray, J.; LeCun, Y.; Paluri, M. A closer look at spatiotemporal convolutions for action recognition. In Proceedings of the IEEE Conference on Computer Vision and Pattern Recognition, Salt Lake City, UT, USA, 18–23 June 2018; pp. 6450–6459.

electronics

Article

Towards Smarter Positioning through Analyzing Raw GNSS and Multi-Sensor Data from Android Devices: A Dataset and an Open-Source Application

Antoine Grenier, Elena Simona Lohan *, Aleksandr Ometov and Jari Nurmi

Electrical Engineering Unit, Tampere University, 33720 Tampere, Finland; antoine.grenier@tuni.fi (A.G.); aleksandr.ometov@tuni.fi (A.O.); jari.nurmi@tuni.fi (J.N.)
* Correspondence: elena-simona.lohan@tuni.fi

Abstract: The state-of-the-art Android environment, available on a major market share of smartphones, provides an open playground for sensor data gathering. Moreover, the rise in new types of devices (e.g., wearables/smartwatches) is further extending the market opportunities with a variety of new sensor types. The existing implementations of biometric/medical sensors can allow the general public to directly access their health measurements, such as Electrocardiogram (ECG) or Oxygen Saturation (SpO2). This access greatly increases the possible applications of these devices with the combination of all the onboard sensors that are broadly in use nowadays. In this study, we look beyond the current state of the art into the positioning capacities of Android smart devices and wearables, with a focus on raw Global Navigation Satellite Systems (GNSS) measurements that are still mostly lacking in the research world. We develop a novel open-source Android application working in both smartphone and smartwatch environments for multi-sensor measurement data logging that also includes GNSS, an Inertial Navigation System (INS) magnetometer, and a barometer. Four smartphones and one smartwatch are used to perform surveys in different scenarios. The extraction of GNSS raw data from a wearable device has not been reported yet in the literature and no open-source app has existed so far for extracting GNSS data from wearables. Not only the developed app but also the results of these measurement surveys are provided as an open-access dataset. We start by defining our methodology and the acquisition protocol, and we dive into the structure of the dataset files. We also propose a first analysis of the data logged and evaluate the data according to several performance metrics. A discussion reviewing the capacities of smart devices for advanced positioning is proposed, as well as the current open challenges.

Keywords: Global Navigation Satellite System (GNSS); Global Positioning System (GPS); low-power electronics; smart devices; smartphone; wearables; measurements; Android

Citation: Grenier, A.; Lohan, E.S.; Ometov, A.; Nurmi, J. Towards Smarter Positioning through Analyzing Raw GNSS and Multi-Sensor Data from Android Devices: A Dataset and an Open-Source Application. *Electronics* **2023**, *12*, 4781. https://doi.org/10.3390/electronics12234781

Academic Editors: Lei Jing, Jiehan Zhou and Zhan Zhang

Received: 6 November 2023
Revised: 20 November 2023
Accepted: 22 November 2023
Published: 25 November 2023

1. Introduction

Interest in using smartphones for sensor logging has been part of the research community for the last decade. Smartphones provide a very interesting environment for data gathering, as they include a variety of sensors (e.g., accelerometers, gyroscopes, camera, Wireless Fidelity (WiFi), Bluetooth Low Energy (BLE)) embedded inside a low-cost platform. The Application Programming Interface (API) centralizes access to all the sensor measurements for Android devices, greatly simplifying the developments. Moreover, synchronization of all data is ensured by a common platform, making it useful for sensor fusion studies [1,2].

More recently, the development of "wearables" provides similar but new data-gathering platforms. On top of providing ultra-compact designs for the same sensors available on smartphones, smartwatches propose new kinds of sensors more related to health monitoring, such as ECG, photoplethysmogram (PPG), Galvanic Sensor Response (GSR), and/or SpO2. The different sensors in Android smart devices are displayed in Figure 1, showing the potential of smartwatches. While some manufacturers still propose their ecosystem (e.g., "ConnectIQ" for Garmin [3], Fitbit OS for Fitbit [4]), many manufacturers (e.g., Samsung, Google) are aligning on a common Android-based operating system with the possibility of common development of Android application on both the wearable and the smartphone platforms. This research will refer to them under the terms "smart devices".

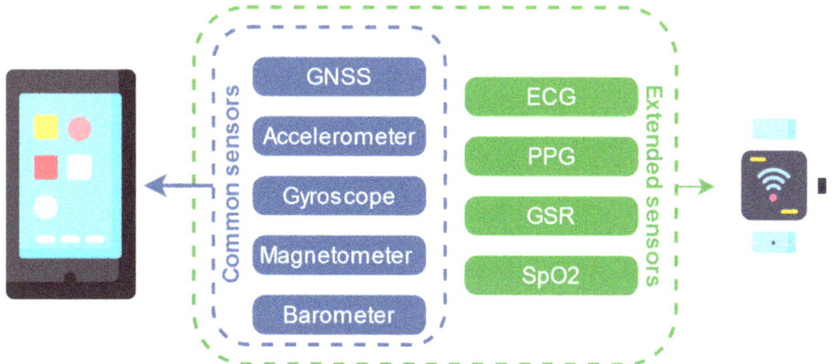

Figure 1. Sensors available in smart devices.

In the context of satellite positioning, all smart devices today include a GNSS receiver, yet the collection of measurements used to be limited to the device coordinates only. In 2016, Google enabled access to the "raw GNSS measurements" directly through the Android API. By "raw GNSS measurements", we understand the measurements made by a receiver to compute a position, such as pseudoranges, carrier phases, Doppler shifts and Signal-to-Noise Ratio (SNR). Since the raw GNSS measurements were made available for smartphone manufacturers, the GNSS research community has been highly interested in the potential of enabling advanced high-precision positioning techniques on such devices, and many research works have been published to assess the receivers' quality for positioning (see Section 2). We note that, as of today, Apple still needs to open access to the raw GNSS measurements in their OS, preventing similar comparisons on Apple devices. Thus, in this paper, we focus only on Android devices.

In the early years of Android GNSS measurements, only a certain number of Android models would provide access to raw data. Since Android 10 (API 29), Google made it mandatory for any Android device to support access to raw GNSS measurements. Discrepancies between device measurements are still present, but access to raw measurements is now open to potentially any Android devices and not only smartphones, including smartwatches.

The main contributions of this paper are as follows:

- Developing a novel open-source sensor logging app called "Mimir", designed for both smartphone and smartwatch environments;
- Providing an open-source multi-sensor dataset acquired on various Android smart devices (four smartphones and one smartwatch), along with a geodesic-grade reference receiver;
- Comparing the data quality and positioning accuracy of the Android smart devices in the context of positioning applications.

To the best of the authors' knowledge, raw GNSS measurements acquisition comparison for smartwatch devices is conducted for the first time in the literature.

This research is performed in the context of the EU-funded projects APROPOS [5] and LEDSOL [6] that aim, among others, to also review low-power/low-cost devices, such as Android smart devices, for navigation purposes. The data acquired consist of four surveys, one static and three pedestrian/dynamic cases, and it was acquired in open-sky and urban areas. The complete dataset is released as an open source under a Creative Commons license CC-BY-4.0 on a Zenodo repository [7].

2. Related Research

2.1. Usage of Android GNSS Measurements

The development of Android GNSS measurements was highly impacted by the release in 2018 of the Xiaomi Mi8, the first device to embed a dual-frequency GNSS chip from Broadcom (BCM4775) [8]. Reception of multiple GNSS frequencies, such as L1 and L5, can enable advanced positioning techniques [9]. Consequently, many publications over the last five years have demonstrated diverse application topics for the usage of Android GNSS measurements: measurement quality [10–13], high-precision positioning [10,14–16], atmospheric error estimation [17,18], sustainable water access through standalone positioning aiding [19,20], spoofing/jamming mitigation [21,22], etc.

While the use of the Android GNSS measurements appears very fruitful and has high potential, acquiring the raw data is not trivial and requires specific apps and access to the device's developer options. In [11], a comprehensive review of the available Android apps for GNSS data acquisition is performed. The paper analyzes the raw measurements available in Android devices and mentions some of the problems associated with the processing. One of the issues pointed out is the measurement quality and unknown hardware characteristics, which prevent more precise positioning as of today. Similarly, as the receiver is greatly influenced by its environment, a lack of details about the acquisition protocol and lack of dataset diversity (e.g., harsh environments) are also highlighted as a limitation with currently available open data.

2.2. Access to Public Datasets

Today, a few public datasets provide access to Android GNSS measurements. The review from [11] was based on a dataset published by Google, in the context of the Google Smartphone Decimeter Challenge (GSDC) that happened in both 2021 and 2022. The data used for the competition is still accessible online and can be downloaded upon acceptance of the competition rules [23,24]. The data from 2021 are composed of dynamic tracks acquired in a car scenario and performed with different Android device models. Additional details can be found in [25].

Other interesting datasets were published by the EUSPA (former GSA) as part of the "GNSS Raw Measurements Task Force" [26]. The data are accessible through a Google Doc, summarizing the campaigns. As of today, two campaigns are accessible through the document. One campaign from 2020 focuses on different wearables devices but provides access only to positions and not to raw GNSS measurements. Another campaign from 2021 contains data acquired in different scenarios (i.e., static, dynamic pedestrian/bike/car) and environments (i.e., open-sky, forest, suburban, urban, highway) from six Android devices. The accessibility and diversity of the data make the EUSPA dataset very interesting for research purposes. However, we did not use this dataset in this research, as we wanted to focus on a newer generation of Android devices including smartwatches, and assess the potential improvements in the GNSS measurement quality.

To summarise, access to raw GNSS measurements in open datasets is still scarce and limited to a few providers. Indeed, all the studies mentioned before that compared several devices have shown the large discrepancies that can exist in terms of measurement quality. Thus, to mitigate the impact of the data over the proposed methods, it is necessary to compare it on a variety of datasets and devices. Given the amount of new Android devices released every year, keeping up with new datasets for every device might seem vain. Yet, crowd-sourcing and publication of the datasets used in studies could enable more device comparison without the need for each research team to purchase a specific device.

2.3. Wearables and Next GNSS Chip Generation

Wearables have the potential to expand the usage of data gathered in smart devices, bringing the health-related sensor dimension. Android smartwatches are based on a slightly different operating system (OS) than smartphones, called WearOS (formerly Android Wear). This OS is based on the same Android API levels, which allows for the common development of Android applications in both smartphones and smartwatches. WearOS 3.0 (Android 11) is the first version based on a higher version than Android 9, which should provide access to raw GNSS measurements. However, to the best of the authors' knowledge, no paper until today has yet investigated the GNSS measurement quality of any Android smartwatch. Nevertheless, medical-oriented studies have shown the potential of wearables for medical research [27,28], as well as Android devices for contact tracing [29]. More specifically, in [30], the usage of indoor/outdoor positioning to retrieve the mobility patterns showed interesting results in recovering mental health assessment for patients. Thus, more accurate positioning and motion estimation could enable further medical research, especially if combined with the health sensors of smartwatches.

In 2019, Broadcom announced the chipset's second generation (BCM4776), a follow-up chipset to the previously mentioned BCM4775 included in the Xiaomi Mi8. According to Broadcom, the new chipset comes with potentially 60% additional L5 signals with the reception of BeiDou-L5 signals [31]. Since 2021, new Android devices have integrated this chipset into their design, such as the Google Pixel 6 and 7, and even the recently released Google Pixel Watch. The release of the new Broadcom chipset could potentially mark the beginning of the second generation of GNSS measurements in Android smart devices.

2.4. Paper Structure

We start by presenting the methodology used for the creation of the dataset. In Section 3, we review the motivation for the device selection, the developed software, and the survey (or measurement) protocol. In Section 4, we provide a deep review of the dataset contents, including the type of sensor logged, the list of the performed surveys, and a description of the scenario and environments. In Section 6, we propose a first analysis of the data, with a comparison of the capacities between the devices. Finally, Section 7 concludes the paper with future development plans for our logging app and perspective research using this dataset.

3. Methodology

In this section, we present the methodology for filtering the different Android devices, in order to select only a subset fitting our research. We proceed with detailing the different software used and developed to acquire and analyze the data, and we finish by presenting our survey protocol.

First, we recall some terminology used in this paper:

- By "survey", we mean the act of surveying and gathering data using a measurement device, as defined in land surveying topics.
- In Differential GNSS (DGNSS), the "base" refer to the static receiver and "rover" refer to the moving receiver.
- In GNSS, an "epoch" is defined a measure of time when a new GNSS measurement is received. A 1 Hz sampling rate would be equivalent to 1 epoch per second.

3.1. Device Selection

Research in Android GNSS measurements have shown that the measurement's quality is not equal among all devices, even under identical measurement scenarios, as the manufacturers can embed different GNSS receivers in each new model [12]. Even access to specific GNSS observables (e.g., carrier phase, navigation message) cannot be ensured for each device. While the impact of these discrepancies can be minimal or unseen by the general user, it can prevent some of the research potential. Therefore, we performed the surveys over multiple Android devices for better assessment of their capacities in this study.

As the number of existing Android smartphones is enormous, the selection of the best devices can prove challenging. Thanks to the research community, a comprehensive list of Android devices with their GNSS capacities has been created through crowdsourcing and it is available online [32]. While it may contain multiple entries and some contradictory information, it still offers a first filter to select the devices with the required GNSS capacities. We note, however, that no smartwatch has been reported in this list so far.

To maximize the potential of each evaluated device and the quality of the dataset, we have decided to select devices under four constraints: (1) support of dual-frequency reception; (2) available carrier-phase measurements; (3) available navigation message; (4) GNSS chipset diversity. Point (3) was the most constraining, as many devices do not retrieve the satellite's ephemeris from the navigation message anymore. Using the aforementioned list, we came down to only 22 devices checking all four requirements. More than half of them used a Broadcom BCM4775 or BCM4776 chip. The BCM4775 has been extensively reviewed in research before given that it was implemented in the Xiaomi Mi8 [10,14,17]; we have decided to focus on evaluating only the new generation BCM4776 instead. The rest of the receivers were chosen based on prices and in order to maximize GNSS chipset and manufacturer diversity. As one smartphone (Samsung A52) was already present in our lab for previous research purposes, we decided to also include it in these surveys for comparison purposes, even though it did not fit the requirements mentioned before. In total, four smartphones were selected and are summarized in Table 1. Note that while the Xiaomi 11T was said to have the navigation message enabled in the online list, we did not receive any during our experiments, and reported it accordingly in the table.

As no previous studies have looked into GNSS measurements in smartwatches, we decided to review the recently released Google Pixel Watch (October 2022) [33]. The watch also embeds a BCM4776 GNSS chip, similar to the one in the Pixel 7, enabling the measurement comparisons between both platforms. For example, while the BCM4776 is supposed to be dual-frequency-enabled, only L1 frequencies were received during the survey. The most logical hypothesis is that the dual-frequency reception was disabled in order to save energy, as it was proven to have a large impact on the energy budget of Android devices [34]. This shows that even when the GNSS chipset model is similar in two devices, large differences in the measurements can still be present.

In the following sections, devices will be referred to by the acronym listed in Table 1 for simplicity.

3.2. Logging Application Developments

In order to log the measurements inside the Android device, an app is used to interact with the Android API. Several apps exist today to log GNSS or sensor data, as reviewed in [11]. The most common one is GNSS logger [35], maintained by Google. It can log GNSS data in multiple formats (e.g., CSV, RINEX, NMEA). Logging other motion sensors is also possible, such as accelerometers, gyroscopes and magnetometers. However, several limitations of the existing applications were found during the first stages of our research. Firstly, only a partial version of the GNSSLogger source code is available on GitHub (old versions of the app), which prevents easy forking to the current version. Secondly, the logging is limited to GNSS and motion sensors logging only (no barometer data), with no possibility of controlling the sampling frequency. Thirdly, GNSSLogger does not exist for smartwatches, and, to the best of the authors' knowledge, there is currently no app available for logging GNSS data from smartwatches.

Table 1. Selected Android devices included in the dataset.

Acronym	Device Name	Type	Release	Android (API)	GNSS Chip	Frequencies	Phase	Navigation Message
GP7	Google Pixel 7	Phone	October 2022	Android 13 (API 33)	Broadcom BCM4776	L1, L5	✓	✓
GPW	Google Pixel Watch	Watch	October 2022	WearOS 3.5 (API 30)	Broadcom BCM4776	L1	✓	✓
ON2	OnePlus Nord 2 5G	Phone	July 2021	Android 12 (API 31)	Unknown	L1, L5	✓	✓
A52	Samsung A52 5G	Phone	March 2021	Android 13 (API 33)	Qualcomm SD 720G	L1	✗	✗
X11	Xiaomi 11T	Phone	September 2021	Android 12 (API 31)	Unknown	L1, L5	✓	✗

Consequently, we decided to develop our own Android app called "Mimir", a name based on the God of Knowledge in Nordic mythology (see https://www.britannica.com/topic/Mimir (accessed on 21 November 2023)). Mimir was developed with the goal of enabling the logging of any sensors present in an Android device. The code is released as open-source, under the Apache 2.0 license and is available on Github [36]. As mentioned earlier, while the Android smartwatch implements different flavors of the WearOS, they are based on a common Android API. Thus, the development of smartwatch apps is similar to smartphones, except for the UI design of the app. It allows the development of a common Android library for sensor logging, implemented in two different apps, for smartphones and smartwatches, respectively. Currently, the app allows logging of GNSS, motion and environment sensors, as defined in the Android developer documentation. The raw data are logged in a Comma Separated Value (CSV) format, similar to GNSSLogger logging format. For compatibility with the former processing software, all measurements related to GNSS have kept the same columns as in GNSSLogger. Other sensors are logged using our own defined format (see Appendix A). Furthermore, the app also allows the logging of health data on the smartwatch version. However, the recording of health data has been tested with the Google Pixel Watch only and is not recorded in the dataset provided for ethical reasons. Future developments planned are discussed in Section 7.

Note that similarly to GNSSLogger, no processing of the sensor is performed in the app. It is only used for sensor data gathering, logging the measurements exactly as provided by the Android API. Any post-processing is left to the computer processing software.

3.3. Analysis Software Developments

Once the data have been logged, it needs to be processed in order to analyze and use the sensor's measurements for specific target applications; in our case, the target applications are based on accurate and/or robust positioning. The aforementioned GNSSLogger app is widely used by the research community because of its companion processing software called "GNSS Analysis App" [35], allowing plotting of the logged data. It is developed in Matlab and the source code can be found on GitHub. During the processing of the data, multiple unknown errors arose that prevented its usage for our analysis goals. Thus, similarly to how we developed the Mimir app [36], we have decided to build our own software to process the measurements.

For processing the data, we developed a Python library called "MimirAnalyzer", with its code also available in open source on Github [37]. The library allows parsing of the CSV data into a Pandas DataFrame format, to be later analyzed. For a quick review of the measurement quality, we also created a local web dashboard. It allows visualization of the trajectory, comparison with a reference and review of the measurements logged into the file. For a deeper analysis, we have created also some Jupyter Notebook templates that allow for visuals and statistical analysis. More details about the performance metrics are provided in Section 5.

3.4. Survey Protocol

Based on the sections above, we can derive the survey protocol used for the acquisition and analysis of the devices, which is summarized in Figure 2.

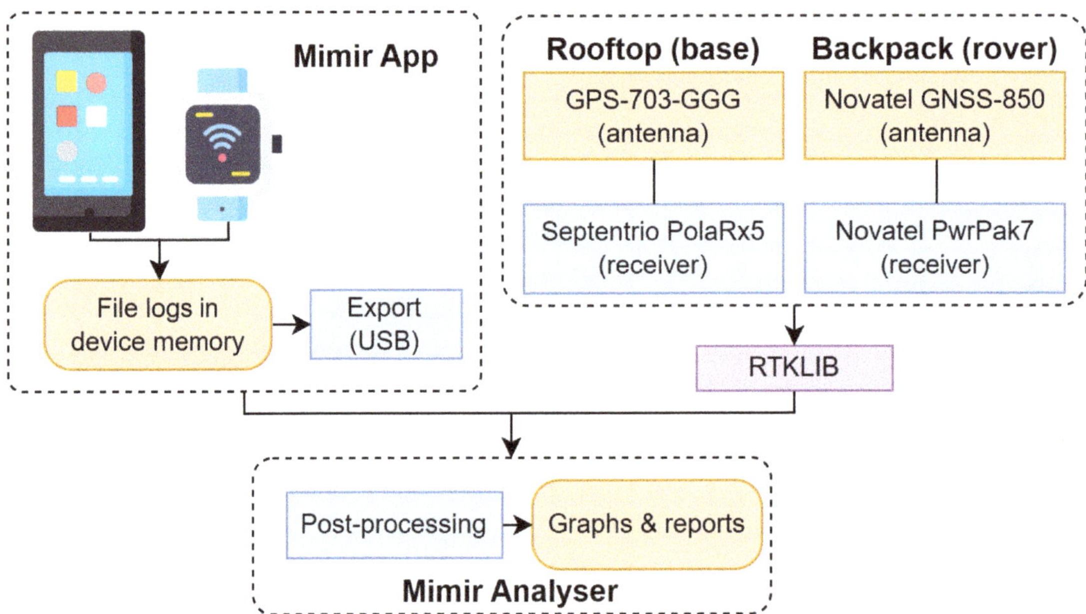

Figure 2. Processing workflow for survey acquisition and processing.

For a proper comparison of the devices' capacities, we define our reference/baseline to compare the measurements. During the process of evaluating the positioning capacities of the receivers, the most obvious benchmarking method is to find the reference coordinates of the surveyed locations. For static surveys, this can be easily defined after or before the survey by surveying the static location with a high-grade GNSS receiver. To minimize device disturbances, we used a tripod as represented in Figure 3a. For dynamic surveys, we used a backpack with an antenna mount on top as represented in Figure 3b. The used receiver was a Novatel PwrPak7 receiver connected to a Novatel GNSS-850 antenna. For the trajectory computations, we opted for Differential GNSS (DGNSS) computations, using the backpack a rover and our local GNSS station located on the university rooftop (Novatel GPS-703-GGG antenna, Septentrio PolaRx5 receiver) as a base. The total baseline between the devices ranges from a few hundred meters to a few kilometres, depending on the scenario. The DGNSS computations were performed using the open-source software RTKLib [38]; as these computations were needed only for the reference dynamic tracks, the DGNSS algorithms are not a part of our MimirAnalyzer software at present.

(**a**)   (**b**)

Figure 3. Description of the acquisition setups. (**a**) Static acquisition setup. (**b**) Backpack setup with reference receiver.

4. Dataset Description

In this section, we provide details on the dataset. We refer to it as a single "dataset" since it is provided at a single Zenodo link [7] but it is to be remembered that it contains several scenarios, both under static and dynamic conditions, and several Android devices. We start by detailing the logged sensors, then the survey scenarios will be presented and we provide a list of the performed surveys. Finally, we review the file structure of the dataset.

4.1. Scenarios and Environments

For our research purposes, we focused on pedestrian scenarios that are highly relevant in a variety of location-based applications such as studying mobility and migration patterns of workers or accessing various fundamental resources such as potable water (see LEDSOL project [6,19]), helping city planners to offer optimized services based on user location, or promoting more sustainable ways of transportation through optimized access to bikes, green lanes, or parks [39].

The university campus area and Hervanta center (a district in Tampere, Finland) offer an interesting and diverse environment for challenging positioning (i.e., open-sky, light forest, lake, suburban). While there are not many high buildings that could represent very harsh urban canyoning conditions, it is possible to recreate a partial blockage within the campus due to campus buildings (having between 3 and 10 floors). We have divided the surveys into three scenarios, all pedestrians with the conditions summarized in Table 2.

Defining the way the device is held or carried is also part of the scenario definition and crucial for future post-processing. Several studies [40,41] have shown how these carry modes determine the positioning models to use [40]. In [41], several carrying modes were assessed (e.g., texting, swinging, jacket pocket) and their effect on the inertial measurements, and consequently on the positioning solution, were reviewed.

Table 2. Summary of the scenarios.

	Motion	Environment	Duration	Device Carrying Mode
S1	Static	Open-sky	45 min	Placed on tripod
S2	Pedestrian	Urban canyoning	10 min	In right-hand, texting *
S3	Pedestrian	Suburban area	30 min	In backpack, pocket *
S4	Pedestrian	Forest & lake	30 min	In backpack, pocket *

* Smartwatch placed on left wrist, texting mode.

In our study, two carry modes were reviewed in the dataset: (1) texting, where the device was held face up in front of the user; (2) backpack pocket, where the device was put in a stood-up orientation. These modes were chosen for their practical aspects, but also to understand how positioning might be affected by partial body blockage, as highlighted in [42]. Note that the smartwatch was placed on the left wrist (except S1). A representation of the scenario's trajectories is displayed in Figure 4.

4.1.1. Static Scenario (S1)

The first scenario is a long static acquisition (45 min), performed in a near-perfect open-sky environment. We took a small light post as a visible landmark to perform the acquisition with each device separately. The devices were placed on a stable tripod as represented in Figure 3a, allowing long acquisition with minimal disturbances. The tripod landmark was surveyed using our reference receiver to obtain its precise location. While the light post is relatively wide, this setup was assessed to be suitable for the purpose of the survey, given the positioning accuracy of the devices. Moreover, the positioning of the antenna phase centre is unknown for all Android devices, so a highly precise benchmark is not required.

The receiver taken as reference for the visibility analysis in this scenario is a base station located on top of the nearest building, about 100 m northeast. Figures and statistics of the survey are provided in Appendix B.

4.1.2. Short Dynamic Scenario with Urban-Canyoning (S2)

The second scenario is a short dynamic pedestrian acquisition (10 min), performed in heavy urban canyoning. The scenario starts and ends with a short (30 s) static acquisition over the landmark previously described in S1. This was performed to allow the convergence of the position and to ensure that every available signal was correctly tracked before the dynamic part started. The trajectory went from south to north. The southern part has relatively good visibility, with only some overpass bridges between the buildings. The northern part presents a very narrow alley between the buildings, with only a quazi-zenital sky view. The data were acquired separately, device by device, and the device was always placed in the right hand of the surveyor. The GPW was placed on the left wrist and acquired at the same time as the GP7, for comparison over the same GNSS chipset model. We used the backpack setup described in Section 3 as a reference receiver for each survey. Figures and statistics of the S2 survey are provided in Appendix C.

4.1.3. Dynamic Scenario in Urban Area (S3)

The third scenario is a dynamic pedestrian acquisition (45 min), performed in a light urban area. Similarly to S2, the S3 scenario starts and ends with a short static acquisition for the usual landmark. The southern part has good open-sky conditions, contrary to the northern part, which has tall buildings around it. Strong noise can be seen in the trajectory due to a tall building with an outside ceiling, which as expected leads to noisier measurements. Finally, the part located on the road shows a small underground tunnel with possible GNSS cutoffs. The data from all the devices were acquired at the same time; all devices were placed in separate side pockets of the backpack. This allows comparison of the devices over the exact same-scenario conditions. Figures and statistics of the S3 survey are provided in Appendix D.

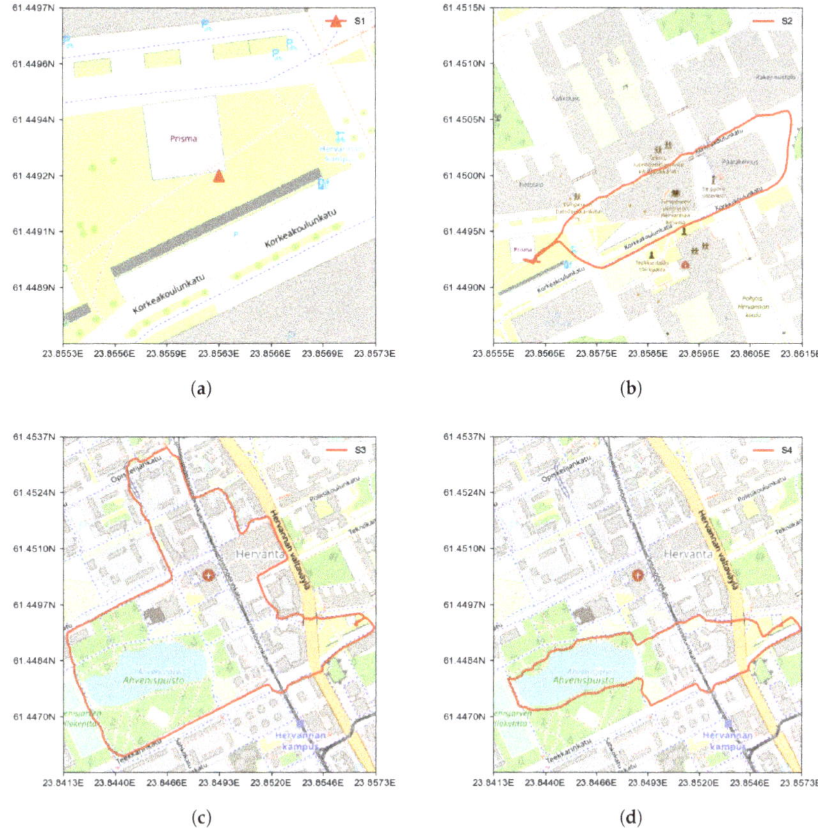

Figure 4. Maps of the scenarios' trajectories. (**a**) S1; (**b**) S2; (**c**) S3; (**d**) S4.

4.1.4. Dynamic Scenario in Forest/Lake Area (S4)

The final scenario is a dynamic pedestrian acquisition (30 min), performed in a light forest area around a lake. Similarly to S2 and S3, the S4 scenario starts and ends with a short static acquisition. The scenario was chosen to evaluate the impact of vegetation and the lake on the measurement quality. Similarly to S3, the data from all the devices were acquired at the same time and all devices were carried along in the backpack. Figures and statistics of the S4 survey are provided in Appendix E.

4.2. List of the Surveys Performed

In total, twelve surveys were performed to acquire the dataset with the devices over all four scenarios. The surveys happened on different days throughout the year, and a summary is available in Table 3. Note that S1 surveys were acquired before the completion of the Mimir app. The measurements were acquired using the GNSSLogger app for smartphones, which explains the lack of barometer data. For S1 and S2, a separate survey was performed for each device, as the device was held in the right hand of the surveyor, thus allowing for only one device per survey. For S3 and S4, all devices were acquired at the same time in different pockets located on the side of the backpack.

Table 3. Summary of the surveys performed and sensors logged.

Scenario	Date	Device	GNSS	INS	Magnetometer	Barometer
S1	17.02.2023	GP7	✓	✓	✓	✗
	03.03.2023	ON2	✓	✓	✓	✗
	03.03.2023	X11	✓	✓	✓	✗
	17.03.2023	A52	✓	✓	✓	✗
	14.08.2023	GPW	✓	✓	✓	✓
S2	01.08.2023	GP7	✓	✓	✓	✓
	01.08.2023	GPW	✓	✓	✓	✓
	01.08.2023	X11	✓	✓	✓	✓
	11.08.2023	A52	✓	✓	✓	✓
	11.08.2023	ON2	✓	✓	✓	✓
S3	11.08.2023	GP7	✓	✓	✓	✓
		ON2	✓	✓	✓	✓
		X11	✓	✓	✓	✓
		A52	✓	✓	✓	✓
		GPW	✓	✗	✗	✗
S4	11.08.2023	GP7	✓	✓	✓	✓
		ON2	✓	✓	✓	✓
		X11	✓	✓	✓	✓
		A52	✓	✓	✓	✓
		GPW	✓	✓	✓	✓

4.3. File Structure

Multiple types of files are present inside the dataset. For the reference data, we provide the observation data in Novatel proprietary log file format, and the RINEX files extracted using Novatel Application Suite software. The observations from the base station are provided in RINEX files.

For Android devices, the raw logs are produced in a simple CSV format. This format was chosen to allow compatibility with the GNSSLogger app (presented in Section 3.3). Each line of the CSV file represents the data from a sensor at a specific timestamp. While multiple clocks might be defined inside the device (e.g., GNSS clock, system clock), all entries provide two timestamps: a UTC time in Unix format based on the global system clock, and a time counter since the last time the smart device was booted. The former is used for temporal indication but it might not be very accurate, as explained in Android documentation [43], and it is recommended to use the latter for more accurate time intervals.

We note that, while it is possible to request a specific sampling frequency through the Android API, our empirical experiments have shown that the accuracy of the sampling clock can be poor. Requesting for a too high or too slow sampling interval might lead to truncated sampling, based on the physical limitation of the sensor. It is recommended to interpolate the values of a log to the wanted sampling frequency before any processing is performed. The time information discussed above should be sufficient for accurate clock alignment between the sensors and with the GNSS clock.

4.4. Sensors Summary

Multiple sensor types were logged during the surveys and are listed below:

- GNSS measurements that can be associated with multiple line entries in the log file: (1) **location measurements** (latitude, longitude, altitude) provided by the phone, from different providers; (2) **raw GNSS measurements**, as described in [11,44]; (3) **navigation messages**, as provided by the Android API and decoded by the GNSS receiver (i.e., not from another network channel).
- INS measurements, composed of two type of sensors: **accelerometers**, providing linear acceleration measurements, and **gyroscopes**, providing rotational acceleration measurements. In total, we have on each mobile device three accelerometers and three

gyroscopes, placed on the three orthogonal axes (X, Y, Z), allowing re-composition of a relative motion in three dimensions.

Additionally, an INS often include **magnetometers**, which measures the magnetic field (similar to a magnetic compass). The magnetometer enables the absolute orientation and estimation of the INS drifts [45]. Similarly to the INS measurements, three magnetometers allow the measurement of the magnetic field in three dimensions. All this information can be put together to form a low-grade INS system to be combined with GNSS measurements [1,2,45]. In the Android documentation, these sensors are regrouped under the term "motion sensors".

- Barometer measurements, related to the **atmospheric pressure** can be converted into altitude. As GNSS is known to have low precision in the up/vertical direction due to the geometry of a GNSS system, a fusion of GNSS/Barometer measurements is often seen in positioning applications [46].

All sensor entries are located in the same file, the first column containing the tag identifying the measurement type. The entry order follows the timeline of the sensor event reception. As different sensors provide different data, a header at the beginning of the files recalls the CSV columns for each sensor type. Motion sensors can also be retrieved "uncalibrated", with their biases provided but not applied. In total, there are 10 different possible tags in the file. Table A1 in Appendix A provides a summary of each logged sensor measurement; additional details can be found directly in the Android developer documentation [43].

5. Performance Metrics

For the analysis of the collected raw data, we have divided our study into three sections for each scenario: (1) positioning, assessing the native capacity of the device; (2) visibility, assessing the constellations/frequencies visible by each device; (3) measurements, assessing the raw GNSS observables' precision, which is relevant to advanced positioning applications. In this section, we explain the performance metrics chosen to evaluate these aspects in each scenario.

5.1. Positioning

The Android API give access to different location providers, with different degrees of precision: NLP, Network Location Provider, related to the mobile network position (coarse precision); GPS, as provided by the GNSS chipset (standard precision); FSL, Fused Location Provider (best precision); in this paper, we focused on the GPS provider locations only, as provided by the phone. We did not perform any post-processing of the positions, nor computed the positions using advanced positioning techniques (e.g., sensor fusion, DGNSS, PPP). Instead, for positioning quality review, we computed the position errors in the east/north/up (ENU) direction, based on the reference receiver used in each scenario (see Section 3). These errors can be reviewed in two graphs, an overall map and ENU error against time, as well as statistic tables.

5.2. Visibility

Throughout our analysis, we realized that the Android API provides an event at a regular sampling rate (i.e., 1 Hz), while the measurements per satellite are not regular. Gaps of several seconds (1–5 s) can be seen for certain satellites, leading to the idea that the device performs some kind of sequential tracking. The signal stops being tracked by the receiver for a certain amount of time to concentrate on the tracking of another signal. The receiver then comes back to the first signal to converge again, and avoids the signal falling out of the tracking loop. The most probable reason is to limit the energy consumption, and possibly to enable more signal tracking than supported by the number of channels available in hardware. This phenomenon has been seen on every device, even with a different number of total visible signals per epoch.

Today, the GNSS tracking is often limited to the four major constellations (GPS, GLONASS, Galileo, BeiDou), with dual-frequency (L1/L5) for the most advanced ones. Some Android devices do support some other satellite-based systems such as IRNSS and QZSS. However, none of the devices reviewed showed the reception of SBAS satellites. Consequently, while the geodesic-grade receivers used as a reference allowed the coverage of almost all constellations and frequency bands, we limited our comparative analysis only to the signal available in the Android devices, for a fairer comparison.

Constellations in the results are named based on their RINEX code, as defined in [47]: G (GPS), R (GLONASS), E (Galileo), C (BeiDou), I (IRNSS) and J (QZSS). Similarly, frequencies and signals are simplified to L1 ($\sim$1575.42 MHz) and L5 ($\sim$1176.45 MHz). For comparison with the reference receiver, we extracted the following signals based on RINEX definition: L1 $\Rightarrow$ (L1C, L1L), L5 $\Rightarrow$ (L5Q, L2I, L5P, L5A).

To best assess the satellite's visibility of the devices, we provide two types of graphs: the number of signals seen at each epoch w.r.t time, and bar graphs with the total amount of unique signals seen during the whole scenario. For the latter, we also add the total amount of signals seen by the reference receiver (the left bar) to compare against each device (the right bar). The bar graphs come in two variants: one per constellation and one per frequency. We also provide a summary table comparing the percentage of visible signals compared to the reference receiver.

5.3. Measurements

Four observations are used in GNSS positioning: pseudorange, carrier phase, Doppler shift, and SNR. The last two are provided straightforwardly in the GNSS event of the Android API; however, the pseudoranges and the carrier-phase data require some post-processing to be usable in a typical GNSS algorithm.

As described in [44], the pseudorange needs to be constructed based on the received and transmitted times given in the event. Several checks of the tracking state should be performed for each constellation to ensure correct measurements. In our analysis, we used the instructions provided by [44], but we only checked for adequate code lock and Time of Week (TOW) decoded. The carrier phases in Android devices are provided as "Accumulated Delta Ranges", which correspond to the regular carrier-phase measurements multiplied by the wavelength of the GNSS signal, thus in meter units and not in cycle units.

For both the pseudoranges and the carrier phases, we performed a double time difference on the measurements, similarly to [10]. The double time difference allows for discarding the residual drift induced by the velocity changes, leaving only the acceleration changes which are assumed too small for both satellites and users in our study cases. It also reduces or cancels out some errors present in the GNSS measurements, such as atmospheric error or satellite orbit/clock error, as two consecutive measurements at a regular small rate (here 1 Hz) should be tainted by similar errors. Consequently, the error left can be assumed to be related mostly to the receiver tracking noise. While other papers have used different error estimations for measurements, such as the Code–Minus–Carrier combination [11,12], the double time difference method showed adequate results for a preliminary analysis. For Doppler measurements, only one time difference is needed as the measurement is already a velocity/rate estimation. For C/N0, it is more interesting to study its value range; thus, there is no difference in performance.

The results are provided as logged, with minimum filtering of the data for a more realistic analysis. However, to minimize the number of outliers and end up with biased statistics, we filtered some of the measurements based on a threshold, as a basic navigation filter would do. The threshold values were chosen empirically, based on the standard deviation and expected precision of each type of data. Consequently, the thresholds were set as follows: pseudorange $\Rightarrow$ 300 m, Doppler $\Rightarrow$ 30 m and phase $\Rightarrow$ 30 m.

For each scenario, a statistic table for each GNSS measurement is provided, along with the percentage of measurements included/removed from the filter. For pseudoranges, Doppler and phase errors, we used box plot graphs to represent the different statistics.

For Carrier-to-Noise ratio (C/N0), we choose to represent the data in "violin plots", which are mostly similar to box plots but provide additional information on the data distribution. It is specifically interesting to study the L1/L5 differences and the range of CN0 values.

6. Results

In this section, we review each scenario and provide a first analysis of the data. We do not review in detail the content of each figure/table provided in the appendices, as they can be redundant for many scenarios. Instead, we summarize the observations from all scenarios in three sections related to positioning, visibility and measurement analysis, and only show the relevant figures within this section. A comprehensive summary of the performance metrics is available for each scenario in Appendix B (S1), Appendix C (S2), Appendix D (S3) and Appendix E (S4). For more details on the data, the results presented in this study are accessible in the Jupyter Notebook file available on MimirAnalyzer's Github.

6.1. Positioning Analysis

All the devices showed signs of heavy filtering (e.g., Kalman Filter) inside their proprietary navigation algorithms. The locations provided by the GNSS chipset inside each Android device followed a stringent user dynamic, which allowed a very smooth trajectory. This heavy filtering is even more apparent during the static survey (S1). In Figures 5 and 6, the GP7 and GPW seem to be converging around the static position, with a looser prediction model. However, other devices such as the ON2 and X11 report a constant position throughout the whole survey, which explains the very small standard deviation. This can probably be explained by a more highly constrained motion model based on user dynamic assessment performed by the receiver internally.

During the dynamic surveys, all the devices provided consistently good positioning with relatively small divergence from the ground truth and within each other. Certain discrepancies can, however, be found with the A52 device during S3 (Figure 7) and S4 (Figure 8), leading to wrong positioning before quickly re-converging to the actual user position. Yet, this was not seen during S2, with a much more challenging signal environment, making it hard to assess the actual cause for such drift in positioning.

6.2. Visibility Analysis

Overall, dual-frequency devices showed great visibility performances, capturing most of the signals compared to the receiver. In certain cases, some devices have even reported more satellites than the reference receiver. In these cases, the satellites were actually flagged as unhealthy and were not included by the reference receiver, contrary to the devices that included any measurement received in the log file. As the positioning solution presented here is computed by the device itself, not all the visible satellites in the raw data are likely to be used by the devices and this is a topic of future investigation; in any case, the effect on navigation should be minimal if not in a highly constrained environment.

The visibility analysis would benefit from additional comparisons that should be performed to compare the C/N0 levels against satellites' elevation w.r.t. the reference receiver. Mono-frequency devices showed very good results in capturing the L1 frequency. Yet, given that all the major constellations today transmit L5 signals, it significantly reduces the amount the total amount of signals seen by the device. An interesting case is the GPW smartwatch, as it should integrate the same GNSS chipset as the GP7 smartphone, but did not show any dual-frequency measurements contrary to GP7. Nevertheless, comparing positioning results between GPW and GP7 showed similar results in all scenario cases.

The most remarkable result seen in smart devices is the capacity to alternate ON/OFF tracking of signals. It allows more signals to be tracked concurrently and optimizes power consumption. Concurrent tracking is not a new concept, as it was used in the very first GNSS receiver with limited hardware channel, and probably is still used today in other receivers. Yet, the effects are directly visible here in the measurements.

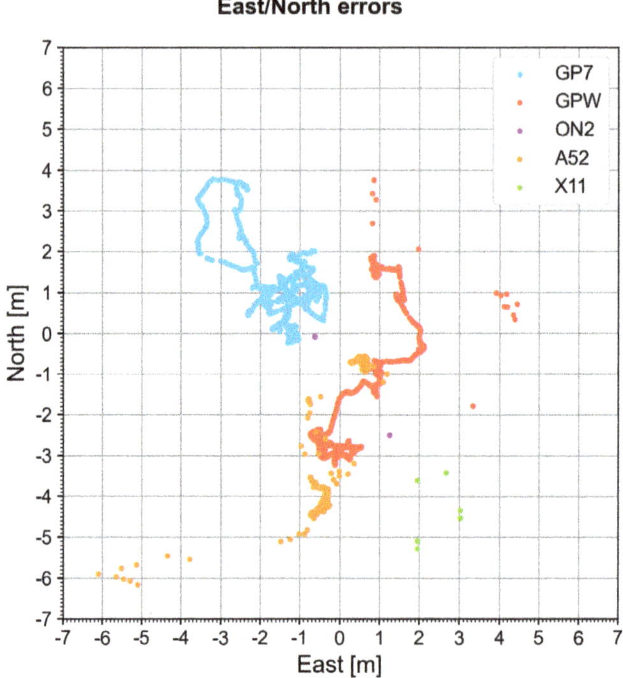

Figure 5. Two−dimensional error in East and North (S1).

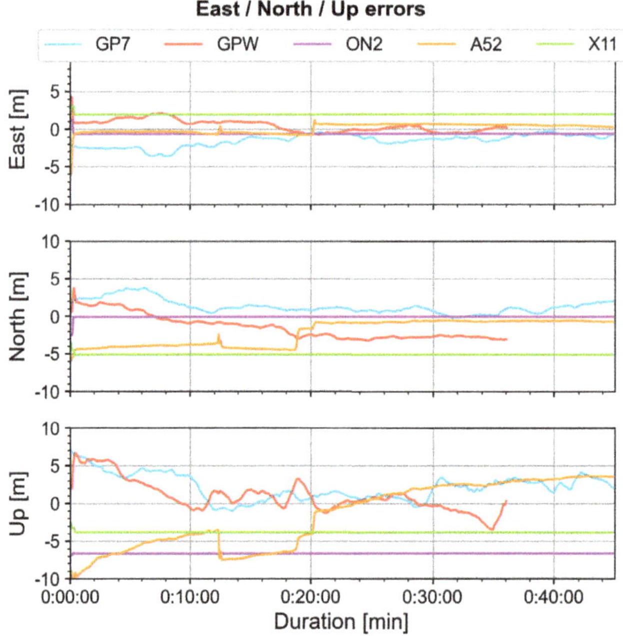

Figure 6. East−North−Up error (S1).

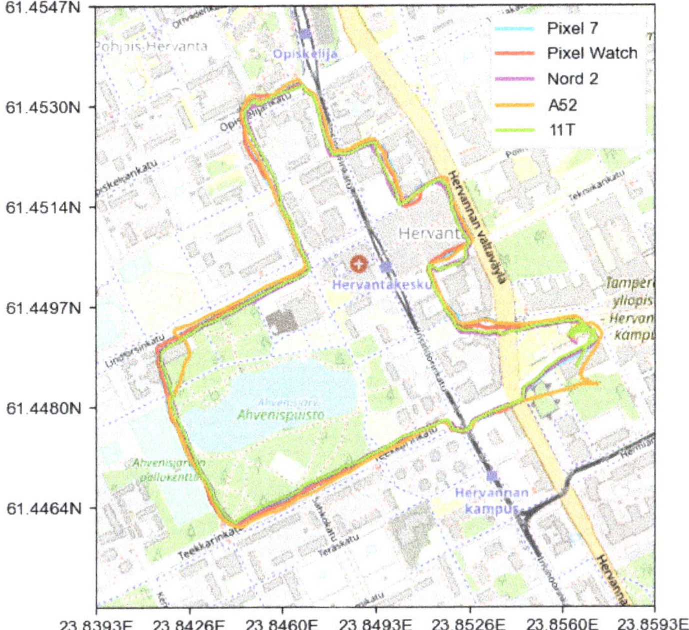

Figure 7. Two–dimensional error in east/north (S3).

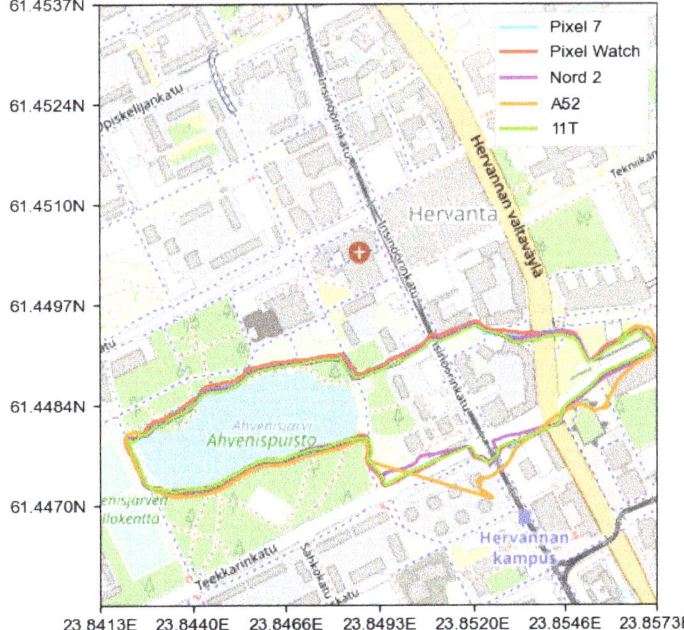

Figure 8. Two–dimensional error in east/north (S4).

The downside of concurrent tracking is it results in many gaps in the measurements. To overcome this issue, one could perform interpolation of the measurements to have a regular rate, as the gaps are relatively short. In our processing, we did not perform any interpolation as it was not needed for our presented analysis. Advanced processing techniques, such as PPP, might indeed require adaptation to their algorithms to account for such measurement discrepancies and should be reviewed for future research.

6.3. Measurements Analysis

All measurement errors were found to be very consistent throughout all the scenarios, with higher noise during dynamic scenarios compared to the static ones. As mentioned before, measurements were filtered to avoid biases in statistics due to outliers. Yet, a more thorough analysis of these outliers would be interesting to better understand the resilience of devices in different environments. Moreover, Android provides a certain amount of flags on code and phase tracking quality, as well as multipath flags, directly within the API. These flags have not been used in this analysis but should be checked for more advanced positioning computations, as assessed in previous studies [48,49].

Pseudoranges errors from certain devices (i.e., ON2 and X11) appeared to have a much smaller dispersion. A possible explanation could be pseudorange smoothing using Doppler or phase measurements, as this technique is often used to reduce pseudorange noise [9].

Regarding Doppler measurements, a relatively small standard deviation was observed for GP7 and GPW, compared to the other devices. Both Doppler and phase measurements showed a tendency for a small positive, regardless of the devices. While the exact reason is not clear, it is believed to be due to the time-differencing operations performed to obtain the error measurements, as described in Section 5.3. Future research should review the effects on measurements, possibly using other techniques such as the Code–Minus–Carrier combination (see Section 5.3).

C/N0 comparison showed a distinct drop in their values between L1 and L5 tracking in all devices. Yet, L5 signals are designed with a higher minimum received power [50], which shows that smart device hardware is not optimized yet for the reception of this additional frequency. Further comparison between frequencies should be performed to further assess the impact on the measurement quality.

7. Discussion and Future Work

7.1. A Suitable Reference Definition

During the processing of dynamic surveys, we realized that the reference trajectory actually looked noisier than the one from the devices. The reference trajectory was computed in DGNSS baser/rover setup, as described in Section 3 and DGNSS computations were performed using the open-source software RTKlib. Even with DGNSS, the errors of the reference rise significantly in challenging environments, due to the limited number of satellites visible and multipath errors. Consequently, the statistics tables of position errors presented in the appendices need to be reviewed with this consideration in mind.

Nevertheless, we do not attribute such error to the receiver, but to the processing itself. As navigation algorithms inside Android devices are proprietary, it is difficult to assess the filtering algorithms and parameters. Consequently, simple DGNSS processing as performed in RTKlib is not sufficient and more advanced processing should probably be performed on the reference data to better assess the positioning capacities of each device. In our next research developments, we are planning to integrate the positioning processing (i.e., Kalman Filter) within the MimirAnalyser software to improve the reference trajectory processing.

Moreover, the dataset would benefit from an INS coupled with the reference receiver setup. It would enable better processing for the reference trajectory, and the INS measurements could be compared against the INS recordings from the smart devices. However, INS methods require alignments of the different devices' frames, which could prove challenging. It is the main reason why using a combined GNSS/INS solution for the reference trajectories was left out of this analysis. Future work should look into the feasibility of enhancing the reference trajectories with INS measurements.

7.2. The Android Platform for Research

Sensor logging and measurement quality assessment on Android come with a set of challenges. During our research, we discovered that while there are numerous apps existing today for sensor logging, there were none suitable to log every sensor available on smartphones/smartwatches. Apps related to GNSS logging (e.g., GNSSlogger) only concentrate on motion-related sensors and do not allow customization on the sensor parameters. Therefore, we had to develop a new application to fit our research needs, which required a certain knowledge of Android development. As this app is open-source, it can be re-used and enhanced by the research community to be used in new applications. Alternatively, the dataset produced through this research can be used as also provided under open-source license.

While the Android platform is not easy to access, it does allow common and easy access to raw measurements from sensors, which would require heavier hardware developments otherwise. Power management inside the device can prove challenging, as the system itself might perform automatic duty-cycling strategies which can impact sensor logging, which can be even trickier on smartwatches. An easy workaround for Mimir was to prevent screen lock and keep the app awake to avoid duty-cycling by the system. Yet, the assessment of power consumption is part of the future work for this research to review the impact of advanced positioning techniques on the battery life of a device.

Regarding the smartwatch, developments have proven to be quite similar to the smartphone case. The common API made the development of a common Android library possible, avoiding the need for redundant code developments specific to each platform. Data extraction was, however, a bit more challenging, as the smartwatch did not have any USB port. Consequently, the data had to be extracted through WiFi and with the help of the Android development software. Future developments on the Mimir app will try to improve the interactivity between smartphone and smartwatch so that the data can be uploaded through a companion phone first before being extracted by USB.

While the smartwatch offers a new playground for health-related sensors, they have not been reviewed here as outside the scope of this study. It has to be noted, however, that extraction of these sensor data is implemented in the Mimir app, and future research will also focus on reviewing their quality. As smartwatches can be very interesting for long-term study of patients, focus will be given on enhancing the energy-consumption of the app for extending the battery-life of the device.

7.3. Android Devices and Scenario Impact

Analysis of a different scenario showed a clear impact on the measurements and positioning quality. S2 was expected to give the largest errors, given the challenging nature of the dense urban canyoning. Yet, the scenario that registered the highest errors is S3. While it also contained challenging sections, the highest error values could be due to the longer survey time, which increases the probability of erroneous measurements in the dataset. Despite the increase in measurement noises, the positioning results stayed consistent for all devices throughout all scenarios and regardless of the device platform (i.e., smartphone/smartwatch). This supports our hypothesis of using smartwatches for research applications in positioning, and possibly other topics (e.g., health applications).

Devices have shown clear positioning abilities to maintain a suitable positioning solution throughout the scenarios, yet with challenges for some devices (see Section 7.4). These discrepancies can partially be explained due to the environment and possible unaccounted multipath, yet would require a deeper analysis to better understand their exact origin. While we focus mainly on the outdoors for this study due to GNSS positioning, the additional indoor scenarios would be interesting to review in a future analysis. More specifically, the transition from outdoor to indoor and/or light indoor environment (e.g., glass ceiling) would be beneficial to assess the sensitivity of the antenna/receiver embedded in smart devices.

Similarly, future research should provide deeper attention to the carry mode effects on the device measurements. While we only performed acquisition texting and pocket mode, new modes like swinging should be considered, especially for the smartwatch review.

7.4. Positioning with Smart Devices

In the results presented, we have seen that all smart devices showed reliability in providing a precise positioning regardless of the environment. While an increase in measurement noise could be seen in more challenging environments, such as S2, the devices maintained a suitable positioning accuracy. This is also the case for C/N0 values, which remained at a similar level throughout the surveys and do not seem to be a real determining factor in this case for positioning precision.

Yet, a comparison of the raw measurement errors between the devices showed large differences from one device to another. With more advanced positioning techniques, this needs to be taken into account, as not all devices might be capable of performing similarly. The most evident discrepancies in measurements seen in our analyses were access to dual-frequency measurements, retrieval of carrier-phase measurements and pseudorange smoothing. Moreover, GP7 and PGW provided interesting comparison points as they had the same GNSS chipset model embedded, which showed that the platform has an impact on the measurements as well. The clearest example is the absence of L5 signals on the GPW.

Another limiting factor in Android devices remains the missing characterizations of the hardware, such as the antenna used for reception of GNSS signal, as already pointed out in [11]. While access to antenna information has been added in the Android API, the information provided by the manufacturer is slim or non-existent. For advanced positioning processing, one has to rely on self-characterization of the hardware.

A similar conclusion can be drawn for INS-related sensors, such as accelerometers and gyroscopes. They have not been reviewed in the analysis even though they are present in the dataset, as we preferred to focus on GNSS capacities, but will be part of the future analysis.

8. Conclusions

In this paper, we presented a new open-source dataset for research on advanced positioning using Android smart devices, with a first review of and the first app for GNSS raw data on a smartwatch platform. We detailed our methodology for creating the dataset by describing the logging software development, the protocol definition and file structure in detail. We also provided a first analysis of the results by comparing five devices (four Android smartphones and a smartwatch) in terms of several performance metrics. It is to be reminded that the paper's focus and main novelty has been to introduce the methodology of the acquisition protocol and the new open-access app and datasets. The results are just examples of what can be achieved with the developed open-access tools, but they do not provide a comprehensive selection of performance metrics; for example, algorithms for user positioning starting from the acquired raw data are still to be implemented and the multi-system multi-frequency GNSS analysis is still a topic for future research. Nevertheless, our results have shown that GNSS raw data collected from mass-market devices is of good quality, with similar results between the smartwatch and smartphones. This strengthens the possibility of using smartwatch sensor data for broader research topics. It can enable significant research in the area, provided that suitable acquisition and processing tools are made available to the research community, such as our developed software "Mimir". Last but not least, we have also reviewed the current challenges associated with Android data processing and highlighted the potential research opportunities for the continuity of this work.

Author Contributions: Conceptualization, A.G. and E.S.L.; methodology, A.G.; software, A.G.; validation, A.G. and E.S.L.; formal analysis, A.G.; investigation, A.G.; resources, A.G.; data curation, A.G.; writing—original draft preparation, A.G. and E.S.L.; writing—review and editing, A.G., E.S.L., A.O. and J.N.; visualization, A.G.; supervision, E.S.L., A.O. and J.N.; project administration, A.G.; funding acquisition, E.S.L., A.O. and J.N. All authors have read and agreed to the published version of the manuscript.

Funding: This research was funded by European Union's Horizon 2020 Research and Innovation Programme under the Marie Skłodowska Curie grant agreement number 956090 (APROPOS: Approximate Computing for Power and Energy Optimisation, http://www.apropos-itn.eu/ (accessed on 21 November 2023)), by the LEAP-RE programme of the European Union's Horizon 2020 Research and Innovation Program, grant agreement number 963530 (LEDSOL project) and the Academy of Finland, project number 352364.

Data Availability Statement: The data presented is published under CC-BY-4.0 on Zenodo repository (https://zenodo.org/records/8340005 (accessed on 21 November 2023). Software developed are published under Apache 2.0 open-source license on GitHub (Mimir App: https://github.com/agrenier-gnss/mimir (accessed on 21 November 2023), Mimir Analyzer: https://github.com/agrenier-gnss/MimirAnalyzer (accessed on 21 November 2023).

Acknowledgments: We would like to thank the following students at Tampere University who have helped with the software developments and with preliminary work on the Android devices: Silja Nahkala, Heini Vesaranta, Henry Andersson, Petrus Jussila, Salla Rouhiainen, Severi Ruusumaa, Ha Chu, and My Nguyen.

Conflicts of Interest: The authors declare no conflict of interest.

Abbreviations

The following abbreviations are used in this manuscript:

API	Application Programming Interface
BLE	Bluetooth Low Energy
CN0	Carrier-to-Noise ratio
CSV	Comma Separated Value
DGNSS	Differential GNSS
ECG	Electrocardiogram
GNSS	Global Navigation Satellite System
ECG	Electrocardiogram
SpO2	Oxygen Saturation
PPG	photoplethysmogram
GSR	Galvanic Sensor Response
INS	Inertial Navigation System
SNR	Signal-to-Noise Ratio
TOW	Time of Week
WiFi	Wireless Fidelity

Appendix A. Android Sensor Measurements

Details on the each entry in files logged by the Mimir app. For details on the measurements content, please refer to the Android documentation [43].

Table A1. Mimir log entries related to GNSS sensors.

Tag	Measurement	Description	Unit
Fix	provider	Origin of the location provided (e.g., GPS, NLP, FLP)	
	latitude	Geodetic latitude (WGS84)	[Dec. Deg.]
	longitude	Geodetic longitude (WGS84)	[Dec. Deg.]
	altitude	Geodetic altitude (WGS84)	[m]
	speed	User velocity	[m/s]
	accuracy	Horizontal uncertainty (1-σ)	[m]
	bearing	Horizontal direction	[Deg.]
	time	UTC time in UNIX	[ms]
	speedAccuracyMetersPerSecond	User velocity uncertainty (1-σ)	[m/s]
	bearingAccuracyDegrees	Horizontal direction velocity (1-σ)	[Deg.]
	elapsedRealtimeNanos	Time since system boot	[ns]
	verticalAccuracyMeters	Vertical uncertainty (1-σ)	[m]
	elapsedRealtimeUncertaintyNanos	Time since system boot uncertainty (1-σ)	[ns]
Raw	utcTimeMillis	UTC time in UNIX	[ms]
	timeNanos	Internal clock from GNSS hardware receiver	[ns]
	leapSecond	Number of leap seconds w.r.t. provided clock	[s]
	timeUncertaintyNanos	Internal clock uncertainty (1-σ)	[ns]
	fullBiasNanos	Full bias between clock and true GPS time	[ns]
	biasNanos	Partial clock bias	[ns]
	biasUncertaintyNanos	Partial clock bias uncertainty (1-σ)	[ns]
	driftNanosPerSecond	Clock drift	[ns/s]
	driftUncertaintyNanosPerSecond	Clock drift uncertainty (1-σ)	[ns/s]
	hardwareClockDiscontinuityCount	Counts of hardware discontinuities	
	svid	Satellite ID	
	timeOffsetNanos	Time offset of the measurements	[ns]
	state	Current tracking state of the signal	
	receivedSvTimeNanos	Received satellite time at measurement time	[ns]
	receivedSvTimeUncertaintyNanos	Received time uncertainty (1-σ)	[ns]
	cnODbHz	Carrier-to-Noise ratio	[dB-Hz]
	pseudorangeRateMetersPerSecond	Pseudorange rate, i.e., Doppler shift	[m/s]
	pseudorangeRateUncertaintyMetersPerSecond	Pseudorange rate uncertainty	[m/s]
	accumulatedDeltaRangeState	Carrier tracking state	
	accumulatedDeltaRangeMeters	Accumulated pseudorange rate, i.e., carrier phase	[m/s]
	accumulatedDeltaRangeUncertaintyMeters	Accumulated pseudorange rate uncertainty	[m/s]
	carrierFrequencyHz	GNSS carrier frequency	[Hz]
	carrierCycles	Number of carrier phase cycles (deprecated)	
	carrierPhase	Carrier phase (deprecated)	
	carrierPhaseUncertainty	Carrier phase uncertainty (deprecated)	
	multipathIndicator	Multipath flag	[Boolean]
	snrInDb	Signal-to-Noise ratio	[dB-Hz]
	constellationType	Constellation ID	
	automaticGainControlLevelDb	Current Automatic Gain Control	[dB]
	basebandCnODbHz	Baseband Carrier-to-Noise ratio	[dB-Hz]
	fullInterSignalBiasNanos	Full GNSS inter-signal bias	[ns]
	fullInterSignalBiasUncertaintyNanos	Full GNSS inter-signal bias uncertainty (1-σ)	[ns]
	satelliteInterSignalBiasNanos	Partial GNSS inter-signal bias	[ns]
	satelliteInterSignalBiasUncertainty	Partial GNSS inter-signal bias uncertainty (1-σ)	[ns]
	codeType	RINEX code type	
	elapsedRealtimeNanos	Time since system boot (1-σ)	[ns]
Nav	utcTimeMillis	UTC time in UNIX	[ms]
	svid	Satellite ID	
	type	Navigation message type	
	status	Parity check status	
	messageId	Message frame ID	
	submessageId	Message sub-frame ID	
	data	Byte array of navigation message	

Table A2. Mimir log entry related to INS sensors. All axis are related to the device frame.

Tag	Measurement	Description	Unit
ACC	x_meterPerSecond2	X-axis acceleration	$[m/s^2]$
	y_meterPerSecond2	Y-axis acceleration	$[m/s^2]$
	z_meterPerSecond2	Z-axis acceleration	$[m/s^2]$
	accuracy	Android accuracy classification	
GYR	x_radPerSecond	X-axis rotation	$[rad/s]$
	y_radPerSecond	Y-axis rotation	$[rad/s]$
	z_radPerSecond	Z-axis rotation	$[rad/s]$
	accuracy	Android accuracy classification	
MAG	x_microTesla	X-axis magnetic field	$[\mu Tesla]$
	y_microTesla	Y-axis magnetic field	$[\mu Tesla]$
	z_microTesla	Z-axis magnetic field	$[\mu Tesla]$
	accuracy	Android accuracy classification	
ACC_UNCAL	x_uncalibrated_meterPerSecond2	Raw X-axis acceleration	$[m/s^2]$
	y_uncalibrated_meterPerSecond2	Raw Y-axis acceleration	$[m/s^2]$
	z_uncalibrated_meterPerSecond2	Raw Z-axis acceleration	$[m/s^2]$
	x_bias_meterPerSecond2	Compensated X-axis acceleration	$[m/s^2]$
	y_bias_meterPerSecond2	Compensated Y-axis acceleration	$[m/s^2]$
	z_bias_meterPerSecond2	Compensated Z-axis acceleration	$[m/s^2]$
	accuracy	Android accuracy classification	
GYR_UNCAL	x_uncalibrated_radPerSecond	Raw X-axis rotation	$[rad/s]$
	y_uncalibrated_radPerSecond	Raw Y-axis rotation	$[rad/s]$
	z_uncalibrated_radPerSecond	Raw Z-axis rotation	$[rad/s]$
	x_bias_radPerSecond	Compensated X-axis rotation	$[rad/s]$
	y_bias_radPerSecond	Compensated Y-axis rotation	$[rad/s]$
	z_bias_radPerSecond	Compensated Z-axis rotation	$[rad/s]$
	accuracy	Android accuracy classification	
MAG_UNCAL	x_uncalibrated_microTesla	Raw X-axis magnetic field	$[\mu Tesla]$
	y_uncalibrated_microTesla	Raw Y-axis magnetic field	$[\mu Tesla]$
	z_uncalibrated_microTesla	Raw Z-axis magnetic field	$[\mu Tesla]$
	x_bias_microTesla	Compensated X-axis magnetic field	$[\mu Tesla]$
	y_bias_microTesla	Compensated Y-axis magnetic field	$[\mu Tesla]$
	z_bias_microTesla	Compensated Z-axis magnetic field	$[\mu Tesla]$
	accuracy	Android accuracy classifications	

Table A3. Mimir log entry related to environment sensors.

Tag	Measurement	Description	Unit
PSR	pressure_hPa	Ambient air pressure	[hPa or mBar]
	accuracy	Android accuracy classification	

Appendix B. Scenario 1—Static Acquisition in Open-Sky Environment

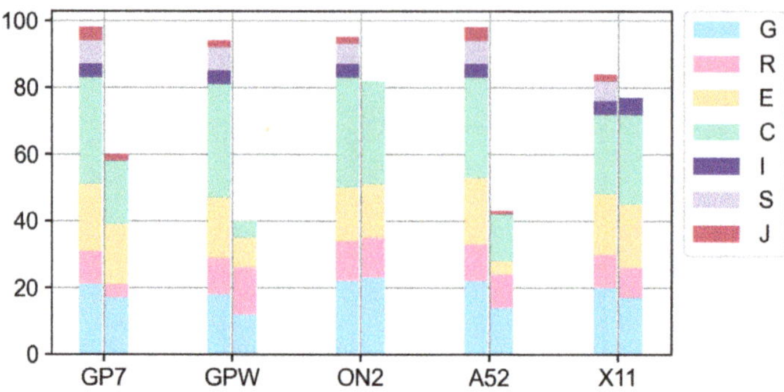

Figure A1. Signals seen per system (S1) (left: reference, right: device).

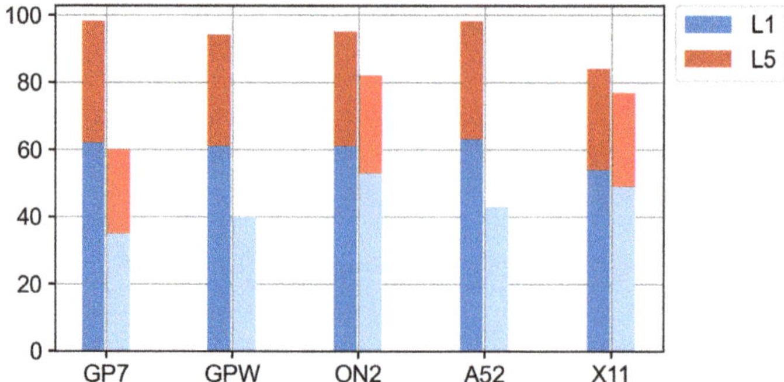

Figure A2. Signals seen per frequency (S1) (left: reference, right: device).

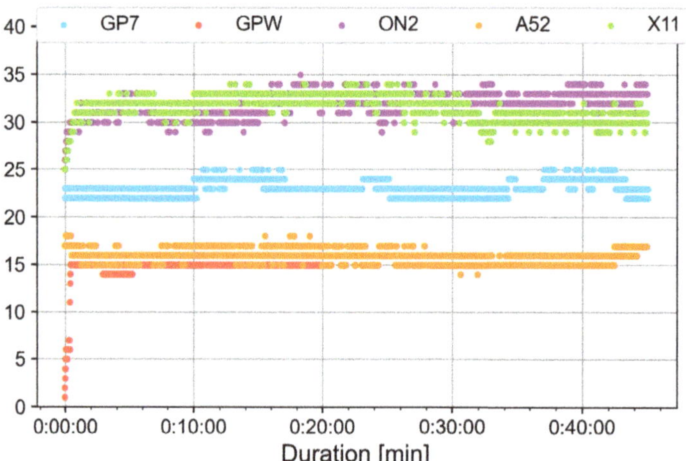

Figure A3. Signals seen per epoch (S1).

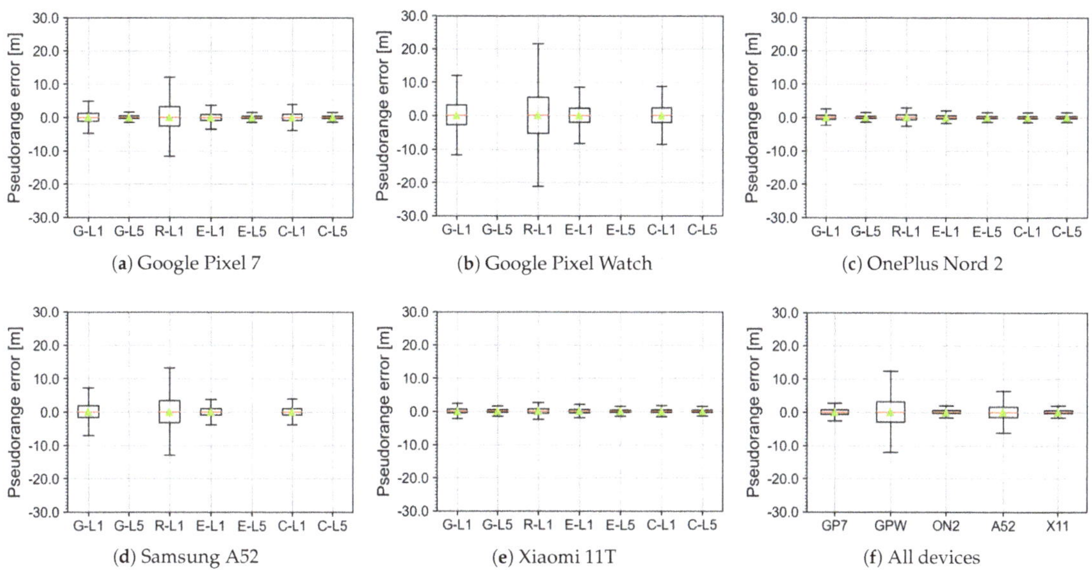

(**a**) Google Pixel 7   (**b**) Google Pixel Watch   (**c**) OnePlus Nord 2

(**d**) Samsung A52   (**e**) Xiaomi 11T   (**f**) All devices

Figure A4. Pseudorange errors (S1), with outlier threshold of 300 m. See Table A6 for details.

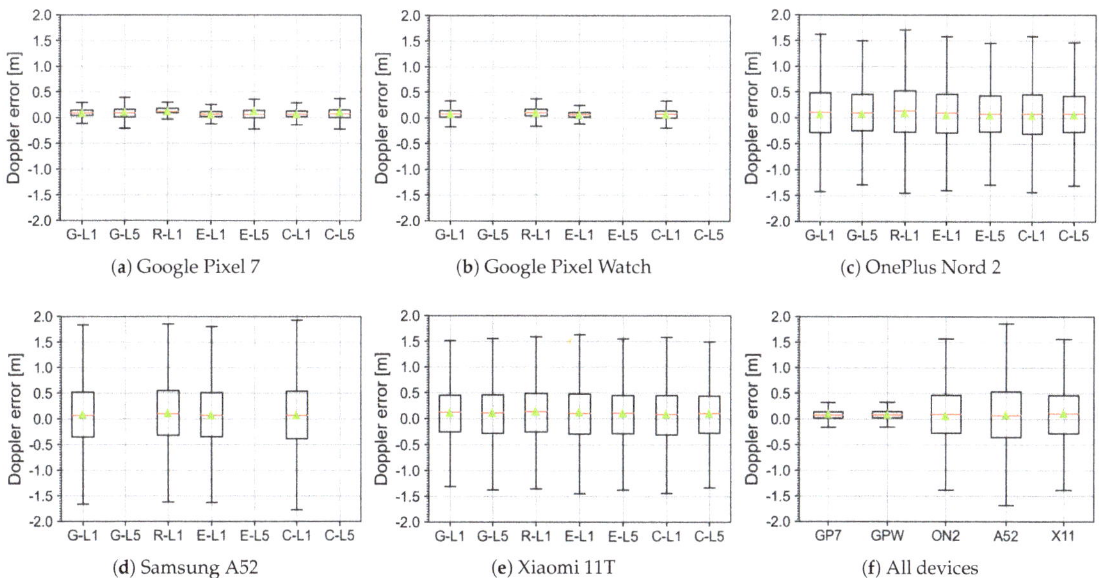

(**a**) Google Pixel 7   (**b**) Google Pixel Watch   (**c**) OnePlus Nord 2

(**d**) Samsung A52   (**e**) Xiaomi 11T   (**f**) All devices

Figure A5. Doppler errors (S1), with outlier threshold of 30 m. See Table A7 for details.

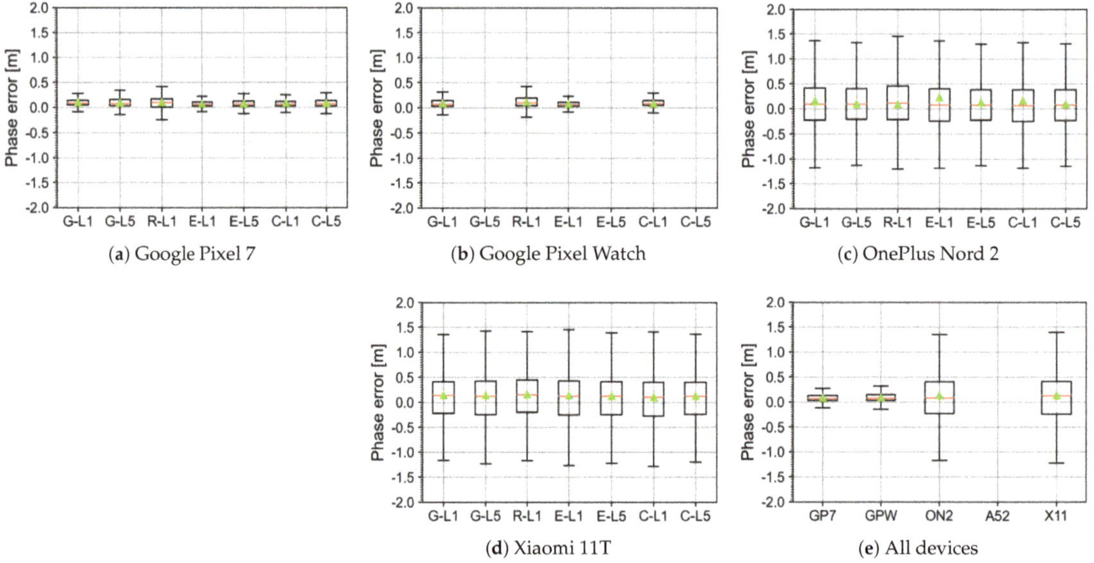

Figure A6. Phase errors (S1), with outlier threshold of 30 m. See Table A8 for details.

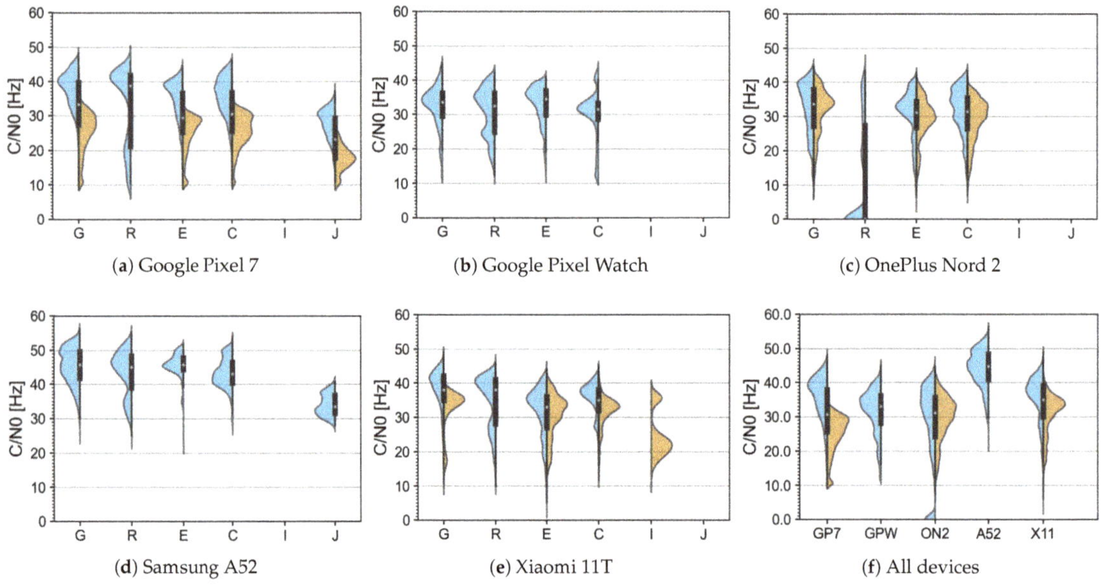

Figure A7. Carrier-to-Noise ratio (C/N0) (S1) [Blue: L1, Orange: L5]. See Table A9 for details.

Table A4. Statistics for position errors (S1).

| Device | Inc. [%] | Mean ± StD (1-σ) [m] | | | RMSE [m] | |
		East	North	Up	2D	3D
GP7	100.00	-1.533 ± 0.815	1.338 ± 0.934	2.298 ± 1.747	2.382	3.742
GPW	100.00	0.513 ± 0.839	-1.396 ± 1.614	1.021 ± 2.117	2.349	3.323
ON2	100.00	-0.596 ± 0.160	-0.097 ± 0.207	-6.612 ± 0.018	0.658	6.645
A52	100.00	0.076 ± 0.685	-2.164 ± 1.681	-1.595 ± 4.312	2.825	5.395
X11	100.00	1.976 ± 0.086	-5.086 ± 0.096	-3.803 ± 0.087	5.457	6.652

Table A5. Signals' percentages visibility w.r.t. the base receiver (S1).

| Device | Freq. [%] | | Constellations [%] | | | | | |
	L1	L5	G	R	E	C	I	J
GP7	56.5	69.4	81.0	40.0	90.0	59.4	—	50.0
GPW	65.6	0.0	66.7	127.3	50.0	14.7	—	—
ON2	86.9	85.3	104.5	100.0	100.0	93.9	—	—
A52	68.3	0.0	63.6	90.9	20.0	46.7	—	25.0
X11	90.7	93.3	85.0	90.0	105.6	112.5	125.0	—

Table A6. Statistics for pseudorange errors [m] (S1).

	Device	Inc. [%]	Mean [m]	StD [m]	Min. [m]	Max. [m]
Pseudorange	GP7	99.90	0.061	1.736	-108.757	108.773
	GPW	99.93	0.058	0.456	-16.099	22.945
	ON2	100.00	0.059	1.545	-50.682	50.699
	A52	99.98	0.072	1.643	-72.456	87.806
	X11	99.96	0.063	0.482	-23.339	23.511

Table A7. Statistics for Doppler errors [m] (S1).

	Device	Inc. [%]	Mean [m]	StD [m]	Min. [m]	Max. [m]
Doppler	GP7	99.97%	0.073	0.252	-17.607	25.736
	GPW	100.00%	0.083	0.160	-4.161	2.966
	ON2	100.00%	0.069	1.308	-18.050	15.844
	A52	100.00%	0.086	1.088	-17.533	14.831
	X11	100.00%	0.112	0.910	-26.402	19.542

Table A8. Statistics for carrier phase errors [m] (S1).

	Device	Inc. [%]	Mean [m]	StD [m]	Min. [m]	Max. [m]
Phase	GP7	99.50%	0.093	0.748	-29.155	29.441
	GPW	99.98%	0.098	0.327	-11.922	11.616
	ON2	99.09%	0.100	1.386	-28.982	29.283
	A52	—	—	—	—	—
	X11	99.45%	0.136	1.092	-29.632	29.967

Table A9. Statistics for C/n0 errors [dB-Hz] (S1).

	Device	Inc. [%]	Mean [dB]	StD [dB]	Min. [dB]	Max. [dB]
C/n0	GP7	100.00%	30.8	8.0	10.6	48.2
	GPW	100.00%	31.5	6.4	12.1	44.9
	ON2	100.00%	28.2	10.6	—	44.7
	A52	100.00%	44.1	5.4	21.4	56.3
	X11	100.00%	33.8	6.7	3.0	49.0

Appendix C. Scenario 2—Pedestrian Dynamic in Urban Canyoning Environment

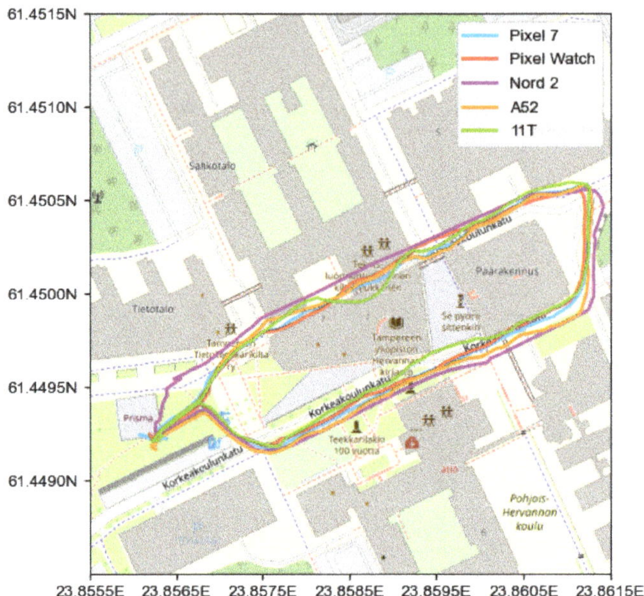

Figure A8. Two-dimensional error in east/north (S2).

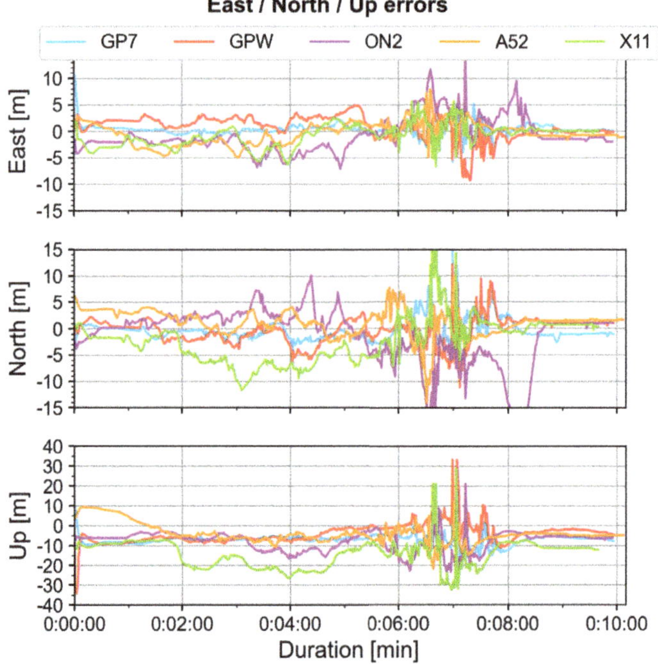

Figure A9. East/north/up error (S2).

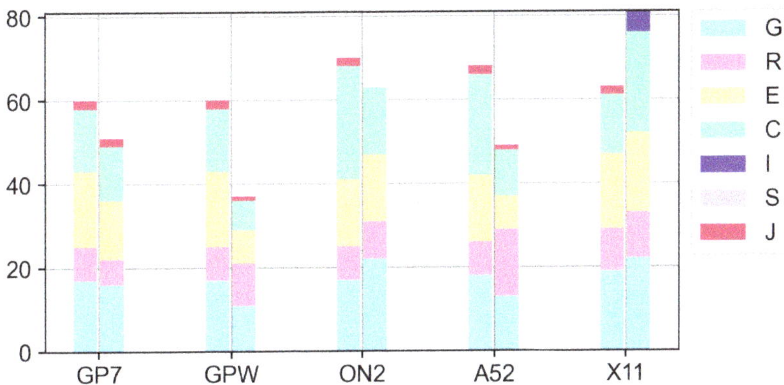

Figure A10. Signals seen per system (S2) (left: reference, right: device).

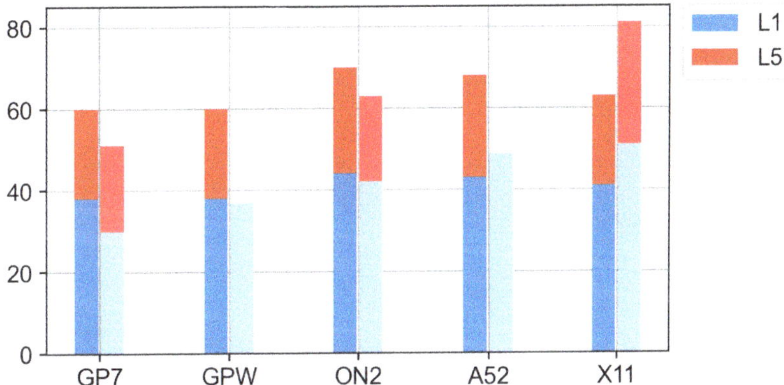

Figure A11. Signals seen per frequency (S2) (left: reference, right: device).

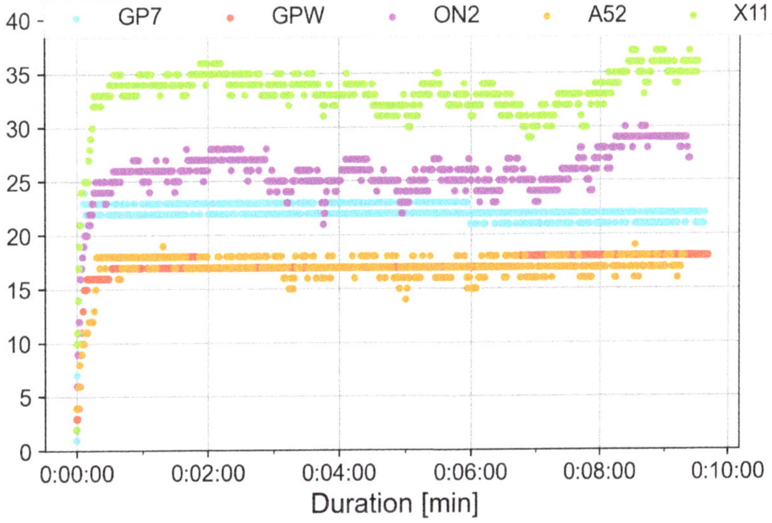

Figure A12. Signals seen per epoch (S2).

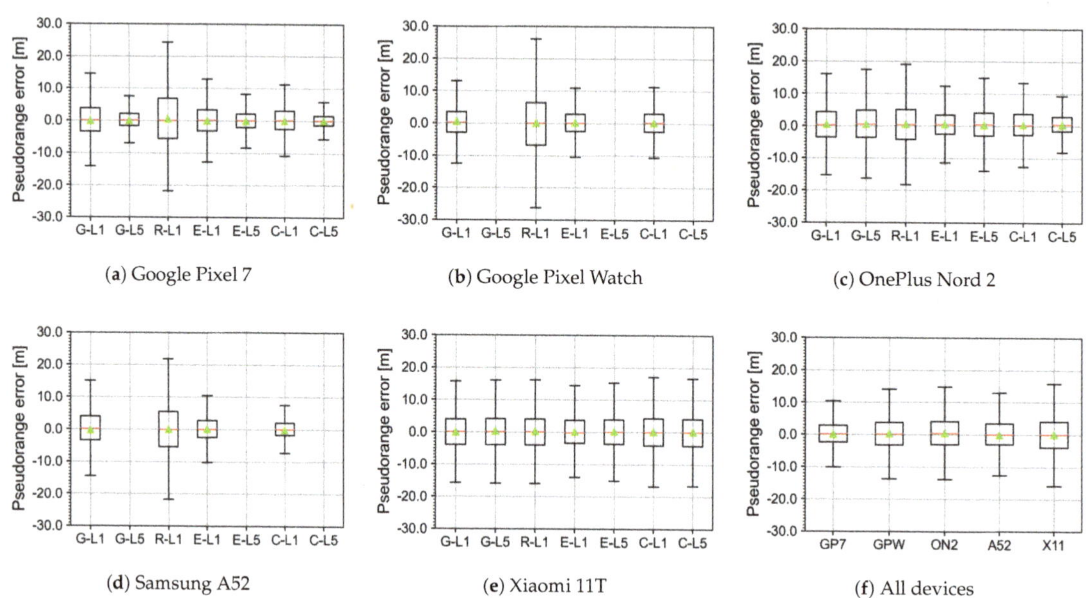

(**a**) Google Pixel 7

(**b**) Google Pixel Watch

(**c**) OnePlus Nord 2

(**d**) Samsung A52

(**e**) Xiaomi 11T

(**f**) All devices

Figure A13. Pseudorange errors (S2), with outlier threshold of 300 m. See Table A12 for details.

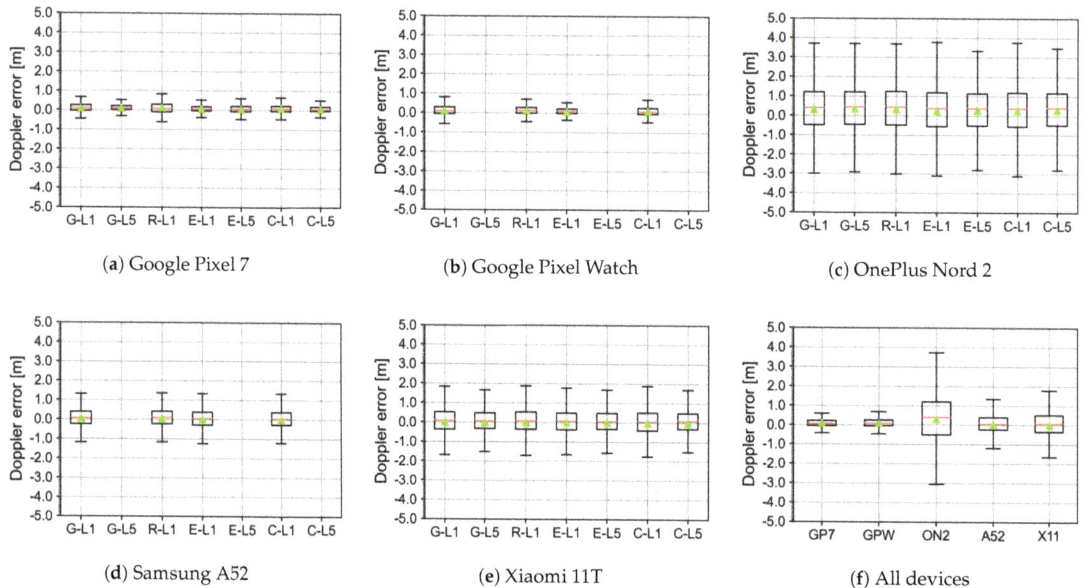

(**a**) Google Pixel 7

(**b**) Google Pixel Watch

(**c**) OnePlus Nord 2

(**d**) Samsung A52

(**e**) Xiaomi 11T

(**f**) All devices

Figure A14. Doppler errors (S2), with outlier threshold of 30 m. See Table A13 for details.

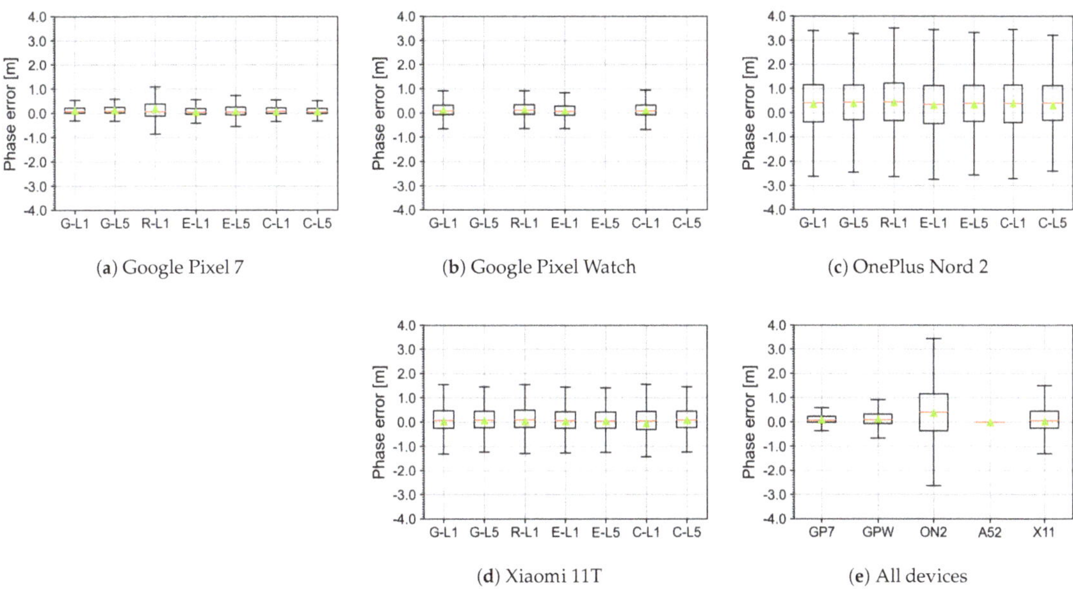

(**a**) Google Pixel 7 (**b**) Google Pixel Watch (**c**) OnePlus Nord 2

(**d**) Xiaomi 11T (**e**) All devices

Figure A15. Phase errors (S2), with outlier threshold of 30 m. See Table A14 for details.

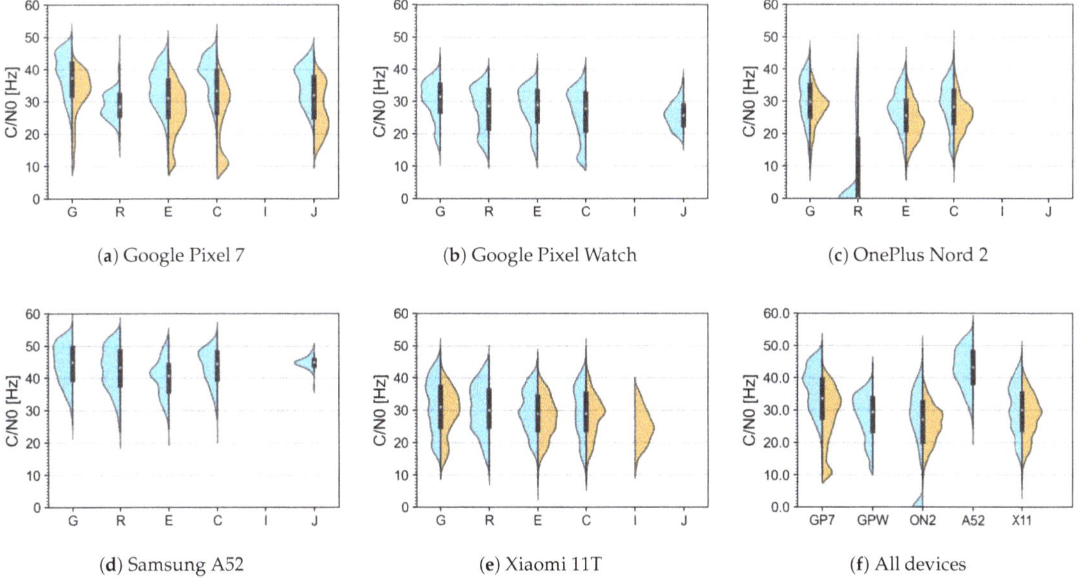

(**a**) Google Pixel 7 (**b**) Google Pixel Watch (**c**) OnePlus Nord 2

(**d**) Samsung A52 (**e**) Xiaomi 11T (**f**) All devices

Figure A16. Carrier-to-Noise ratio (C/N0) (S2) [Blue: L1, Orange: L5]. See Table A15 for details.

Table A10. Statistics for position errors (S2).

| Device | Inc. [%] | Mean ± StD (1-σ) [m] | | | RMSE [m] | |
		East	North	Up	2D	3D
GP7	100.00	0.372 ± 1.301	−0.445 ± 1.969	−6.928 ± 3.090	2.428	7.964
GPW	100.00	0.935 ± 2.034	−0.329 ± 2.444	−4.135 ± 5.342	3.329	7.528
ON2	100.00	−0.835 ± 3.150	−1.301 ± 5.496	−7.707 ± 5.009	6.515	11.265
A52	—	−0.381 ± 1.980	1.212 ± 2.459	−3.867 ± 5.071	3.400	7.224
X11	100.00	−0.802 ± 2.076	−2.360 ± 4.233	−14.201 ± 6.892	5.329	16.657

Table A11. Signals' percentages visibility w.r.t. the base receiver (S2).

| Device | Freq. [%] | | Constellations [%] | | | | | |
	L1	L5	G	R	E	C	I	J
GP7	78.9	95.5	94.1	75.0	77.8	86.7	—	100.0
GPW	97.4	—	64.7	125.0	44.4	46.7	—	50.0
ON2	95.5	80.8	129.4	112.5	100.0	59.3	—	—
A52	114.0	—	72.2	200.0	50.0	45.8	—	50.0
X11	124.4	136.4	115.8	110.0	105.6	171.4	500.0	0.0

Table A12. Statistics for pseudorange errors [m] (S2).

	Device	Inc. [%]	Mean [m]	StD [m]	Min. [m]	Max. [m]
Pseudorange	GP7	99.94%	0.124	9.752	−153.536	246.959
	GPW	99.83%	0.413	18.186	−296.118	226.721
	ON2	99.35%	0.553	13.840	−128.524	138.471
	A52	99.98%	0.090	15.080	−273.992	195.191
	X11	99.27%	0.281	13.038	−99.589	143.735

Table A13. Statistics for Doppler errors [m] (S2).

	Device	Inc. [%]	Mean [m]	StD [m]	Min. [m]	Max. [m]
Doppler	GP7	100.00%	0.076	0.395	−5.145	10.060
	GPW	100.00%	0.098	0.374	−6.700	6.036
	ON2	99.99%	0.317	2.073	−22.044	13.691
	A52	100.00%	0.058	0.651	−12.502	7.768
	X11	100.00%	0.007	1.085	−19.211	21.977

Table A14. Statistics for carrier phase errors [m] (S2).

	Device	Inc. [%]	Mean [m]	StD [m]	Min. [m]	Max. [m]
Phase	GP7	99.91%	0.096	1.004	−29.357	22.052
	GPW	99.98%	0.119	0.798	−11.452	18.423
	ON2	98.35%	0.383	2.542	−26.589	29.935
	A52	100.00%	0.000	0.000	0.000	0.000
	X11	98.04%	0.039	1.335	−29.847	28.787

Table A15. Statistics for C/n0 errors [dB-Hz] (S2).

	Device	Inc. [%]	Mean [dB]	StD [dB]	Min. [dB]	Max. [dB]
C/n0	GP7	100.00%	32.742	8.964	10.600	51.452
	GPW	100.00%	28.139	6.772	12.100	44.307
	ON2	100.00%	24.779	11.000	0.000	49.000
	A52	100.00%	42.822	6.292	21.300	57.300
	X11	100.00%	29.477	7.533	5.000	49.000

Appendix D. Scenario 3—Pedestrian Dynamic in Urban Environment

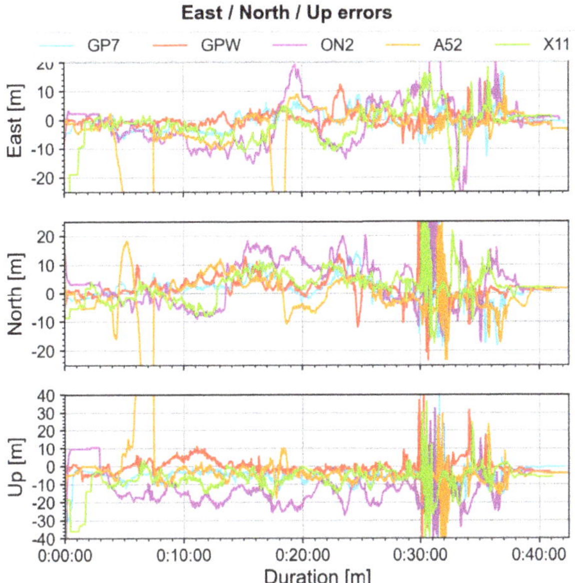

Figure A17. East/north/up error (S3).

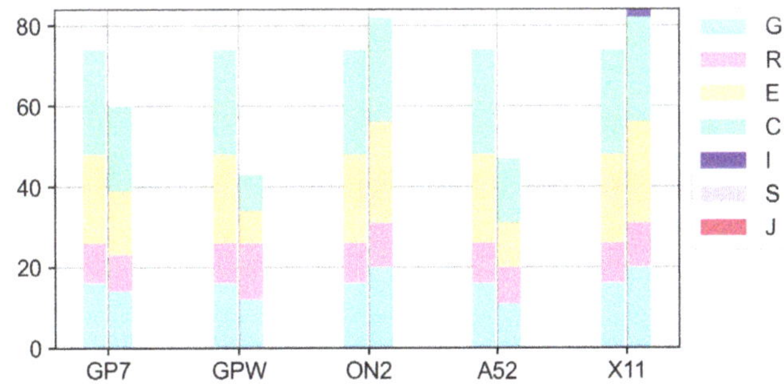

Figure A18. Signals seen per system (S3) (left: reference, right: device).

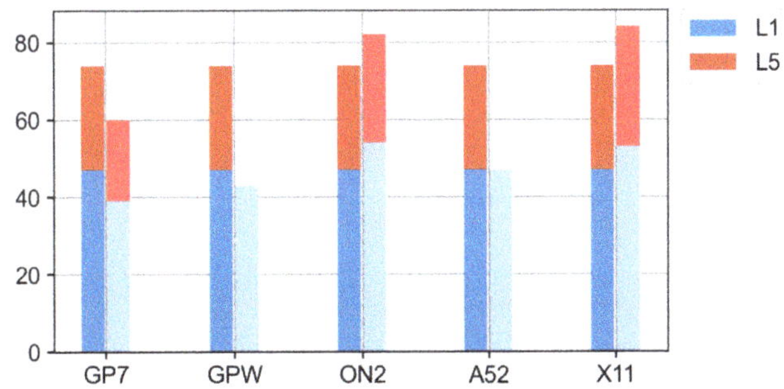

Figure A19. Signals seen per frequency (S3) (left: reference, right: device).

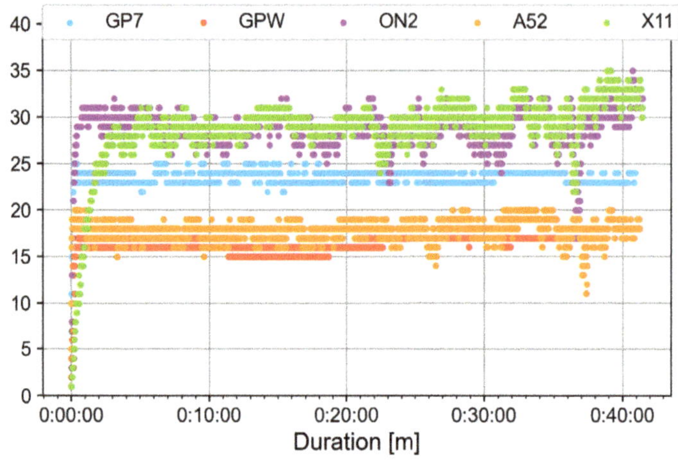

Figure A20. Signals seen per epoch (S3).

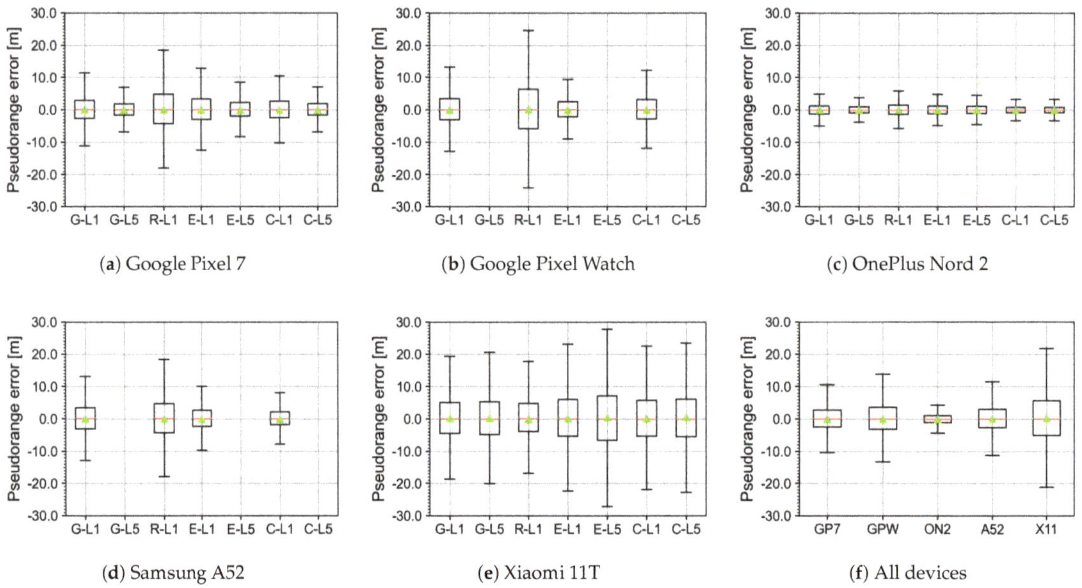

(**a**) Google Pixel 7

(**b**) Google Pixel Watch

(**c**) OnePlus Nord 2

(**d**) Samsung A52

(**e**) Xiaomi 11T

(**f**) All devices

Figure A21. Pseudorange errors (S3), with outlier threshold of 300 m. See Table A18 for details.

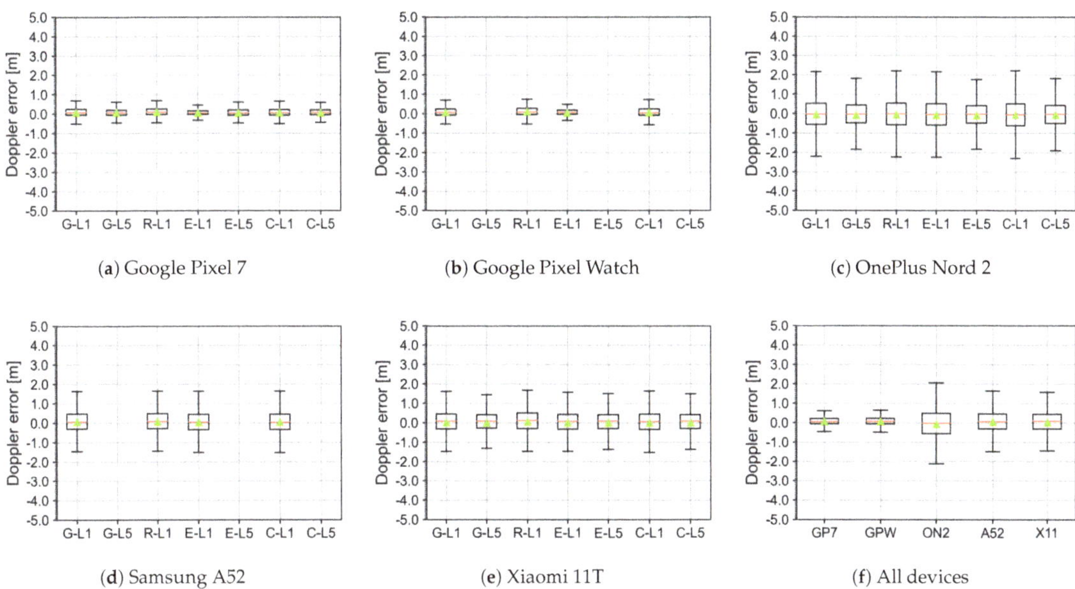

(**a**) Google Pixel 7

(**b**) Google Pixel Watch

(**c**) OnePlus Nord 2

(**d**) Samsung A52

(**e**) Xiaomi 11T

(**f**) All devices

Figure A22. Doppler errors (S3), with outlier threshold of 30 m. See Table A19 for details.

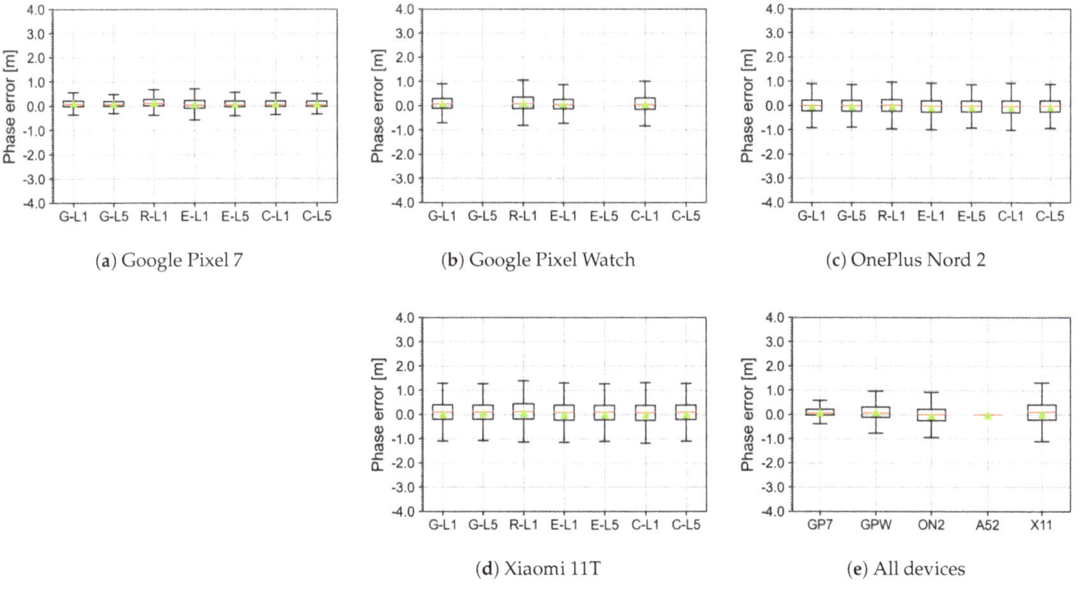

(**a**) Google Pixel 7

(**b**) Google Pixel Watch

(**c**) OnePlus Nord 2

(**d**) Xiaomi 11T

(**e**) All devices

Figure A23. Phase errors (S3), with outlier threshold of 30 m. See Table A20 for details.

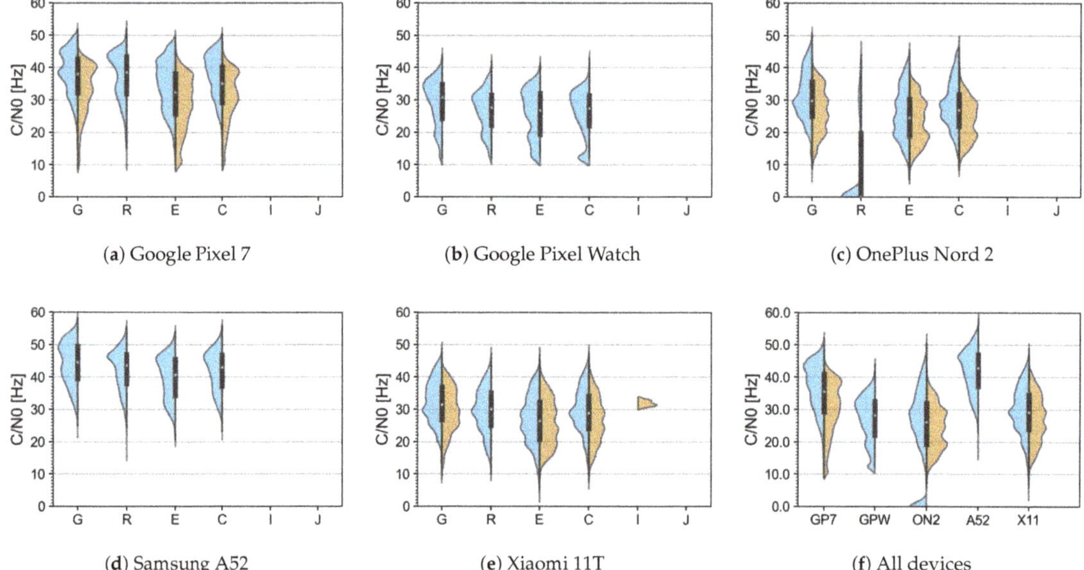

(a) Google Pixel 7 **(b)** Google Pixel Watch **(c)** OnePlus Nord 2

(d) Samsung A52 **(e)** Xiaomi 11T **(f)** All devices

Figure A24. Carrier-to-Noise ratio (C/N0) (S3) [Blue: L1, Orange: L5]. See Table A21 for details.

Table A16. Statistics for position errors (S3).

| Device | Inc. [%] | Mean $\pm$ StD (1-σ) [m] | | | RMSE [m] | |
		East	North	Up	2D	3D
GP7	100.00	-0.434 ± 3.359	1.040 ± 4.977	-5.256 ± 5.615	6.109	9.822
GPW	100.00	-0.024 ± 2.290	1.810 ± 4.551	-0.661 ± 5.498	5.405	7.738
ON2	100.00	-0.677 ± 8.928	3.771 ± 8.038	-12.473 ± 9.570	12.607	20.151
A52	100.00	-8.516 ± 26.865	-0.529 ± 7.447	-1.653 ± 13.730	29.149	32.263
X11	100.00	-2.150 ± 7.576	2.359 ± 5.934	-7.482 ± 8.829	10.137	15.384

Table A17. Signals' percentages visibility w.r.t. the base receiver (S3).

| Device | Freq. [%] | | Constellations [%] | | | | | |
	L1	L5	G	R	E	C	I	J
GP7	83.0	77.8	87.5	90.0	72.7	80.8	—	—
GPW	91.5	—	75.0	140.0	36.4	34.6	—	—
ON2	114.9	103.7	125.0	110.0	113.6	100.0	—	—
A52	100.0	—	68.8	90.0	50.0	61.5	—	—
X11	112.8	114.8	125.0	110.0	113.6	100.0	200.0%	—

Table A18. Statistics for pseudorange errors [m] (S3).

	Device	Inc. [%]	Mean [m]	StD [m]	Min. [m]	Max. [m]
Pseudorange	GP7	99.98%	0.106	10.293	-243.124	264.025
	GPW	99.81%	0.095	16.888	-287.794	277.366
	ON2	99.94%	-0.023	4.986	-155.842	137.541
	A52	99.99%	0.123	14.754	-282.783	297.405
	X11	99.99%	0.394	22.109	-280.577	269.135

Table A19. Statistics for pseudorange errors (S3).

	Device	Inc. [%]	Mean [m]	StD [m]	Min. [m]	Max. [m]
Pseudorange	GP7	99.98%	0.106	10.293	−243.124	264.025
	GPW	99.81%	0.095	16.888	−287.794	277.366
	ON2	99.94%	−0.023	4.986	−155.842	137.541
	A52	99.99%	0.123	14.754	−282.783	297.405
	X11	99.99%	0.394	22.109	−280.577	269.135

Table A20. Statistics for carrier phase errors [m] (S3).

	Device	Inc. [%]	Mean [m]	StD [m]	Min. [m]	Max. [m]
Phase	GP7	99.62%	0.096	0.972	−26.408	29.383
	GPW	99.95%	0.098	0.821	−26.192	19.709
	ON2	98.43%	−0.042	1.335	−29.954	29.333
	A52	100.00%	0.000	0.000	0.000	0.000
	X11	98.78%	0.046	1.061	−29.611	29.073

Table A21. Statistics for C/n0 errors [dB-Hz] (S3).

	Device	Inc. [%]	Mean [dB]	StD [dB]	Min. [dB]	Max. [dB]
C/n0	GP7	100.00%	34.5	8.2	10.6	52.0
	GPW	100.00%	27.0	7.0	12.1	43.9
	ON2	100.00%	24.5	11.1	0.0	50.4
	A52	100.00%	42.0	6.5	16.1	58.1
	X11	100.00%	29.0	7.3	3.6	48.8

Appendix E. Scenario 4—Pedestrian Dynamic in Light Forest and Lake Environment

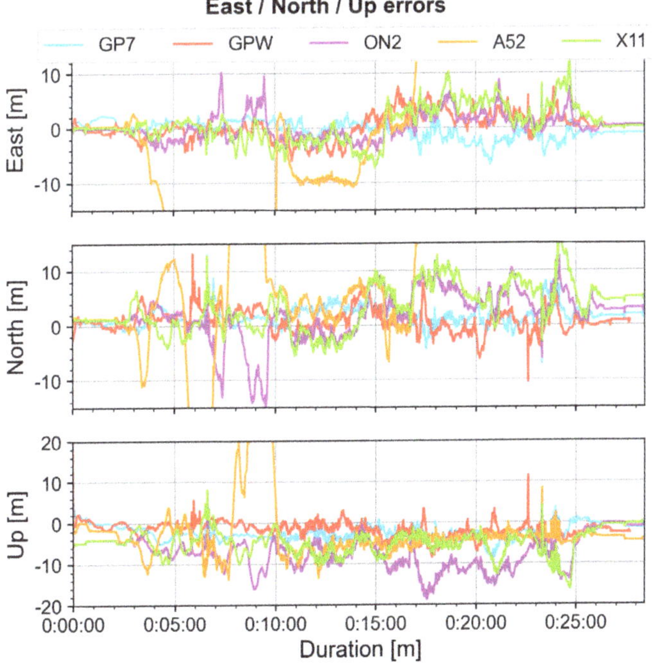

Figure A25. East/north/up error (S4).

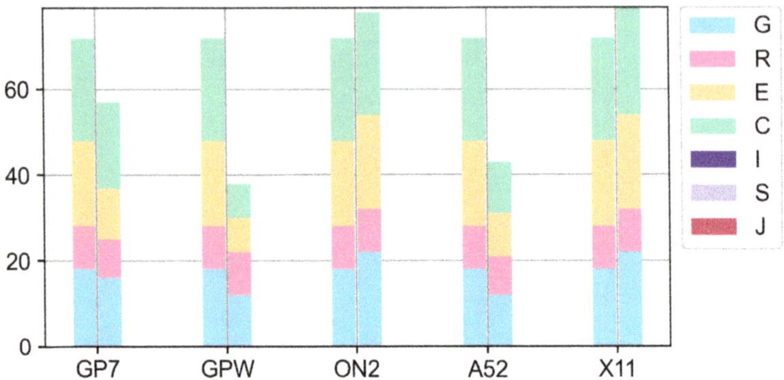

Figure A26. Signals seen per system (S4) (left: reference, right: device).

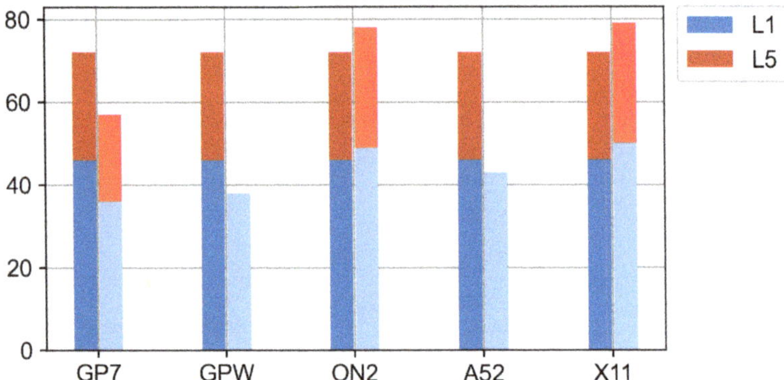

Figure A27. Signals seen per frequency (S4) (left: reference, right: device).

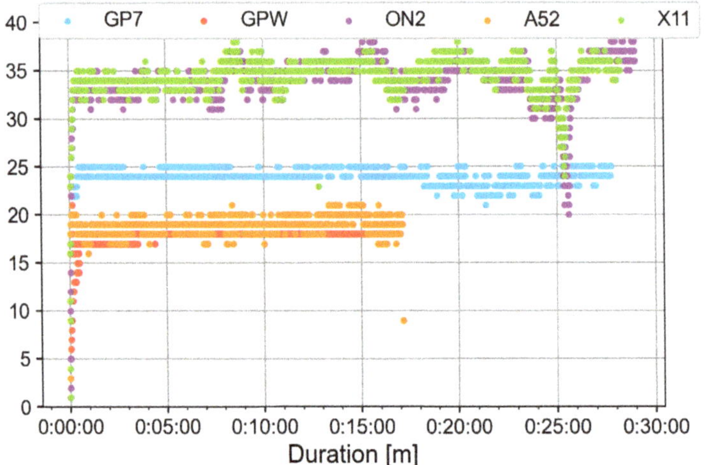

Figure A28. Signals seen per epoch (S4).

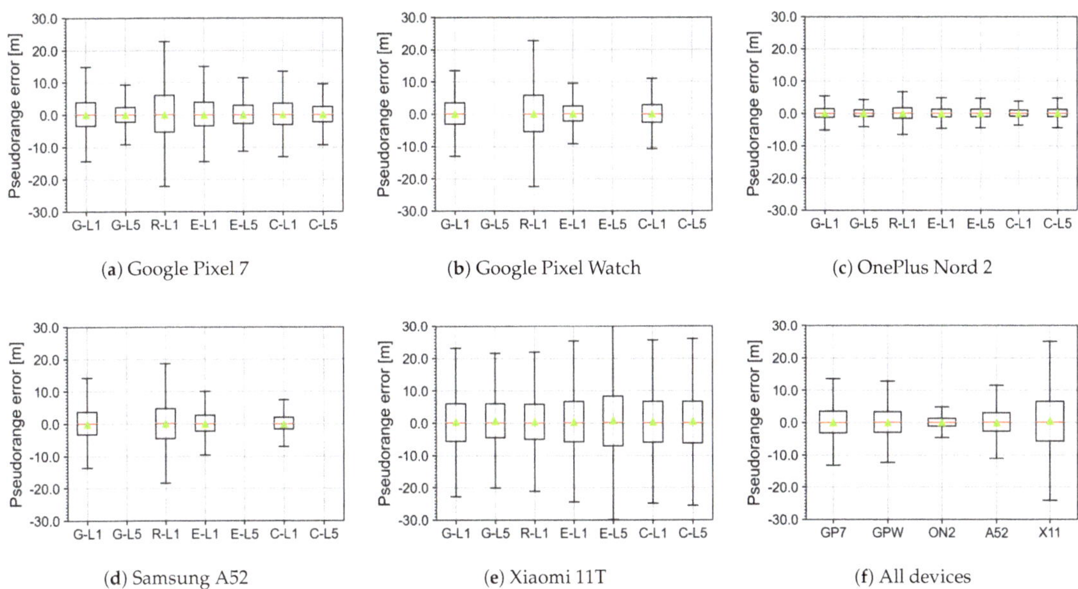

(**a**) Google Pixel 7

(**b**) Google Pixel Watch

(**c**) OnePlus Nord 2

(**d**) Samsung A52

(**e**) Xiaomi 11T

(**f**) All devices

Figure A29. Pseudorange errors (S4), with outlier threshold of 300 m. See Table A24 for details.

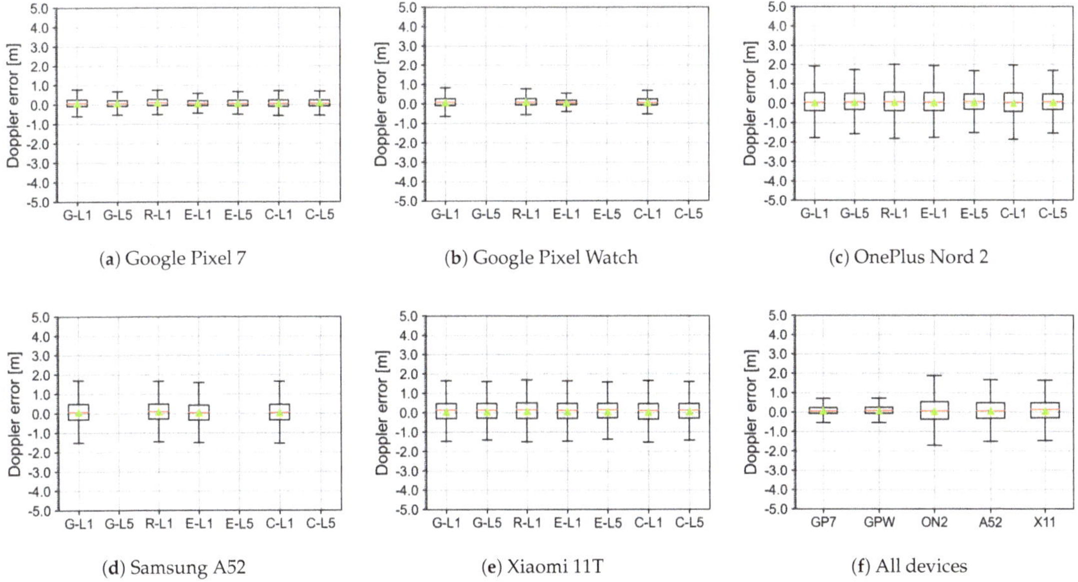

(**a**) Google Pixel 7

(**b**) Google Pixel Watch

(**c**) OnePlus Nord 2

(**d**) Samsung A52

(**e**) Xiaomi 11T

(**f**) All devices

Figure A30. Doppler errors (S4), with outlier threshold of 30 m. See Table A25 for details.

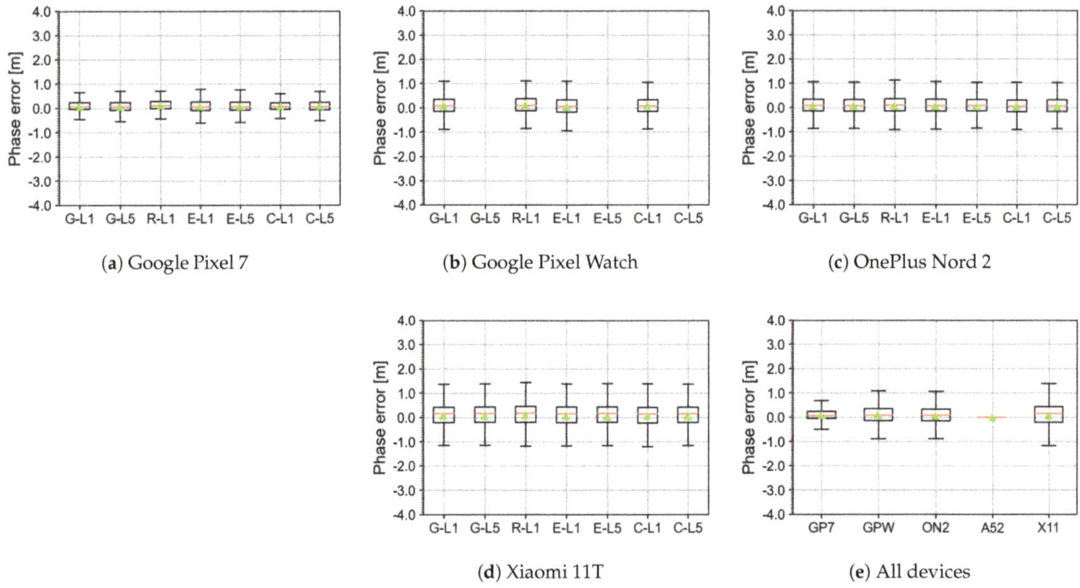

(**a**) Google Pixel 7

(**b**) Google Pixel Watch

(**c**) OnePlus Nord 2

(**d**) Xiaomi 11T

(**e**) All devices

Figure A31. Phase errors (S4), with outlier threshold of 30 m. See Table A26 for details.

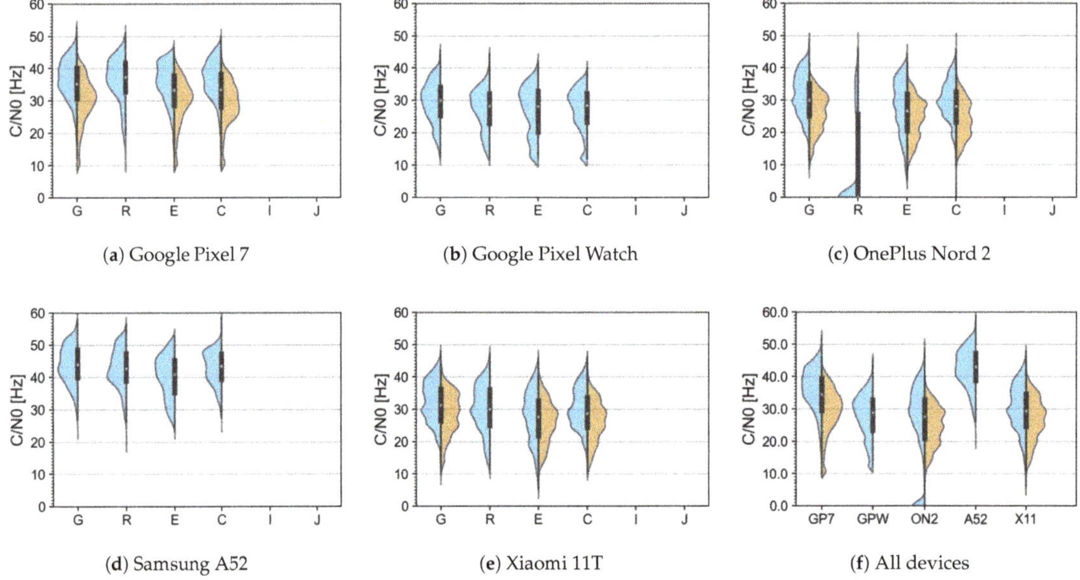

(**a**) Google Pixel 7

(**b**) Google Pixel Watch

(**c**) OnePlus Nord 2

(**d**) Samsung A52

(**e**) Xiaomi 11T

(**f**) All devices

Figure A32. Carrier-to-Noise ratio (C/N0) (S4) [Blue: L1, Orange: L5]. See Table A27 for details.

Table A22. Statistics for position errors (S4).

Device	Inc. [%]	Mean ± StD (1-σ) [m] East	North	Up	RMSE [m] 2D	3D
GP7	100.00	-0.600 ± 1.741	1.734 ± 1.738	-2.449 ± 1.964	3.069	4.390
GPW	100.00	0.166 ± 2.286	0.881 ± 2.121	-1.709 ± 2.330	3.244	4.344
ON2	100.00	0.180 ± 2.241	1.829 ± 4.352	-7.407 ± 3.991	5.228	9.905
A52	—	—	—	—	—	—
X11	100.00	0.593 ± 3.188	3.145 ± 4.068	-4.926 ± 3.044	6.078	8.394

Table A23. Signals' percentages visibility w.r.t. the base receiver (S4).

Device	Freq. [%] L1	L5	Constellations [%] G	R	E	C	I	J
GP7	78.2	80.7	88.9	90.0	60.0	83.3	—	—
GPW	82.6	—	66.7	100.0	40.0	33.3	—	—
ON2	106.5	111.5	122.2	100.0	110.0	100.0	—	—
A52	93.4	—	66.7	90.0	50.0	50.0	—	—
X11	108.6	111.5	122.2	100.0	110.0	104.2	—	—

Table A24. Statistics for pseudorange errors [m] (S4).

	Device	Inc. [%]	Mean [m]	StD [m]	Min. [m]	Max. [m]
Pseudorange	GP7	99.95%	0.141	12.020	-224.975	154.113
	GPW	99.73%	0.168	14.831	-261.697	274.383
	ON2	99.87%	0.064	5.215	-102.422	115.495
	A52	99.98%	0.054	13.572	-216.617	176.996
	X11	99.81%	0.603	25.354	-261.007	271.240

Table A25. Statistics for Doppler errors [m] (S4).

	Device	Inc. [%]	Mean [m]	StD [m]	Min. [m]	Max. [m]
Doppler	GP7	99.91%	0.085	0.610	-19.214	29.231
	GPW	100.00%	0.088	0.395	-6.177	6.730
	ON2	100.00%	0.055	0.958	-21.951	22.330
	A52	100.00%	0.077	0.755	-8.661	6.299
	X11	100.00%	0.064	0.846	-18.461	25.563

Table A26. Statistics for carrier phase errors [m] (S4).

	Device	Inc. [%]	Mean [m]	StD [m]	Min. [m]	Max. [m]
Phase	GP7	98.97%	0.090	1.127	-29.219	29.975
	GPW	99.89%	0.110	0.895	-22.653	27.966
	ON2	98.92%	0.065	0.888	-29.929	28.932
	A52	100.00%	0.000	0.000	0.000	0.000
	X11	99.10%	0.079	0.931	-29.507	29.919

Table A27. Statistics for C/n0 errors [dB-Hz] (S4).

	Device	Inc. [%]	Mean [dB]	StD [dB]	Min. [dB]	Max. [dB]
C/n0	GP7	100.00%	33.8	7.4	10.6	52.7
	GPW	100.00%	27.8	6.7	12.1	45.3
	ON2	100.00%	25.6	10.5	0.0	48.8
	A52	100.00%	42.5	6.1	19.3	58.7
	X11	100.00%	29.2	7.1	5.0	47.8

References

1. Zhu, F.; Tao, X.; Liu, W.; Shi, X.; Wang, F.; Zhang, X. Walker: Continuous and Precise Navigation by Fusing GNSS and MEMS in Smartphone Chipsets for Pedestrians. *Remote Sens.* **2019**, *11*, 139. [CrossRef]
2. Harke, K.; O'Keefe, K. Gyroscope Drift Estimation of a GPS/MEMSINS Smartphone Sensor Integration Navigation System for Kayaking. In Proceedings of the 35th International Technical Meeting of the Satellite Division of The Institute of Navigation (ION GNSS+ 2022), Denver, CO, USA, 19–23 September 2022; pp. 1413–1427.
3. Garmin. Garmin Developers. Available online: https://developer.garmin.com/connect-iq/overview/ (accessed on 12 September 2023).
4. Fitbit. Fitbit Developers. Available online: https://dev.fitbit.com/ (accessed on 12 September 2023).
5. APROPOS. Approximate Computing for Power and Energy Optimisation. Available online: https://projects.tuni.fi/apropos/ (accessed on 12 September 2023).
6. LEDSOL. Enabling Clean and Sustainable Water through Smart UV/LED Disinfection and SOLar Energy Utilization. Available online: https://www.leap-re.eu/ledsol/ (accessed on 12 September 2023).
7. Grenier, A.; Lohan, E.S.; Ometov, A.; Nurmi, J. *Multi-Sensor Dataset From Android Smart Devices*; Zenodo: Geneva, Switzerland, 2023. [CrossRef]
8. EUSPA. World's First Dual-Frequency GNSS Smartphone Hits the Market. Available online: https://www.euspa.europa.eu/newsroom/news/world-s-first-dual-frequency-gnss-smartphone-hits-market (accessed on 12 September 2023).
9. Subirana, J.; Zornoza, J.; Hernández-Pajares, M. *GNSS Data Processing Volume I: Fundamentals and Algorithms*; ESA Communications; European Space Agency: Paris, France, 2013.
10. Grenier, A. Development of a GNSS Positioning Application under Android OS Using GALILEO Signals. Master's Thesis, Ecole Nationale de Sciences Geographiques, Champs-sur-Marne, France, 2019.
11. Zangenehnejad, F.; Gao, Y. GNSS Smartphones Positioning: Advances, Challenges, Opportunities, and Future Perspectives. *Satell. Navig.* **2021**, *2*, 1–23. [CrossRef] [PubMed]
12. Paziewski, J.; Fortunato, M.; Mazzoni, A.; Odolinski, R. An Analysis of Multi-GNSS Observations Tracked by Recent Android Smartphones and Smartphone-Only Relative Positioning Results. *Measurement* **2021**, *175*, 109162. [CrossRef]
13. Magiera, W.; Vārna, I.; Mitrofanovs, I.; Silabrieds, G.; Krawczyk, A.; Skorupa, B.; Apollo, M.; Maciuk, K. Accuracy of Code GNSS Receivers under Various Conditions. *Remote Sens.* **2022**, *14*, 2615. [CrossRef]
14. Wen, Q.; Geng, J.; Li, G.; Guo, J. Precise Point Positioning with Ambiguity Resolution Using an External Survey-Grade Antenna Enhanced Dual-Frequency Android GNSS Data. *Measurement* **2020**, *157*, 107634. [CrossRef]
15. Wang, L.; Li, Z.; Wang, N.; Wang, Z. Real-Time GNSS Precise Point Positioning for Low-Cost Smart Devices. *GPS Solut.* **2021**, *25*, 1–13. [CrossRef]
16. Li, M.; Huang, G.; Wang, L.; Xie, W. BDS/GPS/Galileo Precise Point Positioning Performance Analysis of Android Smartphones Based on Real-Time Stream Data. *Remote Sens.* **2023**, *15*, 2983. [CrossRef]
17. Fortunato, M.; Ravanelli, M.; Mazzoni, A. Real-Time Geophysical Applications with Android GNSS Raw Measurements. *Remote Sens.* **2019**, *11*, 2113. [CrossRef]
18. Stauffer, R.; Hohensinn, R.; Herrera-Pinzón, I.D.; Pan, Y.; Moeller, G.; Kłopotek, G.; Soja, B.; Brockmann, E.; Rothacher, M. Estimation of Tropospheric Parameters with GNSS Smartphones in a Differential Approach. *Meas. Sci. Technol.* **2023**, *34*, 095126. [CrossRef]
19. Lohan, E.S.; Bierwirth, K.; Kodom, T.; Ganciu, M.; Lebik, H.; Elhadi, R.; Cramariuc, O.; Mocanu, I. Standalone Solutions for Clean and Sustainable Water Access in Africa through Smart UV/LED Disinfection, Solar Energy Utilization, and Wireless Positioning Support. *IEEE Access* **2023**, *11*, 81882–81899. [CrossRef]
20. Lohan, E.S.; Kodom, T.; Lebik, H.; Grenier, A.; Zhang, X.; Cramariuc, O.; Mocanu, I.; Bierwirth, K.; Nurmi, J. Raw GNSS Data Analysis for the LEDSOL Project—Preliminary Results and Way Ahead. In Proceedings of the WiP in Hardware and Software for Location Computation (WIPHAL 2023), Castellon, Spain, 6–8 June 2023; Volume 3434.
21. Lo, S.; Chen, Y.H.; Akos, D.; Cotts, B.; Miralles, D. Test of Crowdsourced Smartphones Measurements to Detect GNSS Spoofing and other Disruptions. In Proceedings of the International Technical Meeting of The Institute of Navigation, Reston, VA, USA, 28–31 January 2019; pp. 373–388.
22. Spens, N.; Lee, D.K.; Nedelkov, F.; Akos, D. Detecting GNSS Jamming and Spoofing on Android Devices. *Navig. J. Inst. Navig.* **2022**, *69*, navi.537. [CrossRef]
23. Orendorff, D.; Van Diggelen, F.; Elliott, J.; Fu, M.; Khider, M.; Dane, S. Google Smartphone Decimeter Challenge. 2021. Available online: https://kaggle.com/competitions/google-smartphone-decimeter-challenge (accessed on 21 November 2023).
24. Howard, A.; Chow, A.; Julian, B.; Orendorff, D.; Fu, M.; Khider, M.; Dane, S. Google Smartphone Decimeter Challenge. 2022. Available online: https://kaggle.com/competitions/smartphone-decimeter-2022 (accessed on 21 November 2023).
25. Fu, G.M.; Khider, M.; van Diggelen, F. Android Raw GNSS Measurement Datasets for Precise Positioning. In Proceedings of the 33rd International Technical Meeting of the Satellite Division of the Institute of Navigation (ION GNSS+ 2020), Online, 22–25 September 2020; pp. 1925–1937.
26. EUSPA. GNSS Raw Measurements Task Force. 2021. Available online: https://www.euspa.europa.eu/euspace-applications/gnss-raw-measurements/gnss-raw-measurements-task-force (accessed on 21 November 2023).

27. Reeder, B.; David, A. Health at Hand: A Systematic Review of Smart Watch Uses for Health and Wellness. *J. Biomed. Inform.* **2016**, *63*, 269–276. [CrossRef] [PubMed]
28. Jat, A.S.; Grønli, T.M. Smart Watch for Smart Health Monitoring: A Literature Review. In Proceedings of the International Work-Conference on Bioinformatics and Biomedical Engineering, Maspalomas, Spain, 27–30 June 2022; Springer: Cham, Switzerland, 2022; pp. 256–268.
29. Hernández-Orallo, E.; Manzoni, P.; Calafate, C.T.; Cano, J.C. Evaluating How Smartphone Contact Tracing Technology Can Reduce the Spread of Infectious Diseases: The Case of COVID-19. *IEEE Access* **2020**, *8*, 99083–99097. [CrossRef] [PubMed]
30. Site, A.; Lohan, E.S.; Jolanki, O.; Valkama, O.; Hernandez, R.R.; Latikka, R.; Alekseeva, D.; Vasudevan, S.; Afolaranmi, S.; Ometov, A.; et al. Managing Perceived Loneliness and Social-Isolation Levels for Older Adults: A Survey with Focus on Wearables-Based Solutions. *Sensors* **2022**, *22*, 1108. [CrossRef] [PubMed]
31. Broadcom. Broadcom Introduces Second Generation Dual-Frequency GNSS. Available online: https://www.broadcom.com/blog/broadcom-introduces-second-generati (accessed on 12 September 2023).
32. Barbeau, S. Crowdsourcing GNSS Features of Android Devices. 2021. Available online: https://barbeau.medium.com/crowdsourcing-gnss-capabilities-of-android-devices-d4228645cf25 (accessed on 21 November 2023)
33. Google. Google Pixel Watch. Available online: https://store.google.com/us/product/google_pixel_watch?hl=en-US (accessed on 12 September 2023).
34. Karki, B.; Won, M. Characterizing Power Consumption of Dual-Frequency GNSS of Smartphone. In Proceedings of the IEEE Global Communications Conference, Taipei, Taiwan, 7–11 December 2020; pp. 1–6.
35. Google. GPS Measurement Tools Github. Available online: https://github.com/google/gps-measurement-tools (accessed on 12 September 2023).
36. University, T. Mimir Github. Available online: https://github.com/agrenier-gnss/mimir (accessed on 12 September 2023).
37. University, T. Mimir Analyzer Github. Available online: https://github.com/agrenier-gnss/MimirAnalyzer (accessed on 12 September 2023).
38. Takasu, T. PPP Ambiguity Resolution Implementation in RTKLIB. *Geophys. J. Int.* **2013**, *194*, 1441–1454.
39. Ferreira, D.L.; Nunes, B.A.A.; Campos, C.A.V.; Obraczka, K. User Community Identification through Fine-Grained Mobility Records for Smart City Applications. *IEEE Trans. Intell. Transp. Syst.* **2022**, *23*, 4387–4401. [CrossRef]
40. Fu, H.; Kone, Y.; Renaudin, V.; Zhu, N. A Survey on Artificial Intelligence for Pedestrian Navigation with Wearable Inertial Sensors. In Proceedings of the IEEE 12th International Conference on Indoor Positioning and Indoor Navigation (IPIN), Beijing, China, 5–8 September 2022; pp. 1–8.
41. Fu, H.; Bonis, T.; Renaudin, V.; Zhu, N. A Computer Vision Approach for Pedestrian Walking Direction Estimation with Wearable Inertial Sensors: PatternNet. In Proceedings of the 2023 IEEE/ION Position, Location and Navigation Symposium (PLANS), Monterey, CA, USA, 24–27 April 2023; pp. 691–699.
42. Zhu, N.; Bouronopoulos, A.; Leduc, T.; Servières, M.; Renaudin, V. Evaluation of the Human Body Mask Effects on GNSS Wearable Devices for Outdoor Pedestrian Navigation Using Fisheye Sky Views. In Proceedings of the 2023 IEEE/ION Position, Location and Navigation Symposium (PLANS), Monterey, CA, USA, 24–27 April 2023; pp. 841–850.
43. Google. Android Developers. Available online: https://developer.android.com/ (accessed on 12 September 2023).
44. EUSPA. *Using GNSS Raw Measurements on Android Devices*; Technical Report; European Agency for the Space Program: Paris, France, 2019.
45. Perul, J.; Renaudin, V. HEAD: SmootH Estimation of wAlking Direction with a handheld device embedding inertial, GNSS, and magnetometer sensors. *Navigation* **2020**, *67*, 713–726. [CrossRef]
46. Chiang, K.W.; Chang, H.W.; Li, Y.H.; Tsai, G.J.; Tseng, C.L.; Tien, Y.C.; Hsu, P.C. Assessment for INS/GNSS/Odometer/Barometer Integration in Loosely Coupled and Tightly Coupled Scheme in a GNSS-degraded Environment. *IEEE Sens. J.* **2019**, *20*, 3057–3069. [CrossRef]
47. Romero, I. *RINEX: The Receiver Independent Exchange Format Version 3.05*; ESA/ESOC/Navigation Support Office: Darmstadt, Germany, 2020.
48. Massarweh, L.; Fortunato, M.; Gioia, C. Assessment of Real-Time Multipath Detection with Android Raw GNSS Measurements by Using a Xiaomi Mi 8 Smartphone. In Proceedings of the IEEE/ION Position, Location and Navigation Symposium (PLANS), Portland, OR, USA, 20–23 April 2020; pp. 1111–1122.
49. Verheyde, T.; Blais, A.; Macabiau, C.; Marmet, F.X. Analyzing Android GNSS Raw Measurements Flags Detection Mechanisms for Collaborative Positioning in Urban Environment. In Proceedings of the International Conference on Localization and GNSS (ICL-GNSS), IEEE, Tampere, Finland, 2–4 June 2020; pp. 1–6.
50. Grenier, A.; Lohan, E.S.; Ometov, A.; Nurmi, J. A Survey on Low-Power GNSS. *IEEE Commun. Surv. Tutor.* **2023**, *25*, 1482–1509. [CrossRef]

Article

VR Drumming Pedagogy: Action Observation, Virtual Co-Embodiment, and Development of Drumming "Halvatar"

James Pinkl * and Michael Cohen *

Spatial Media Group, University of Aizu, Tsuruga, Ikki-Machi, Aizu-Wakamatsu 965-8580, Fukushima, Japan
* Correspondence: jamesjpinkl@gmail.com (J.P.); xilehence@gmail.com (M.C.)

Abstract: Virtual Co-embodiment (VC) is a relatively new field of VR, enabling a user to share control of an avatar with other users or entities. According to a recent study, VC was shown to have the highest motor skill learning efficiency out of three VR-based methods. This contribution expands on these findings, as well as previous work relating to Action Observation (AO) and drumming, to realize a new concept to teach drumming. Users "duet" with an exemplar half in a virtual scene with concurrent feedback to learn rudiments and polyrhythms. We call this puppet avatar controlled by both a user and separate processes a "halvatar". The development is based on body-part-segmented VC techniques and uses programmed animation, electromechanical drum strike detection, and optical bimanual hand-tracking informed by head-tracking. A pilot study was conducted with primarily non-musicians showing the potential effectiveness of this tool and approach.

Keywords: personal sensing; Virtual Co-embodiment (VC); virtual reality; mixed reality; music pedagogy; rhythm training; Action Observation (AO); motor skill learning; sound and music computing; information interfaces and presentation; sense of agency

1. Introduction

In this introduction, an overview of drumming and general motor skill learning is presented. Polyrhythms and rudimental drumming, musical concepts referenced throughout this research, are described. Virtual reality (VR) fields relevant to this research, such as Action Observation (AO) and virtual co-embodiment (VC), are reviewed. Lastly, rhythm games and other technology-based musical education tools are also surveyed.

1.1. Motor Skill Learning

Motor skill learning is defined as the process of increasing the accuracy of movement in the space and time domains through practice. This learning can be further broken down into cognitive, associative, and autonomous phases [1]. The cognitive phase consists of learners striving to understand movements correctly. The associative phase consists of conscious effort to gain the skill to perform movements correctly. The autonomous phase is the process of gaining the ability to perform movements unconsciously.

1.2. Technology-Accelerated Motor Skill Learning

Recently developed systems have been designed to help teach movement and gestures in new ways. YouMove is a novel augmented reality mirror and system that helps users learn sequences of physical movements [2]. The system displays guidance and feedback at varying levels to help train users. A study using YouMove showed that the system significantly improved learning and short-term retention compared to learning performed via video.

Similarly positive results were found with OctoPocus, a dynamic guide for the teaching of specific gestures with visual feedforward and feedback functionalities [3]. The system helps users achieve proficiency and was observed to teach significantly faster than that of standard Help menus.

Citation: Pinkl, J.; Cohen, M. VR Drumming Pedagogy: Action Observation, Virtual Co-Embodiment, and Development of Drumming "Halvatar". *Electronics* **2023**, *12*, 3708. https://doi.org/10.3390/electronics12173708

Academic Editor: George A. Papakostas

Received: 21 June 2023
Revised: 2 August 2023
Accepted: 18 August 2023
Published: 1 September 2023

1.3. Drumming and Rhythm

Drumming is defined as the action of expressing rhythm by striking a membranophone (such as a drum), idiophone (such as cymbal), or other objects using drum sticks or by directly striking with the hands or fingers. Rhythm, despite having been described countlessly, eludes a comprehensive definition [4]. The book *"The Geometry of Rhythm"* compiles many interpretations. "A measuring of time by means of some kind of movement" was declared by Baccheios the Elder. Didymus defined it as "a schematic arrangement of sounds", and R. Parncutt said, "a musicial rhythm is an acoustic sequence evoking a sensation of pulse" [4].

When considered from a physical perspective, drumming, like any motor skill, requires practice before successful execution [5]. It entails both fine and gross motor skills, as small muscles in the fingers are used for stick control, while the movement of larger muscles such as those in one's arms and torso are stretched to reach various parts of the drummer's kit.

From a motor control perspective, the varied motions of drumming can be classified as serial, discrete, or continuous [6]. A challenging piece of music may demand combinations of all three movements within a relatively short amount of time.

1.4. Musical Practice Tools

The metronome is a musical tool used since the 19th century, designed to improve musical performance of practicing musicians [7]. The metronome expresses a steady pulse, atop which musicians can practice. It prevents unwanted modulations of tempo and helps a player understand time-related issues with performance. Since then, many electronic and digital versions of metronomes have been released, giving players many more options. PolyNome [8] is a mobile application that lets users program desired rhythms, including drum set grooves and accent patterns. PocketDrum [9] and HyperDrum [10] are examples of portable virtual drum set products that detect drumming motions via motion sensing, rendering air drum-like gestures into audible drum sounds. They feature volume dynamics and haptic feedback capabilities, allowing drummers to practice anywhere.

1.5. Musical Practice

Practice can be described as a "systematic activity with predictable stages and activities" [11]. Musical practice can fall into one of four categories: technical practice, automation and memorization, detail-focused review, and maintenance of a piece [12]. Practice with a metronome is usually considered technical practice. For example, a common technique is to start a metronome at a relatively slow tempo, practice an excerpt until it can be naturally performed, and then, gradually increase the tempo in increments until it reaches a desired speed and level of execution.

1.6. Rudiments

In 1933, drummers in the United States organized a meeting to discuss drum instruction. Considering control, coordination, and endurance, thirteen rudiments were selected and deemed essential for any drummer to know [13]. They remain an important part of the musical language of drumming in most contexts and cultures, regardless of genre. They also remain staples in most drummers' practice routines, as complete mastery of a rudiment is never fully achieved. The speed, clarity, and technique involved in the performance of a rudiment can be improved endlessly. While some rudiments have rhythmic dynamics, others, such as those shown in Figure 1, focus on sticking or the specific assignment of notes to the left or right hand. These exercises build coordination and help players play specific accent patterns or phrases smoothly.

Figure 1. Western notation for two drumming rudiments: triple paradiddle and paradiddle-diddle.

1.7. Polyrhythms

A polyrhythm is "a set of simultaneous rhythms, where each rhythm strongly implies an entirely different meter" [14]. In other words, it is the simultaneous use of mathematically related divisions of pulse in music. For example, fitting a group of three notes in the same amount of time as another musical line fitting two notes creates a 3:2 polyrhythm. More generally, a musical phrase with m subdivisions played at the same time as a musical phrase with n subdivisions (where $n \neq x * m$ and n, m, and x are whole numbers) results in a $m{:}n$ polyrhythm. Although polyrhythms are often written as different subdivisions of the same tempo when scored or transcribed using Western notation, as shown in Figure 2, polyrhythms can also be thought of as rhythms played in different, but mathematically related tempos.

Peter Magadini defines the elements of a polyrhythm as the juxtaposition of basic pulse (or primary tempo) and counterrhythm(s), the accompanying pulse(s) that references a different division of time relative to the basic pulse [15]. A polyrhythmic piece or musical exercise can consist of a primary tempo overlaid by one or more counterrhythms. The simultaneous awareness of the primary tempo and counterrhythm, an often difficult task to perform just as a listener, has been called "metronome sense" [16].

Figure 2. Western notation for 3:2 and 5:3 polyrhythms, with corresponding verbal mnemonics.

Gilbert Rouget said of polyrhythms: "When reduced to a single instrument, the rhythmic pattern is always simple. But the combination of these patterns in an interwoven texture, whose principles remain to be discovered, produces a complicated rhythm that is apparently incomprehensible" [17]. This trait makes learning polyrhythms, particularly by ear, relatively difficult. Due to brain-related limitations of listening to music, humans may have trouble focusing on multiple tempi at once [14]. Therefore, slowing music down with the aid of a metronome while learning, a longstanding learning strategy for simpler rhythms, is not so effective for polyrhythmic practice.

Alternative methods for polyrhythmic learning involves metric patterns in language [18]. For example, the phrase "hot cup of tea" can rhythmically represent the 3:2 polyrhythm as in the first example of Figure 2; "what atrocious weather" can represent the 3:4 polyrhythm; "I'm looking for a home to buy" can represent the 5:4 polyrhythm [19]. Figure 2 shows compatible phrases below the two written polyrhythms. However, these too are limited as there does not exist a helpful mnemonic phrase for every $m{:}n$ polyrhythm.

1.8. Action Observation (AO)

Action Observation (AO) is an application of the phenomenon in which observing the behavior of another person or avatar produces neural activity similar to that when performed by one's self. Such observational learning can be an effective approach for learning specific movements, including surgical skills [20] or for movements related to physical therapy [21,22].

AO can be implemented via the observation and mimicry of actions by another human, an avatar animated via motion-capture, or a virtual avatar with a combination of programmed animations, rigging, constraints, and models for motion. For example, in the authors' previous pedagogic tool [23], three different 1st-person VR AO perspectives were utilized in the learning phase of a pedagogic tool teaching rhythm: Photospherical Video, Programmed Animation, and Mocap Animation [23].

In a recent study [24], subjects who learned via 1st-person VR exhibited significantly better prosthetic limb control skills than those in a control group. In another study involving Tai Chi novices [25], positive results, specifically limb position error reduction, were observed while practicing using a system that rendered the movement of an instructor's avatar atop the users'.

Limitations of AO have been alluded to or observed in previous studies. For example, in the previously mentioned prosthetic-limb-training application [24], the benefits of AO training were observed for a bow knot test (asking a subject to use an unaccompanied hand and prosthetic arm to hold and tie shoelaces), but not for a simpler-to-conduct test that entailed picking up and moving blocks with the prosthetic limb. It was, therefore, inferred that VR-based AO is likely to be effective in relatively difficult, bimanual tasks of relatively high coordination [24], a description that fits the action of drumming complicated rhythms.

Sense of agency (SoA), the subjective feeling of initiating and controlling an action, is also an area where AO-based approaches may not be adequate [26]. The body schema is the unconscious representation of the self that supports all types of motor actions, and it can be updated via repeated body movements with a high SoA [26]. AO-based learning has a generally lower level of SoA, as it consists solely of following the body movements of another person or avatar. This, in turn, lowers the potential for a user to acquire an optimal body schema, preventing motor skill learning from achieving maximum efficiency.

Embodied simulation is a concept related to Action Observation and explores the possibility of sharing experiential realities. Paint with Me [27] is an application with which art students can embody the experience of a teacher to learn via movement-based instruction in mixed reality. Much like 1st-person AO-based motion learning, the user learns painting techniques by following another's movements via hand-tracking.

1.9. Virtual Co-Embodiment (VC)

Avatar embodiment is most conventionally implemented via a humanoid avatar with limbs controlled by a single user. Virtual Co-embodiment (VC), a relatively new concept and field of research, describes applications of embodiment with multiple users collaboratively controlling "joint actions" of a single avatar [28]. Studies show that the sensation of agency regarding actions not performed by oneself can arise [29], and VC is a means of achieving that effect.

Real-life analogs for joint actions include two-player pinball, when each player controls one flipper of a shared machine, and three-legged racing, when the inner legs of side-by-side runners are lashed together. Also related is the personal psychological attachment to a body part that is not truly attached to one's body, such as a fake limb. A body part assimilated into one's notional self can be called a body ownership illusion.

One implementation of VC includes weighted average-based VC, whereby the positions of multiple users' limbs are tracked and averaged to control an avatar. In a recent study [26], a weighted average-based VC application was developed for motor skill training. When used in a subjective experiment, learning efficiency was found to be higher than that of AO (perspective sharing with a more-skilled user) and solitary practice.

Allowing multiple users to control different sections of an avatar's limbs is a separate approach to VC [30]. By allowing two users to collaborate and control a robotic arm, more rotational freedom and inhuman movements were realized. Similar to weighted average-based VC, the adjustable mixing of movements allows an expert to scalably support a beginner trying to practice stable movements [30].

Another approach is called body-part-segmented VC, whereby multiple users control separate limbs as a single avatar. In a body-part-segmented VC study [28], SoA and ownership towards the avatar arm controlled by a partner were significantly higher when both subjects could see each others' targets, demonstrating the importance of visual information in terms of enhancing embodiment towards limbs controlled by others.

A focus of our project entails actualization of body-part-segmented VC with an autonomous entity. Rather than multiple users sharing control of a virtual avatar, control is shared between a single user and a program. This concept was a part of VC's original definition [31], but to the authors' knowledge, the realization of this concept for motor learning pedagogy did not heretofore exist.

An example of VC where a human shares an avatar with an autonomous agent is in a contribution exploring social presence for robots and conversational agents [32]. Pondering whether or not autonomous entities and robots should have a human-like social presence, that work experimented with social presentation bound to a single body and more extraordinary capabilities where social presence could be migrated across avatars. A study showed that participants felt comfortable with such capable agents and with the comparatively seamless and efficient experiences that resulted.

1.10. Rhythm Games

A genre of music-based video games, so-called "rhythm games", allows players to play along to music on drum pads or other instrument-like interfaces. Using a dynamic "note highway" or "notefall" visual display, players interpret and perform rhythms to play, as running and final scores are displayed and reported. Representative instances include Parappa The Rappa (1996) [33], Guitar Hero Live (2015) [34], and Beat Saber (2018) [35].

While these games might provide experience somewhat similar to physical and mental musical practice, the lack of authentic musical instruments and formal musical notation prevent the genre from being widely considered a serious educational tool. A previous study [36] suggested that players skilled at rhythm games improve performance in terms of perception and action rather than sensitivity to metrical structure. Other music video games such as Rocksmith [37] teach players how to play the guitar, but their interface is based on notefall and guitar tablature and does not incorporate AO or VC techniques.

NeuralDrum is an extended reality interactive game that measures and analyzes the brain signals of two users as they play drums together [38]. Brain synchronization allows people to better work together [39], occurring when two individuals perform the same activity simultaneously. The Phase Locking Value is calculated to measure brain synchronicity, and it is used to alter the audio and visuals within the game in real-time. While subjects reported relatively high enjoyment, synchronization was not easily perceivable between pairs of subjects [38].

2. System

2.1. Hardware and Sensors

We developed a system to explore AO- and VC-informed drumming pedagogy. Our system uses a Roland TD-25 Drum Sound Module sensing user performance on Roland PD-7 electronic drum pads, the integrated sensors on the Meta Quest 2 head-mounted display (HMD), and a workstation or laptop computer running the Unity 2023.1.0b6 game engine. The components and connections are shown in Figure 3. Each PD-7 dual-trigger drum pad is a 7.5″ rubber pad with two piezo-electric transducers embedded in the pad's rim and center [40]. When a pad is struck in either area, an analog, impulse-like signal is output via a stereo 1/4″ cable to the sound module via the trigger input.

A workstation runs the virtual scene developed in the Unity editor while connected to the sound module and a Quest 2 VR helmet. A high-speed USB 3 cable is used for the HMD connection, which allows it to be used as a thin client display.

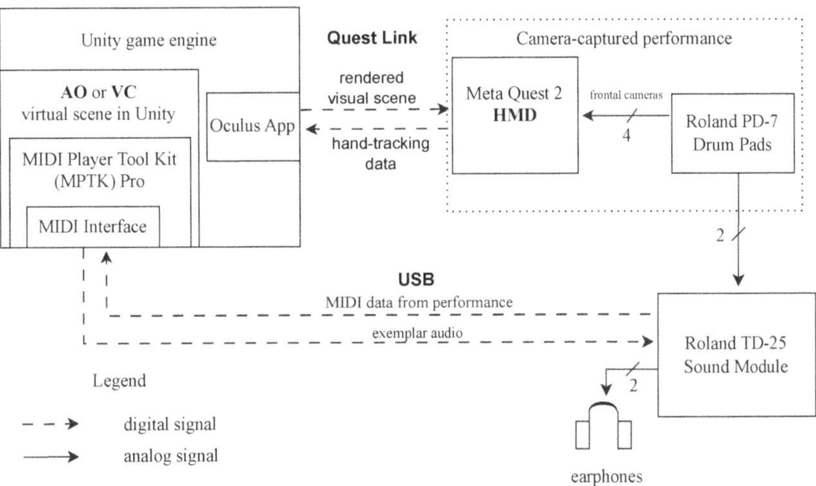

Figure 3. System schematic.

2.2. Development

The Unity virtual environment (VE) for our system utilizes assets including Maestro Midi Player Tool Kit Pro (for real-time MIDI capability) [41] and Open XR (for VR/AR development) [42]. The Oculus desktop application and the Quest Link function on the Quest 2 are used to display the scene through the HMD.

Each virtual scene is populated with drum and drum stick elements. The positions and sizes of these GameObjects correspond to the physical arrangement of the Roland drum pads and the physical drum sticks held by the user. Using programmed keyframe animation, the drum heads atop the drums within the scene momentarily dilate when struck. The exemplar drum sticks similarly use programmed animations that the user is expected to look at while practicing. A short video summary of the system is publicly available: https://www.youtube.com/watch?v=d-pw7aZpvME.

The binaural auditory soundscape (a mix of the user performance on the drum pads and the virtual scene's audio) is displayed to the user via stereo earphones driven by the audio output of the Roland sound module. Via built-in MIDI soundfonts, the module can express performance on the drum pads with a variety of virtual percussive digital instruments. The module sends the MIDI data of the user's performance to the computer over USB while also receiving the virtual scene's rendered audio stream from the workstation. Audio displayed to the user via earphones is a perceptually balanced mix of the exemplar avatar's performance in the virtual scene and the user's own performance.

The computer reads the MIDI data of the performance to control concurrent feedback elements in the scene for the user. The timing of the notes from the exemplar avatar's performance and the user's performance are compared in real-time, determining whether each note is played too late, too early, or accurately timed (within a tolerance). These data are used to control illuminating light objects within the scene. Temporary illumination of a red light above the virtual drum positioned towards the drum head indicates the last strike was too late, while a green light signifies the note was played too early. If the pad is struck within the timing tolerance, neither light object is switched on. This is similar to practice with a metronome: a click track can be heard when successive notes are played on either side of the beat, but often becomes nearly inaudibly masked when a player "locks in" to the metronome's expressed tempo. Tolerances for this are currently set such that, if a note is played within 25% of its duration, either ahead of or behind the beat, it is considered on-time.

Another aspect of the real-time feedback is the user's hands being tracked. Through this visual cue, timing of strokes is better understood and potentially easier to compare with those of the exemplar performance, encouraging microadjustments. The user's perspective is entirely virtual, but hand-tracking allows a virtual representation of the user's hands to be seen.

2.3. Action Observation (AO) Scene

In the AO scene, the user wears an HMD and experiences a virtual scene while drumming, as shown in Figure 4. The user is expected to drum in-sync with the exemplar sticks of the virtual scene as the concurrent feedback is conveyed. Because of the lack of SoA, it is expected that this virtual scene is not as beneficial as a VC-based scene, but it is predicted to be more effective than a video. The AO scene may not be optimized for the cognitive phase of learning, but it could excel in the associative and autonomous phases.The user's point of view of the virtual scene set up for AO practice can be seen in Figure 5.

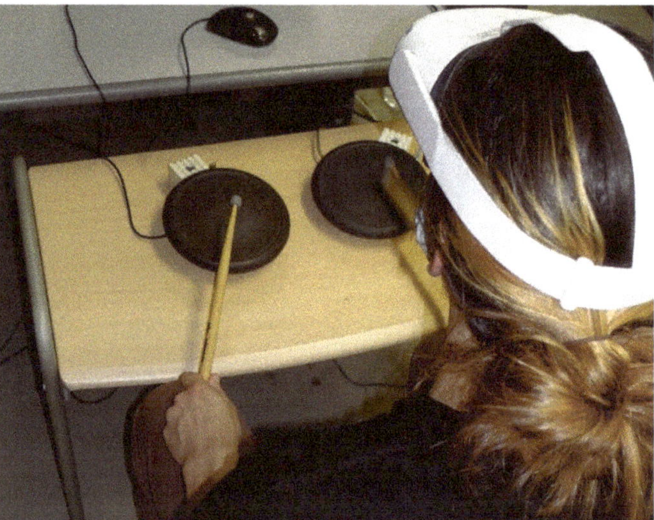

Figure 4. A user interacting with the tool in AO polyrhythm mode: the user performs a polyrhythm on two drum pads in sync with an exemplar avatar observed in the virtual scene.

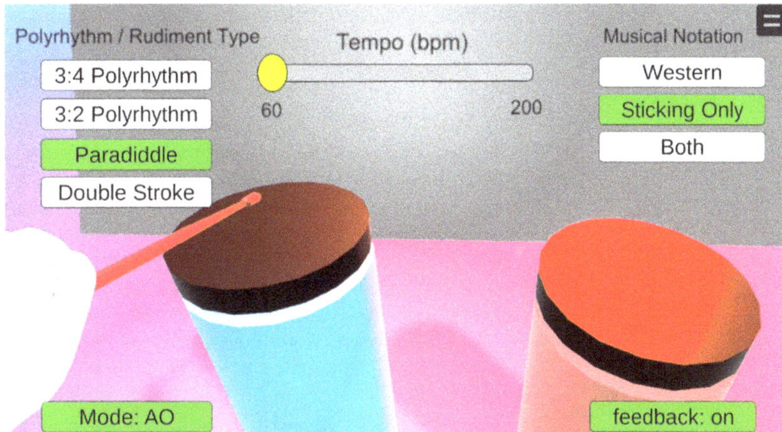

Figure 5. The developed VE can host practice sessions for different rudiments and polyrhythms via AO or VC with real-time hand-tracking. The UI includes controls to achieve a wide range of tempos, feedback options, and other visual aids. This screenshot shows the AO virtual scene at a moment when the user is playing slightly ahead of the beat, as indicated by the red overhead light GameObjects.

2.4. Virtual Co-Embodiment (VC) Scene: The "Halvatar"

While weighted average-based VR-based VC has had observable previous success, foreseeable problems arise when considering this approach for an application purposed for the teaching of drumming. The trajectory of a drum stick in use is inherently full of sharp turns and direction changes that can occur almost instantaneously. While weighted average-based VC techniques have had success in slower actions with smoother trajectories of motion, applying these same techniques to a drumming tool could result in double hits or glitchy motion when multiple users controlling an avatar are at different phases of a drum stroke. Therefore, an approach similar to body-part-segmented VC, as opposed to a method based on weighted averaging, was implemented for this tool.

We call the variation on the body-part-segmented VC concept proposed by this work a "halvatar." Rather than delegating control of an avatar's disparate limbs to human users, a portion of the avatar is controlled by a single user, while another portion (in our approach, the opposing hand and corresponding stick) is controlled exemplarily through programmed animations. This is opposite to body ownership illusions, as parts of a self-identified avatar are shared with an external agent. Embodying a halvatar in practice session results in a duet with an idealized half. This approach avoids complications that would otherwise arise through a weighted average-based VC approach while still maintaining higher levels of SoA.

This approach is particularly suited for polyrhythmic applications. Since the primary pulse and counterrhythm elements of a polyrhythm can be thought of as two musical phrases referencing different tempos, the halvatar is similarly split, the primary pulse expressed on one side and its counterrhythm expressed on the other. VC compartmentalizes the polyrhythm to increase focus on its simpler elements, seemingly helping most with the cognitive phase of learning.

The practicality of the halvatar extends beyond polyrhythms as well. Drum lines in marching bands often consist of multiple-sized bass drums, each carried and played by a different drummer. Sometimes, the bass drum rhythm is "split" amongst members playing different parts of the line to take on melodic properties [43]. Only when an entire line's members are successfully performing their disparate parts together can the bass drum part be heard in full. Hocketing is a similar and more-generalized concept that involves instruments (not only drums) playing interlaced portions of a single musical line [44]. The halvatar's self-duo concept seems to work well for such call-and-response arrangement styles in addition to polyrhythmic phrases.

3. Pilot Study

3.1. Pilot Study Summary

We conducted a pilot study to broadly explore the quantitative effects of our VE and to gather qualitative feedback from subjects. Specifically, we addressed the two following questions. Do practice sessions for drumming using VR-based AO or VC tools offer a measurable benefit over standard practice methods? How do users respond to our development and to practicing drumming in VR in general?

Based on the previously discussed findings, VR-based AO and VC tools have the potential to yield observable positive effects in motor learning pedagogy applications. When considering these techniques in drumming applications, we hypothesized that practicing with either modality would yield better results than practicing with video demonstrations.

To test this hypothesis and to gather baseline data and user feedback moving forward, we conducted a pilot study in which subjects' execution of drumming exercises was recorded before and after three different types of short practice sessions: Video, VR-based AO, and VR-based VC. Subjects were randomly assigned to a group and used one of these three practice methods between an initial and a final recording. To gather subjective feedback, subjects were asked to complete a user experience questionnaire (UEQ) shortly after experiencing his/her assigned practice mode.

3.2. Participants

There were 15 subjects (12 male, 3 female), students recruited from the university, who participated in the pilot study (age: $M = 23 \pm 4(SD)$). Before the start of the experiment, subjects self-reported their age, dominant hand, whether or not they were able to read Western musical notation, musical experience, and primary and secondary instruments (where applicable). Of these, 86% were right-handed, 40% had some prior experience with musical instrument practice, and 20% were currently active performing members of university music clubs. Subjects were randomly assigned to one of three groups corresponding to three modalities they were to use when practicing two rudiments and two polyrhythms throughout the experimental procedure, as shown in Figure 6. The group assignments (Video, AO, and VC) came into effect during the practice session phase of each section (tutorial, rudimental, polyrhythmic) of the experiment. Subjects were given information and instructions via a script read aloud before they were given an informed consent form to sign. The experiment then proceeded from the tutorial.

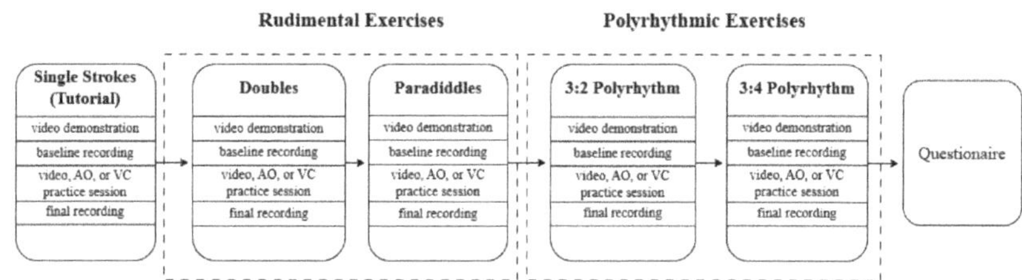

Figure 6. Procedure for the pilot study.

3.3. Tutorial and the Four-Phase Procedure of the Experiment

All subjects first completed a four-phase introductory tutorial to help familiarize themselves with the flow of the experiment and their assigned system. Subjects were told that they could ask questions or adjust the height of the seat and drums any time throughout this tutorial. Simple single-stroke eighth notes were used as the tutorial's lesson, as shown in Figure 7.

Figure 7. Western notation representation for single-stroke eighth notes, the exercise used for the tutorial.

The first phase of the tutorial consisted of showing a video demonstration to the subject via a computer monitor and stereo loudspeakers. The video was a performance of single-stroke eighth notes filmed with an overhead camera placed behind a drummer, as shown in Figure 8. The video started with a metronome for one measure before the demonstration of the exercise began. The exercise was played for four measures, a 22 s video in total. The audio of the performance in the video was quantized and panned, to rectify timing imprecision and to achieve stereo separation, respectively.

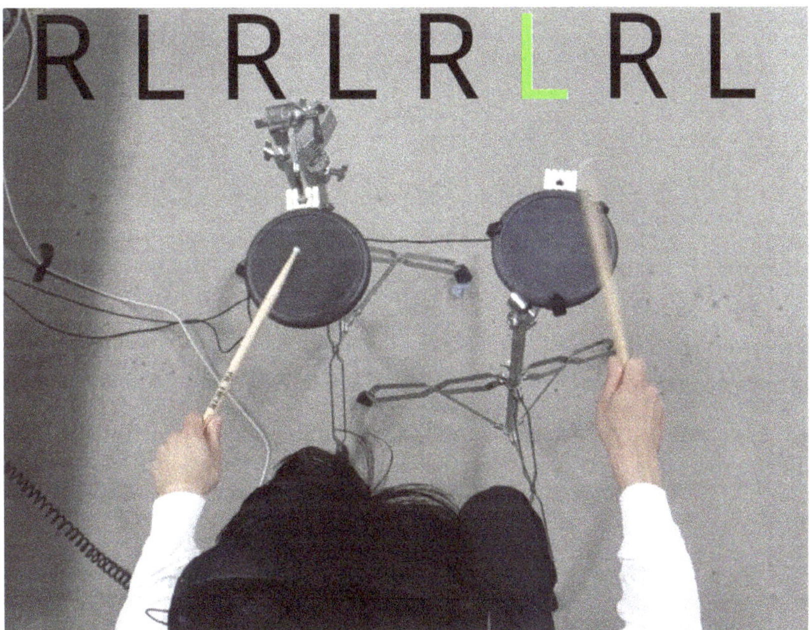

Figure 8. The first phase of all sections of the pilot study was a 3rd-person (overhead view) video demonstration of the corresponding exercise.

Subjects were asked to watch the video attentively before receiving a pair of medium-weight drum sticks and recording their own four-measure take of the previously observed single-stroke eighth note rhythm. As in the overhead demonstration video, the metronome played for the subject at eighth note = 120 beats per minute (bpm), where the metronome expressed every eighth note beat and added a downbeat accent on every fourth beat, or the "1" and "3" of each four-beat measure.

After the baseline recording, subjects were asked to conduct a short practice session with their assigned mode of practice: Video, AO, or VC. If the subject was in either VR group, they were helped with the set-up and fitting of the HMD. Regardless of the assigned group, each practice session had a duration of about two minutes, six repetitions of the exercise, with a small break after each repetition.

In the case of subjects in the Video or AO groups, they were asked to practice along with their assigned exemplar. In the VC case, subjects were given slightly different instructions. Rather than simply playing along with the exemplar, a third of the session was for practicing the left-hand rhythm while the halvatar's right hand drummed autonomously. Another third was to practice with the right hand while the halvatar's left hand autonomously drummed. A final third was for practicing both hands simultaneously.

The last phase of the tutorial was a final recording of the single-stroke eighth notes. Each subject was instructed to once again record a four-measure performance of the previously practiced eighth note exercise. Though this performance was recorded, the data were not analyzed as this portion of the experiment was only a tutorial, intended to get the subject accustomed to the flow of the following four sections. For the proceeding exercises, the pre- and post-practice session recordings were both recorded and analyzed, to measure the improvement of each subject.

3.4. Rudiments

After the tutorial, subjects were asked to complete the same four-phase procedure for four musical exercises, the first two of which were rudiments, specifically doubles and paradiddles, shown in Figure 9. Rudiments were chosen as they are the building blocks of drumming, practiced by players of all levels. Beginners may practice them slowly to learn the sticking, while advanced players may practice them at challenging tempos or with heightened focus on timing accuracy.

Figure 9. Western notation representation for eighth note doubles and paradiddles, the exercises used for the pilot study's rudimental section.

Much like the tutorial, each subject first viewed an overhead video (3rd-person perspective) of a doubles performance before baseline performance recording. Then, the assigned group-dependent practice session and final recording were conducted. After this, the process repeated for paradiddles.

3.5. Polyrhythms

Much like the doubles and paradiddles sections, the same four-phase process was repeated for two polyrhythms. The first learned polyrhythm was the 3:2 polyrhythm, as it is the most common and may have been heard or performed without necessarily having been recognized as a polyrhythm. Following that was the 3:4 polyrhythm, which, while less common, can also be found in pop music.

It is likely that both of these polyrhythms were considered to be more difficult than the double-stroke eighth notes or paradiddles by most subjects. This is because the interval between the exercise's notes varies, and because some notes of the counterrhythm (bottom lines on Figure 10) fall between divisions of the beat expressed by the metronome.

Figure 10. Western notation representation for 3:2 and 3:4 polyrhythms, the exercises used for the polyrhythmic section.

3.6. Practice Session Phase

As previously outlined, each exercise, including the tutorial, contained four phases, corresponding to the rows of the blocks in Figure 6. The third phase was the practice phase and was the only phase dependent on the subjects' assigned group. The first group was the group assigned to use videos shot in 1st-person perspective to practice. Within the videos' frames were the exemplar drummer's hands, sticks, and the two pads being performed on, as shown in Figure 11. Like the overhead 3rd-person perspective demonstration videos, these videos started with one measure of the metronome followed by four measures of performance. Afterward, there was a short pause and the video was looped five additional times.

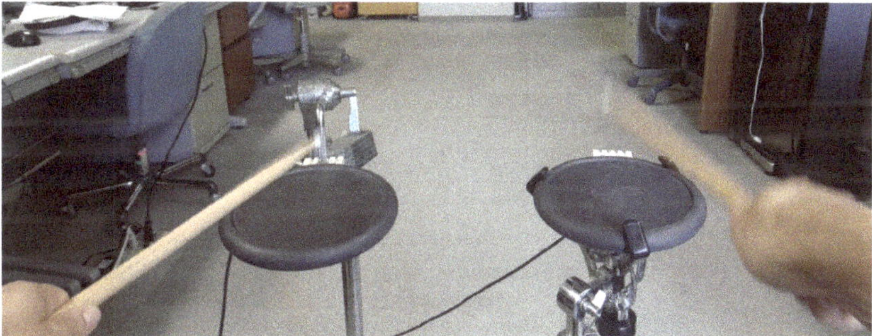

Figure 11. For subjects of the Video group, the third phase of all sections of the pilot study was a 1st-person ("PoV") video demonstration of the corresponding exercise looped 6 times.

The other two groups, VR-based AO and VR-based VC, similarly presented the subject with a 1st-person perspective and a performance of the same length, but within a virtual scene displayed to the user via HMD. Shown in Figures 12 and 13, the interface of the virtual scene was simplified from that of Figure 5. Rather than present the user with options and drop down menus, the virtual scene featured only the sticking cues, drums, and drum sticks. The real-time feedback lighting GameObjects were similarly disabled, as users of the Video group did not have such aids, and it was desired to keep the variables as consistent as possible across the three groups.

Figure 12. Polyrhythmic AO virtual scene: The user's hands are tracked and displayed in a scene with automated drum sticks performing exemplarily.

Figure 13. Polyrhythmic VC scene: The left hand of the halvatar is controlled by the user, and movements of the right hand and stick are controlled via programmed animation.

There was no verbal instruction in the video, AO, or VC practice session media. It was also decided to not display Western musical notation of the exercise due to varying familiarity with it among subjects. However, an animated sticking diagram was displayed in all practice modes, as shown in Figure 14. This helped subjects keep track of where they were within the exercise.

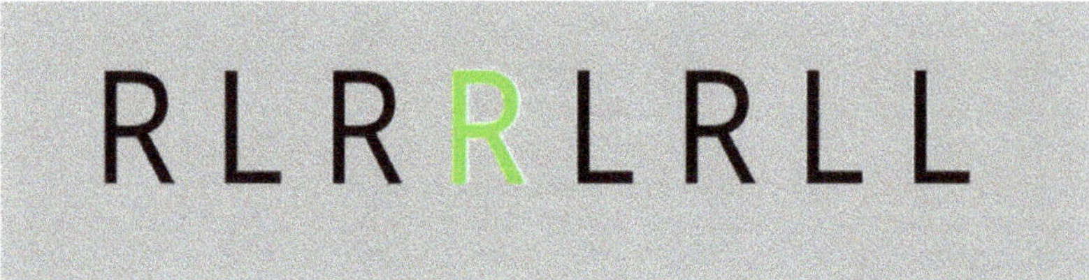

Figure 14. In the video and AO and VC virtual scenes, a visual aid to help subjects learn the sticking during practice sessions is included. Similar to the sticking diagram in Figure 8's 3rd-person video, the sticking is displayed at the top of the video screen and placed behind and above the drums in the virtual scenes. This shows the entire sticking pattern while highlighting the current note.

In all the practice session versions, the musical exercise, tempo (eighth note = 120 bpm), and expression of the metronome were kept consistent. The right drum pad corresponded to a floor tom sound, and the left corresponded to a snares-off snare sound. This was chosen for both sonic and practical reasons. The floor tom and snare occupy significantly different frequency ranges, and the set-up used in the experiment was similar to common placements of a snare and floor tom pair.

Both drum samples were stereo .WAV recordings, where the sustain of the drums' resonance was kept short through acoustic treatment and audio processing. Damper pads and transient shaper audio plug-ins were used to shorten the length of the note, as this articulated each hit and better clarified to the subject the timing of the strike.

For the video playback and AO groups, subjects were asked to study or play along with the teacher or exemplar avatar for all six repetitions of the exercise, whereas in the VC group, subjects performed the right and left hands' rhythm individually twice before combining them in simultaneous performance for an additional two repetitions.

3.7. Post-Experiment Questionnaire

After completing the four-phase procedure for the 3:4 polyrhythm, subjects were asked to complete the user experience questionnaire (UEQ) [45] to gauge the quality of experience of the video, AO, and VC modalities. The scales of the questionnaire covered a comprehensive impression of user experience. Both classical usability aspects (efficiency, perspicuity, dependability) and user experience aspects (attractiveness, stimulation, novelty) were measured. Within the survey, users were prompted with a spectrum between bipolar adjectives. Using a discrete Likert scale from 1 to 7, subjects' evaluations were made. Per the UEQs' standardized methods, scores were later normalized to a -3 to $+3$ scale for final analysis.

4. Results

4.1. Results of Rudimental Exercises

All subjects of the pilot study successfully performed doubles during both the second and fourth phases, before and after the practice session. Three subjects, one subject in each of the video, AO, and VC groups, were not able to play the paradiddle at the time of the initial recording, but were able to after the practice session. One subject in the VC group was not able to play the paradiddle before or after the Phase 3 practice session. These results are shown in Figure 15.

	Number of subjects (out of 5) who successfully performed Doubles	
	Before practice session	After practice session
Video	5	5
AO	5	5
VC	5	5

	Number of subjects (out of 5) who successfully performed Paradiddles	
	Before practice session	After practice session
Video	4	5
AO	4	5
VC	3	4

Figure 15. Number of subjects who could perform the correct sticking for the rudimental exercises before and after their assigned practice session.

For subjects who successfully played the exercise before and after the practice session, jitter was calculated to determine improvement in each subject's rhythmic accuracy. As shown in Figure 16, unquantized performance's timing nonidealities can be seen and heard in a digital audio workstation, particularly when there is a metronome click track present for reference.

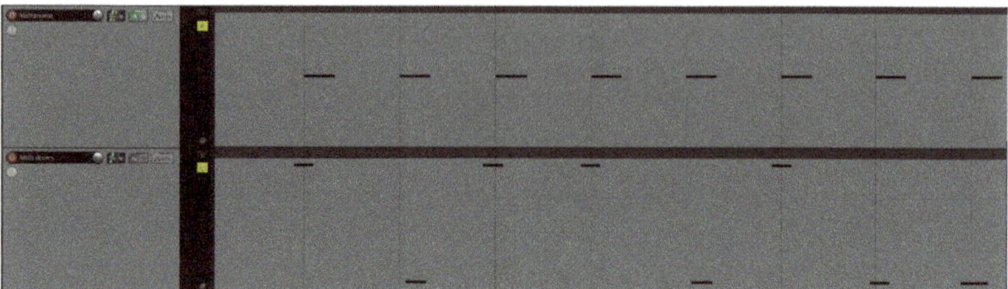

Figure 16. When viewed in a DAW, jitter can be observed between the uniform timing of the metronome (above) and the recording of a paradiddle (below). Relative to the gridline, some strikes of the paradiddle are early or late, whereas the metronome's timing is uniform throughout.

Recordings made by the subjects were exported as MIDI files and converted into csv files with an online tool [46]. The portion of the MIDI files containing the timing of each stroke was used and compared against the timing of the metronome. Time differentials (in fractions of a beat, relative to the tempo of eighth note = 120 bpm) between each subject's every stroke and the metronome were averaged for this calculation.

For doubles, all five subjects in each group had before and after data to compare. VC yielded an average decrease of a quarter of a beat (about a thirty-second note at an eighth note = 120 bpm, or about 120 ms), and video had an increase in jitter of about 80 ms. Conversely, in the paradiddle case, subjects had an increase in jitter of about 130 ms after practicing with VC, whereas the average jitter after a practice session with video had a near-zero decrease. Despite a significant decrease in jitter for VC subjects learning doubles, an improvement on the subjects' averaging timing accuracy, there was an increase in jitter during the paradiddles case, alluding to inconsistency. Changes in jitter for all rhythmic exercises and modes of practice are plotted in Figure 17.

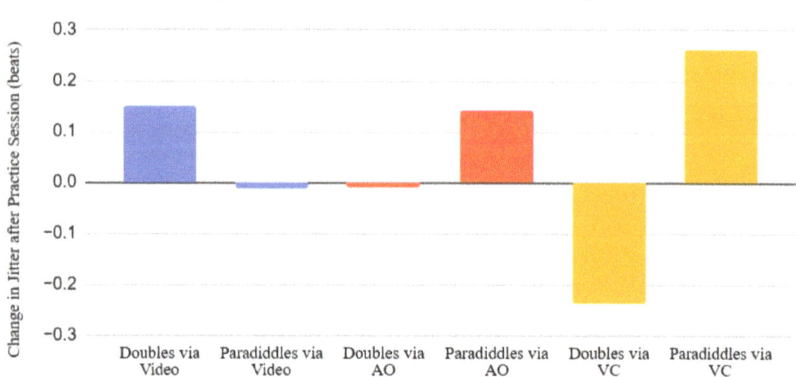

Figure 17. Plot to visualize increases and decreases in average jitter for the two pilot study rudiments after video, AO, and VC practice sessions, respectively.

In many cases, subjects' technique and grip on the stick resulted in double triggering, meaning one intended stroke would be captured as two hits. Such double triggering was ignored, and only the first note of subjects' strokes were considered for jitter calculations. It was also common that subjects' continued to play the rhythmic exercise after the instructed four measures during recording. In those cases, notes after this duration were ignored. Some subjects stopped the exercise a few notes early. In those cases, the notes not played were left out of the averaging calculations.

4.2. Results of Polyrhythmic Exercises

As expected, subjects were not able to perform the two polyrhythmic exercises as proficiently as rudiments. It was decided before the experiment to measure the improvement of polyrhythms via the number of consecutive successful polyrhythmic strikes by a subject. Most subjects were not able to maintain a polyrhythm, especially the 3:4 polyrhythm, for the entirety of the recording. When analyzing the performances, the maximum number of consecutive hits per two-measure phrase (an 8-note phrase for the 3:2 polyrhythm and a 12-note phrase for the 3:4 polyrhythm) during the entire performance was determined for each subject.

In Figure 18, the average numbers of consecutive successful hits for the 3:2 polyrhythm before and after the practice sessions are displayed, with respect to practice mode. All subjects were able to perform eight notes of this polyrhythm correctly with the exception of one subject of the Video group, who scored 5 out of 8 hits.

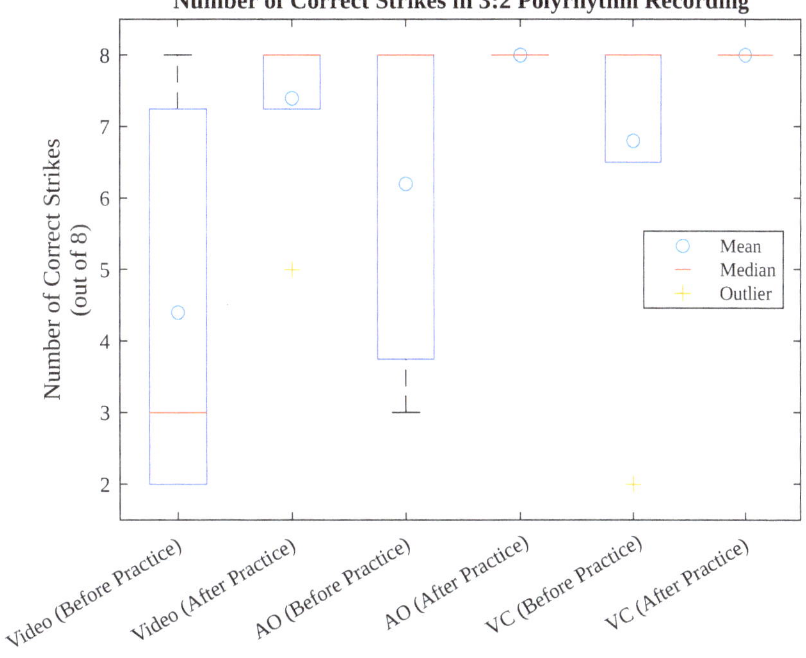

Figure 18. Comparison of successful strikes in 3:2 polyrhythm performance before and after practice session for Video, AO, and VC groups.

In Figure 19, the before and after data for the 3:4 polyrhythm are shown. In both figures, all practice modes show improvement, with some cases reaching the corresponding maximum number of hits (eight or twelve). In terms of the median and mean, the VC group's after-practice session results outperformed the Video group's results, whereas only the AO group's median was higher then the Video group's. The Video group's mean result after the practice session was higher than the AO's after-practice session mean.

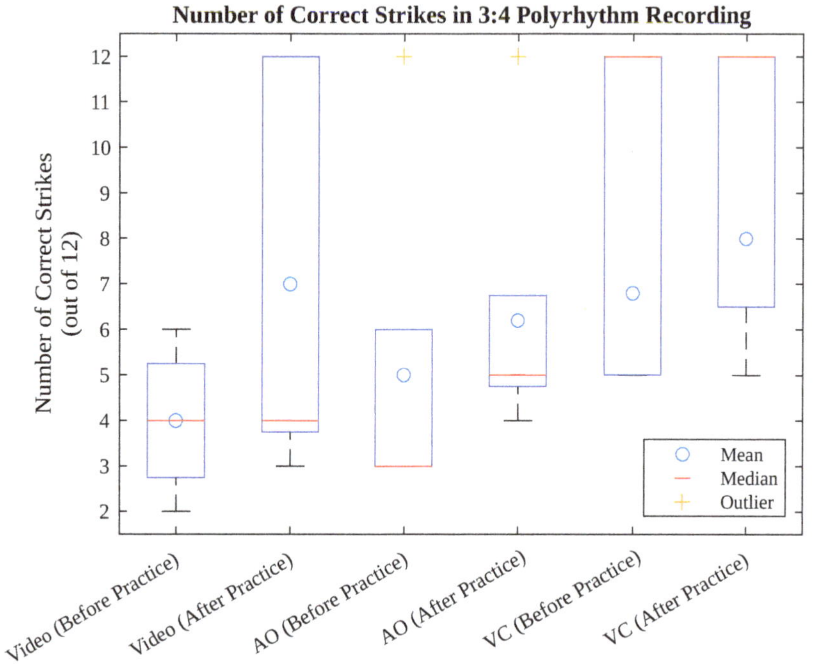

Figure 19. Comparison of successful strikes in 3:4 polyrhythm performance before and after practice session for video, AO, and VC groups.

One unanticipated complication not accounted for before the experiment was subjects' tendencies to invert (mirror-reflect) the sticking, which happened for 10 subjects (3 in video condition, 4 in AO condition, and 3 in VC condition) in the initial recording phase of the 3:4 polyrhythm. After the practice session, only 2 subjects (1 video and 1 VC) performed their final recording with inverted sticking.

4.3. UEQ Questionnaire

The UEQ survey was used to to measure each subject's evaluation of the user experience. After each subject used the electronic drum pads and either the video, AO, or VC tools, an experience that lasted about 30 min, the 26-question survey was administered. Through UEQ data analysis tools, the results were organized into six factors, and a high-level comparison can be seen in Figures 20–22. The data's range was scaled from −3 to +3, and as shown, the subjects' response to the video was mostly comparable with the two VR experiences. Exceptions included perceived novelty, where the VC group's mean achieved a score 1.3 points higher than that of subjects using video, and efficiency, where the Video group's mean achieved a score 0.55 points higher than the VC group.

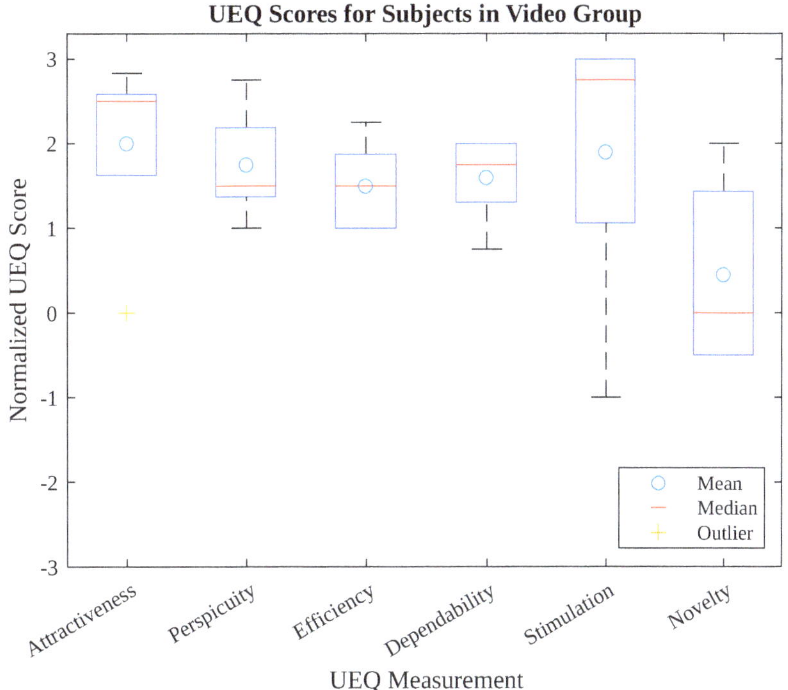

Figure 20. Questionnaire results from Video subjects after the experiment.

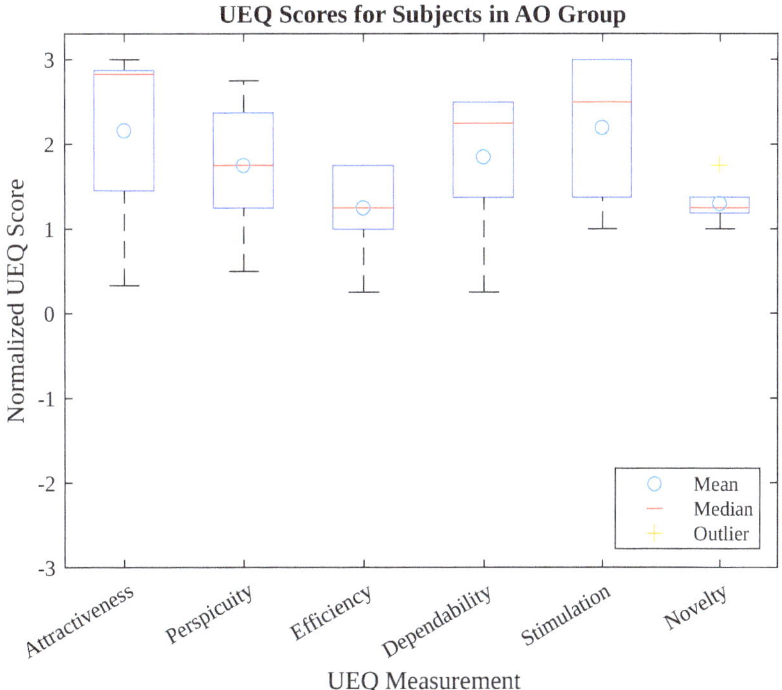

Figure 21. Questionnaire results from AO subjects after the experiment.

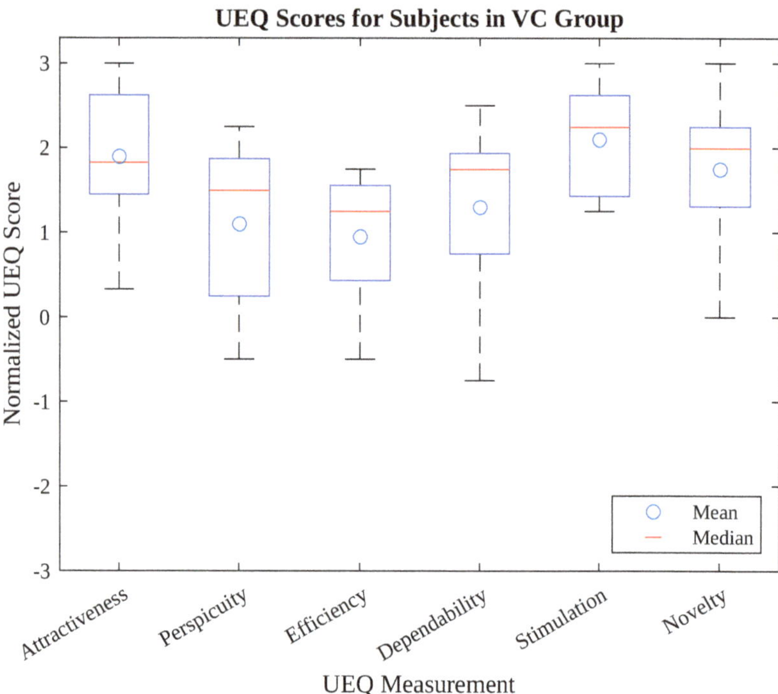

Figure 22. Questionnaire results from VC subjects after the experiment.

4.4. Other Observations

Due to his familiarity with the system, one of the authors conducted the pilot study, including recording the subjects and setting up the demonstration videos and HMD used by the subjects. During the pilot study, several behaviors were observed.

Some of the students within the AO and VC groups did not directly view or interact with the drum stick animations in the practice session. Instead, they were entirely focused on the dynamic sticking charts in front of the drums. Although these were deliberately placed in the VR scene such that both could be viewed simultaneously and subjects were instructed to focus on the movement of the animated sticks, some students left the moving sticks mostly out of their field of view. In previous AO and VC studies, allowing the subjects to clearly see such movements has been correlated with successful results [24,28]; so it is possible that this unanticipated complication had an effect on the AO and VC results in this pilot study.

One student performed a 3:2 polyrhythm perfectly after an AO practice session where he did not perform it correctly once, which was surprising. In most cases, successful performances following the practice session meant that the subject was able to follow the exemplar and show some comfort with the rhythm before final recording.

Feedback regarding general excitement and enjoyment of the pedagogical technique seemed high. Students who learned even with just the video expressed positive final comments in the questionnaire. For subjects with and without musical experience, the process of learning how to drum was perceived as fun.

5. Discussion

This development was successful in creating a tool for rhythmic pedagogy that incorporates both AO and VC techniques. The pilot study, while successful in some respects, did not show promising results in others. The VC group achieved a significantly higher novelty score on the UEQ survey relative to the Video group, but received a lower score in the efficiency category. Similarly, the average jitter results for subjects using VC to practice doubles showed drastic improvement compared to the other groups, but also a significant increase in average jitter relative to the other groups after practicing paradiddles. The VC condition taught a paradiddle to one of two subjects who could not successfully perform the rudiment prior to the practice session, whereas the other two conditions taught one out of one subject.

The skill and experience level of recruited subjects had wide variation, with many subjects having no musical experience and some being performing drummers in bands. In addition, the exemplar avatar in the AO scene and the autonomous portion of the halvatar in the VC scene were designed with programmed animations rather than imported Mocap data. This could have had an effect on subjects' perception and experience compared with the video, as that practice session modality was based on a human performance. Goals and ideas for future work follow.

5.1. Further Quantitative Experimenting

The pilot study was successful in obtaining initial data and subjective feedback while various, mostly musically novice, users interacted with our system. In the near future, we would like to perform more rigorous testing with subjects.

This includes advanced polyrhythm pedagogy trials with experienced drummers. In past AO studies [24], it was presumed that VR-based AO is likely to be effective in complex, bimanual tasks. Having an experienced drummer learn a seldom-performed, technically difficult pattern such as a 13:9 polyrhythm could yield interesting discoveries.

Retention of learning was also not explored within the scope of the pilot study. VC excels in motor learning because of its ability to update the body schema of users, as opposed to limiting users' involvement to following a teacher. Because a relationship between the body schema and long-term retention of motor skills has been observed in previous studies [26], it seems possible that the VC mode of our developed system has unexplored potential in cases of retention, as opposed to immediate results. Such analysis, however, was beyond the scope of the pilot study. A future study across multiple days might show more-positive effects of VC-based drumming practice.

The pilot study was also limited to a single tempo for all groups. Modulation of tempo, however, is a commonly practiced rehearsal technique [47] and worth considering in future experiments. Comfortable tempos are subject-dependent, so granting subjects of future studies some control of tempo is desirable. Real-time feedback cues comprise another feature of the VE that was not utilized in the pilot study. Including this functionality in a build used in an experiment would similarly be of interest.

5.2. Drumming with Brushes

Brushes started to be used by drummers in the early 1900s as an alternative to sticks. Although still used in contemporary performance, brushes were often used to maintain quieter volumes on drums in acoustic musical settings of the last century. In addition to the staccato articulation that results from playing brushes in the same overhead manner as drum sticks, brushes could also achieve more legato, or smooth-sounding, notes by using lateral strokes. Brushes can also achieve infinite sustain by swirling bristles around a drum head in circular motion, very unlike the dynamic volumes that drums usually produce. This technique, or the combination of this technique with lateral or overhead strokes, makes up the bulk of the musical language of drumming with brushes.

Building off the successes seen during the experimental phase of motor learning as taught by weighted average-based VC [26], a next step of interest is developing a tool to teach the motions of brushes. Because brushes comprise a style of drumming that consists of smoother movement with fewer sharp direction changes, it could match well with a weighted average-based VC approach.

5.3. User-Shared Body-Part-Segmented Virtual Co-Embodiment Drumming Application

As the development of a body-part-segmented VC drumming application that uses programmed animations has been validated, a next step is to develop a tool that allows two drumming users to simultaneously occupy the same body-part-segmented avatar. In a previous motor-learning VC study [26], it was thought to be important that the teacher be a human who could make quick adjustments to adapt to the student's inexperience with the system and motion being taught. If this approach were to be realized in a drumming application, a human teacher could potentially correct for microtiming issues of a student and could provide alternative benefits. Comparing the results and subject feedback from the halvatar approach and this standard body-part-segmented approach could provide interesting findings.

6. Conclusions

The developed VE offers a new way to teach rhythmic exercises and rhythmic techniques. Technology-accelerated drum instruction often entails watching professional teachers' or musicians' demonstrations from a distance or from a 3rd-person perspective. Our project gives students the opportunity to learn through an exemplar drumming avatar via two VR-based 1st-person perspectives: AO and body-part segmented VC.

Based on the quantitative data of the pilot study, our VR-based AO and VC virtual scenes did not result in the observation of a measurable significant benefit over video-based drumming practice in our experiment. Subjects in all three groups were able to learn new rhythms, and the improvements in jitter for subjects in the VC group lacked consistency. The results of the survey showed VC to have a significantly higher perceived novelty score than the other two modes. However, VC scored slightly lower than AO and video in the other rating scales. Aside from its quantitative and qualitative results, this work also inspired ideas for future extensions and experiments.

This project represents a "softening" or generalization of the VC concept. Rather than rigidly configuring the control:display mapping between human pilots and digital puppets, as instantiated by humanoid avatars, we exposed the rigging to not only humans, but also surrogates, namely programmed animation for our teaching system, but generally any kind of exemplar (keyframe animation, IK-informed events, AI-driven dynamics, etc.), prepared offline (as in our system) or derived from live events determined and delivered at runtime. In ordinary scenarios, a single human controls each avatar "solo", but more-flexible arrangements can articulate an avatar, so that its control can be shared by multiple humans or algorithmic agents. Ultimately, mapping between human users and virtual or robotic expression is flexible, and a single human can "fan-out" to multiple displays, and multiple users in conjunction with computational agents can control arbitrarily assigned "fan-in" inputs.

Author Contributions: Conceptualization, J.P.; methodology, J.P. and M.C.; software, J.P.; validation, J.P. and M.C.; writing—original draft preparation, J.P.; writing—review and editing, M.C.; visualization, J.P. and M.C.; supervision, M.C.; project administration, M.C. and J.P.; funding acquisition, M.C. All authors have read and agreed to the published version of the manuscript.

Funding: The Spatial Media Group at University of Aizu provided some financial support for this project.

Informed Consent Statement: Prior to their participation, the experimental participants provided written consent after being fully informed about the contents of this study. Additionally, explicit written consent was obtained from each subject regarding publication of any potentially identifiable data within this article.

Data Availability Statement: Data acquired and used in this manuscript can be found online: https://github.com/jPinkl/VRDrumPedagogy.

Acknowledgments: We thank the anonymous Referees for their useful and thoughtful suggestions. We also thank the subjects of the experiment and fellow members of the Spatial Media Group, especially Julián Villegas, Peter Kudry, and Juanca Arevalo.

Conflicts of Interest: The authors declare no conflict of interest.

References

1. Shuell, T.J. Phases of meaningful learning. *Rev. Educ. Res.* **1990**, *60*, 531–547. [CrossRef]
2. Anderson, F.; Grossman, T.; Matejka, J.; Fitzmaurice, G. YouMove: Enhancing movement training with an augmented reality mirror. In Proceedings of the ACM Symposium on User Interface Software and Technology, St. Andrews, UK, 8–11 October 2013; pp. 311–320. [CrossRef]
3. Bau, O.; Mackay, W.E. OctoPocus: A dynamic guide for learning gesture-based command sets. In Proceedings of the 21st Annual ACM Symposium on User Interface Software and Technology, Monterey, CA, USA, 19–22 October 2008; pp. 37–46. [CrossRef]
4. Toussaint, G.T. *The Geometry of Musical Rhythm: What Makes a "Good" Rhythm Good?*; CRC Press: Boca Raton, FL, USA, 2019.
5. Magill, R.A.; Lee, T.D. *Motor Learning: Concepts and Applications: Laboratory Manual*; WCB McGraw-Hill: New York, NY, USA, 1998.
6. Schmidt, R.A.; Lee, T.D.; Winstein, C.; Wulf, G.; Zelaznik, H.N. *Motor Control and Learning: A Behavioral Emphasis*; Human Kinetics: Champaign, IL, USA, 2018.
7. Caracciolo, M. *The Experientiality of Narrative: An Enactivist Approach*; Walter de Gruyter GmbH & Co KG: Berlin, Germany, 2014; Volume 43.
8. Crabtree, J. PolyNome. Available online: https://polynome.net (accessed on 15 June 2023).
9. AeroBand. PocketDrum 2 Plus. Available online: https://www.aeroband.net/collections/all-products/products/pocketdrum2-plus (accessed on 15 July 2023).
10. Theodots. HyperDrum. Available online: https://www.theodots.com (accessed on 17 July 2023).
11. Ericsson, K.A.; Hoffman, R.R.; Kozbelt, A. *The Cambridge Handbook of Expertise and Expert Performance*; Cambridge University Press: Cambridge, UK, 2018.
12. Chaffin, R.; Imreh, G.; Crawford, M. *Practicing Perfection: Memory and Piano Performance*; Psychology Press: London, UK, 2005.
13. Carson, R.; Wanamaker, J.A. *International Drum Rudiments*; Alfred Music Publishing: New York, NY, USA, 1984.
14. Frishkopf, M. West African Polyrhythm: Culture, theory, and graphical representation. In Proceedings of the ETLTC: 3rd ACM Chapter Conference on Educational Technology, Language and Technical Communication, Aizu-Wakamatsu, Japan, 27–30 January 2021. [CrossRef]
15. Magadini, P. *Polyrhythms: The Musician's Guide*; Hal Leonard Corporation: Milwaukee, WI, USA, 2001.
16. Waterman, R.A.; Tax, S. African Influence on the Music of the Americas. In *Mother Wit from the Laughing Barrel: Readings in the Interpretation of Afro-American Folklore*; Wiley: Hoboken, NJ, USA, 1973; pp. 81–94.
17. Arom, S. *African Polyphony and Polyrhythm: Musical Structure and Methodology*; Cambridge University Press: Cambridge, UK, 2004.
18. Neely, A. 7:11 Polyrhythms. Available online: https://youtube.com/watch?v=U9CgR2Y6XO4 (accessed on 15 June 2023).
19. Huang, A. Polyrhythms vs. Polymeters. Available online: https://youtube.com/watch?v=htbRx2jgF-E (accessed on 12 June 2023).
20. Harris, D.; Vine, S.; Wilson, M.; McGrath, J.S.; LeBel, M.; Buckingham, G. Action observation for sensorimotor learning in surgery. *J. Br. Surg.* **2018**, *105*, 1713–1720. [CrossRef] [PubMed]
21. Buccino, G.; Arisi, D.; Gough, P.; Aprile, D.; Ferri, C.; Serotti, L.; Tiberti, A.; Fazzi, E. Improving upper limb motor functions through action observation treatment: A pilot study in children with cerebral palsy. *Dev. Med. Child Neurol.* **2012**, *54*, 822–828. [CrossRef] [PubMed]
22. Ertelt, D.; Small, S.; Solodkin, A.; Dettmers, C.; McNamara, A.; Binkofski, F.; Buccino, G. Action observation has a positive impact on rehabilitation of motor deficits after stroke. *Neuroimage* **2007**, *36*, T164–T173. [CrossRef] [PubMed]
23. Pinkl, J.; Cohen, M. Design of a VR Action Observation Tool for Rhythmic Coordination Training. In Proceedings of the IEEE Conference on Virtual Reality and 3D User Interfaces Abstracts and Workshops, Christchurch, New Zealand, 12–16 March 2022; pp. 762–763. [CrossRef]
24. Yoshimura, M.; Kurumadani, H.; Hirata, J.; Osaka, H.; Senoo, K.; Date, S.; Ueda, A.; Ishii, Y.; Kinoshita, S.; Hanayama, K.; et al. Virtual reality-based action observation facilitates the acquisition of body-powered prosthetic control skills. *J. NeuroEng. Rehabil.* **2020**, *17*, 113. [CrossRef] [PubMed]
25. Hoang, T.N.; Reinoso, M.; Vetere, F.; Tanin, E. Onebody: Remote posture guidance system using first person view in virtual environment. In Proceedings of the Nordic Conference on Human-Computer Interaction, Gothenburg, Sweden, 23–27 October 2016; pp. 1–10.
26. Kodama, D.; Mizuho, T.; Hatada, Y.; Narumi, T.; Hirose, M. Effects of Collaborative Training Using Virtual Co-embodiment on Motor Skill Learning. *IEEE Trans. Vis. Comput. Graph.* **2023**, *29*, 2304–2314. [CrossRef] [PubMed]
27. Gerry, L.J. Paint with me: Stimulating creativity and empathy while painting with a painter in virtual reality. *IEEE Trans. Vis. Comput. Graph.* **2017**, *23*, 1418–1426. [CrossRef] [PubMed]
28. Hapuarachchi, H.; Kitazaki, M. Knowing the intention behind limb movements of a partner increases embodiment towards the limb of joint avatar. *Sci. Rep.* **2022**, *12*, 11453. [CrossRef] [PubMed]

29. Wegner, D.M.; Sparrow, B.; Winerman, L. Vicarious agency: Experiencing control over the movements of others. *J. Personal. Soc. Psychol.* **2004**, *86*, 838. [CrossRef] [PubMed]
30. Hagiwara, T.; Katagiri, T.; Yukawa, H.; Ogura, I.; Tanada, R.; Nishimura, T.; Tanaka, Y.; Minamizawa, K. Collaborative avatar platform for collective human expertise. In Proceedings of the SIGGRAPH Asia Emerging Technologies, Tokyo, Japan, 14–17 December 2021; pp. 1–2. [CrossRef]
31. Fribourg, R.; Ogawa, N.; Hoyet, L.; Argelaguet, F.; Narumi, T.; Hirose, M.; Lécuyer, A. Virtual co-embodiment: Evaluation of the sense of agency while sharing the control of a virtual body among two individuals. *IEEE Trans. Vis. Comput. Graph.* **2020**, *27*, 4023–4038. [CrossRef] [PubMed]
32. Luria, M.; Reig, S.; Tan, X.Z.; Steinfeld, A.; Forlizzi, J.; Zimmerman, J. Re-Embodiment and Co-Embodiment: Exploration of social presence for robots and conversational agents. In Proceedings of the Designing Interactive Systems Conference, San Diego, CA, USA, 23–28 June 2019; pp. 633–644.
33. Parappa The Rappa Remastered. Available online: https://rodneyfun.com/parappa (accessed on 13 June 2023).
34. Activision. Guitar Hero Live. Available online: https://support.activision.com/guitar-hero-live (accessed on 16 July 2023).
35. Beat Saber. Available online: https://beatsaber.com (accessed on 17 June 2023).
36. Gaydos, M. Rhythm Games and Learning. In Proceedings of the International Society of the Learning Sciences (ISLS), Chicago, IL, USA, 1–2 July 2010.
37. Rocksmith Plus. Available online: https://ubisoft.com/en-gb/game/rocksmith/plus (accessed on 11 June 2023).
38. Pai, Y.S.; Hajika, R.; Gupta, K.; Sasikumar, P.; Billinghurst, M. NeuralDrum: Perceiving brain synchronicity in XR drumming. In Proceedings of the SIGGRAPH Asia Technical Communications, Virtual Event, Republic of Korea, 1–9 December 2020; pp. 1–4. [CrossRef]
39. Toppi, J.; Borghini, G.; Petti, M.; He, E.J.; De Giusti, V.; He, B.; Astolfi, L.; Babiloni, F. Investigating cooperative behavior in ecological settings: An EEG hyperscanning study. *PLoS ONE* **2016**, *11*, e0154236. [CrossRef] [PubMed]
40. PD-7 Dual Trig. Drum Pad-7″. Available online: https://roland.com/us/products/pd-7 (accessed on 29 May 2023).
41. Maestro—Midi Player Tool Kit (MPTK). Available online: https://paxstellar.fr (accessed on 17 June 2023).
42. OpenXR Plugin. Available online: https://docs.unity3d.com/Packages/com.unity.xr.openxr@1.7/manual/index.html (accessed on 21 May 2023).
43. Weiss, L.V. Marching Percussion in the 20th Century. In *A History of Drum and Bugle Corps*; Sights and Sounds, Inc.: Madison, WI, USA, 2002; pp. 91–96.
44. Ellison, B.; Bailey, T.B.W. *Sonic Phantoms: Composition with Auditory Phantasmatic Presence*; Bloomsbury Publishing: New York, NY, USA, 2020.
45. UEQ. User Experience Questionnaire. Available online: https://www.ueq-online.org (accessed on 15 July 2023).
46. Rex Rainbow's Github. Available online: https://rexrainbow.github.io/C2Demo/MIDI%20to%20CSV/ (accessed on 17 June 2023).
47. Donald, L.S. *The Organization of Rehearsal Tempos and Efficiency of Motor Skill Acquisition in Piano Performance*; University of Texas at Austin: Austin, TX, USA, 1997.

 *electronics*

MDPI

Article

Enhanced Wearable Force-Feedback Mechanism for Free-Range Haptic Experience Extended by Pass-Through Mixed Reality

Peter Kudry * and Michael Cohen *

Spatial Media Group, University of Aizu, Tsuruga, Ikki–Machi, Aizu-Wakamatsu 965-8580, Fukushima, Japan
* Correspondence: peterkudry@protonmail.com (P.K.); xilehence@gmail.com (M.C.)

Abstract: We present an extended prototype of a wearable force-feedback mechanism coupled with a Meta Quest 2 head-mounted display to enhance immersion in virtual environments. Our study focuses on the development of devices and virtual experiences that place significant emphasis on personal sensing capabilities, such as precise inside-out optical hand, head, and controller tracking, as well as lifelike haptic feedback utilizing servos and vibration rumble motors, among others. The new prototype addresses various limitations and deficiencies identified in previous stages of development, resulting in significant user performance improvements. Key enhancements include weight reduction, wireless connectivity, optimized power delivery, refined haptic feedback intensity, improved stylus alignment, and smooth transitions between stylus use and hand-tracking. Furthermore, the integration of a mixed reality pass-through feature enables users to experience a comprehensive and immersive environment that blends physical and virtual worlds. These advancements pave the way for future exploration of mixed reality applications, opening up new possibilities for immersive and interactive experiences that combine useful aspects of real and virtual environments.

Keywords: haptic interface; wearable computer; virtual reality (VR); augmented reality (AR); mixed reality (MR); extended reality (XR); human–computer interaction (HCI); ambulatory applications; force-feedback; perceptual overlay; tangible user interface (TUI)

Citation: Kudry, P.; Cohen, M. Enhanced Wearable Force-Feedback Mechanism for Free-Range Haptic Experience Extended by Pass-Through Mixed Reality. *Electronics* **2023**, *12*, 3659. https://doi.org/10.3390/electronics12173659

Academic Editor: Stefanos Kollias

Received: 20 June 2023
Revised: 4 August 2023
Accepted: 20 August 2023
Published: 30 August 2023

1. Introduction

In recent years, virtual reality (VR), augmented reality (AR), and mixed reality (MR) have seen rapid advancements, leading to exciting new trends in these immersive technologies. One significant trend is the convergence of VR and AR, giving rise to the concept of MR. Mixed reality combines virtual elements with the real-world environment, allowing users to interact with both physical and digital objects [1]. In the field of virtual reality, research has focused on improving the hardware and user experience. Development of stand-alone VR headsets, such as Meta Quest 2, has eliminated the need for external sensors or cables, enhancing mobility and accessibility. Moreover, advancements in tracking technologies, such as inside-out tracking, have improved accuracy of hand and body movement tracking within virtual environments [2]. Augmented reality has experienced significant growth, especially in industrial applications. The use of AR in fields such as manufacturing, maintenance, and training has shown promising results. By overlaying digital information upon the real world, AR enhances situational awareness, reduces errors, and improves efficiency [3,4]. MR has emerged as a promising area, combining the best aspects of VR and AR. It enables users to interact with virtual objects while maintaining a sense of presence in the real world. Research in this domain has focused on seamless integration of digital content with the physical environment, precise alignment of virtual objects, and realistic occlusion effects. Furthermore, advancements in haptic feedback technologies have contributed to immersive experiences in VR, AR, and MR. Researchers have explored various haptic feedback modalities, such as vibration, texture rendering, and force-feedback, to enhance user interactions and improve the sense of presence [5–7].

For example, vibrotactile feedback is commonly employed through handheld controllers, providing users with tactile sensations corresponding to virtual interactions. Moreover, researchers have investigated the integration of wearable haptic devices, such as gloves or suits, to enhance realism and immersion in virtual environments [8,9]. In AR and MR, haptic feedback is important for creating compelling experiences. Haptic feedback can be utilized to enhance the perception of physical objects and interactions in the augmented space. For instance, researchers have explored the use of devices with force-feedback capabilities to simulate the feeling of touching virtual objects in AR [10]. Wearable haptic devices, such as finger-mounted actuators or arm exoskeletons, provide users with force-feedback and simulate the resistance and interaction forces of virtual objects [11,12]. These advancements in haptic feedback technology contribute to creating more immersive, accurate, and realistic mixed reality experiences. In VR experiences, even pseudo-haptic feedback has been found to improve object recognition and manipulation tasks [13]. In AR, haptic feedback has been shown to enhance the perception of virtual objects' properties, such as texture and stiffness [14]. As haptic feedback continues to evolve, one particularly notable trend is the development of lightweight and wearable haptic devices that provide high-fidelity feedback while ensuring user comfort and freedom of movement.

2. Materials and Methods

2.1. Problem Description

While existing haptic feedback devices have made significant progress, there remain limitations that hinder the full realization of accurate and realistic haptic experiences. Current wearable or handheld force-feedback devices lack sufficient accuracy, simulation of textures and stiffness, support for positional translation, or a combination of these features, thus impacting the fidelity of user interactions. For instance, handheld controllers with vibrotactile feedback allow tactile sensation that corresponds to virtual interactions but lack the ability to constrain the movement of fingers, limiting the realism of grasping and virtual object stiffness simulation. Similarly, haptic suits provide vibrotactile feedback or other force-feedback sensations but fail to arrest the movement of a user's arms, compromising immersive experience. Arm exoskeletons or fixtures can affect limb movement, but they often lack fine precision, simulation of grasping, and vibrotactile feedback. To address these challenges, our research projects have focused on developing (designing, prototyping, refining) a wearable haptic interface derived from a force-feedback stylus initially intended for desktop use. Our primary objective has been to build a device that can offer untethered six degrees of freedom in virtual reality while ensuring accurate tracking and force-feedback capabilities, specifically tailored for computer-aided design (CAD) applications such as 3D modeling and sculpting. In addition to these functionalities, our prototype aims to simulate textures and object stiffness, and provide vibrotactile feedback. Development of this prototype has presented various additional challenges and complexities, including the ergonomic design of a wearable harness, considerations regarding weight, and the device's overall form factor [15]. The capabilities of this prototype had been previously demonstrated exclusively within VR environments, with no exploration in the context of AR or the intermediate spectrum of MR experiences.

2.2. Antecedent Prototype Description

2.2.1. Hardware Overview

In our previous work, we presented a prototype for immersive and realistic interaction with virtual objects. The ambulatory prototype includes an adjustable platform with a 3D Systems Touch force-feedback device mounted in front of the user, providing lifelike tactile sensations. The force-feedback device, operated through a stylus affordance held in the user's hand like a pen, can exert forces up to 3.3 N and features a force-feedback workspace size of 431 W × 348 H × 165 D mm. It can simulate stiffness on all three of its force-feedback-providing axes: X (1.26 N/mm), Y (2.31 N/mm), and Z (1.02 N/mm), as well as inertia (mass at the stylus tip) of up to 45 g [16]. This platform is integrated into a versatile tactical vest with aluminum plates on its front and back. The backplate holds a laptop that streams content to a Meta Quest 2 HMD, while the frontplate serves as the haptic device base. This cantilevered design allows adjustments for varying user heights and arm lengths. The stylus assembly underwent modifications, including weight removal and relocation of the main board. The wearable assembly comprises three devices with distinct power sources. The HMD and laptop use internal batteries, while the device's servomotors rely on an external power bank conveniently strapped to the user's waist. These hardware changes enable a more immersive and realistic haptic feedback experience with virtual objects [15].

2.2.2. Software Overview

In the antecedent prototype's software implementation, Meta Quest 2, an Android-based VR headset, was utilized. However, to overcome compatibility issues with Open Haptics SDK, it operated in Link mode, which tethered the "thin client" device to a laptop. The VR environment was implemented in the Unity game engine, using a combination of an Oculus XR plugin and a Mixed Reality Toolkit (MRTK). The Oculus XR plugin handled lower-level tasks, including stereoscopic rendering, Quest Link functionality, input subsystems for controller support, and HMD tracking [17]. Additional features—such as hand-tracking, gesture-operated teleportation, ray-cast reticle operation, and physics-enabled hand models—were extended through the MRTK [18]. For controlling the 3D Systems stylus, the Open Haptics for Unity plugin is employed, allowing integration of various 3D haptic interactions within Unity. This plugin comprises the Quick Haptics micro API, Haptic Device API (HDAPI), Haptic Library API (HLAPI), Geomagic Touch Device Drivers (GTDD), and additional utilities [19]. However, structure of the Open Haptics plugin for Unity differs from the native edition. Instead of using Quick Haptics, it employs the OHToUnityBridge dynamic-link library (DLL) to establish communication between the Unity controller, the HapticPlugin script written in C#, and the HD/HL APIs written in the C programming language. The `OHToUnityBridge.dll` library directly invokes the HD, HL, and OpenGL libraries without relying on Quick Haptics [15,20].

2.2.3. Pilot Experience Overview

The virtual environment for the pilot study, as shown in Figure 1, immersed users in a range of haptic simulations that demonstrated various functionalities of the haptic mechanism. The experience began with a shape-sorting activity, wherein users interacted with a virtual table containing diverse shapes. Using the physical haptic stylus, users could manipulate virtual shapes with realistic feedback. When the stylus touched a virtual object, it responded by mechanically locking or constraining movement and rotation, creating a lifelike sensation of the object's massive presence. The HMD also displayed spatialized binaural audio cues to the user indicating that an object had been touched or grabbed. Another simulation involved two sculptures. One sculpture featured outer and inner spheres, the outer layer being deformable and the inner layer rigid. Users experienced the tactile sensation of piercing deformable material, like to popping a balloon with a pin. The second sculpture virtually attracted the stylus to its surface and limited movement, providing the feeling of dragging a magnetic stick across a metal table. Next, users encountered two boards made of different materials, glass and wood, to showcase the haptic system's ability to simulate textures and friction smoothness effectively. The difference between these materials deepened the user's perception of texture roughness or smoothness, complemented by audio cues such as the sound of scratching on wood or the squeaking when dragging the stylus across the glass surface. The final segment presented a table with five colored capsules, each representing a distinct tactile effect: elasticity, viscosity, vibration, constant force, and friction. Users could explore these effects to enhance their interactive experience. Furthermore, users had the option to create basic 3D forms (cube, cylinder, sphere, or plane) and manipulate them regarding scale, location and orientation using bimanual manipulation through hand-tracking. Interaction with objects could be achieved through ray-casting selection, collision-based interaction, or direct hand-grabbing, depending on their proximity. These capabilities highlighted the creative potential of free-range haptic and tactile interfaces in digital content creation and CAD applications, resembling the functionality of the desktop edition of the Geomagic Freeform modeling tool [21].

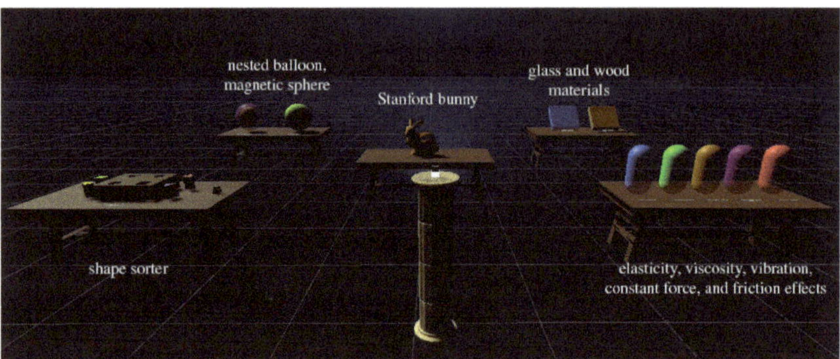

Figure 1. Virtual environment showcasing haptic simulation. Shape-sorting activity, two sculptures (nested balloon and a magnetic sphere), model of a Stanford bunny, glass and wood material simulation, and five distinct tactile effects.

2.3. Preliminary Validation

2.3.1. Summary of a Pilot Experiment

A pilot experiment compared two conditions of experiencing haptic feedback: seated with a normal desktop monitor, and mobile VR with a head-mounted display (HMD). The baseline condition involved playing a Jenga game on a desktop computer with a haptic device, while the room-scale condition allowed participants to explore an immersive haptic sandbox using a harness and VR headset. After experiencing both setups, participants completed a questionnaire, rating their impressions on a Likert scale. The results showed that users had neutral to slightly positive feelings of immersion during the desktop experience. Feedback regarding the wearable harness was mixed, with concerns about weight and comfort, but participants provided strong positive feedback on immersiveness, despite the crudeness of the harness. The combination of virtual reality and haptic force-feedback contributed to overall immersion. Users found the combination of inputs usable but somewhat tricky, with some issues related to avatar misalignment, drift, and hand-tracking transitions.

Overall, participants believed this type of device arrangement could be used for computer-aided design (CAD) applications. The participants' overall experience manifested positive feedback, with an average score of 8.5 out of 10 for the proof of concept. The experiment provided valuable feedback for future refinement and expansion of the system [21].

2.3.2. Summary of a Performance Experiment

The performance experiment aimed to confirm whether the ambulatory condition, utilizing mobile VR with an HMD, could match or surpass performance of the seated condition with a regular monitor. The experiment included block-sorting and Jenga games presented to separate groups of participants for the ambulatory and desktop conditions to avoid bias from learning effects. During the Jenga game, participants were instructed to remove as many blocks as possible from the tower within a 4-min time limit without causing it to collapse. The block-sorting game required participants to successfully sort a variety of shapes within the same time frame. The student participants had diverse levels of VR experience, and all were right-handed, with academic backgrounds in Computer Science or Software Engineering. Data collection involved observation, stopwatch timing, and questionnaires eliciting user experience and subjective impressions. The results indicated that there were no significant differences between the desktop and ambulatory conditions in terms of user experience dimensions (attractiveness, dependability, efficiency, novelty, perspicuity, stimulation) and subjective experience factors (effort, importance, interest, enjoyment, perceived choice, perceived competence, pressure, tension, value, usefulness). Likewise, the performance measurements, including the maximum number of removed and stacked Jenga blocks, number of Jenga trials, and number of filled shape-sorter boards, did not exhibit any notable variances between the two conditions.

The experiment demonstrated that the ambulatory condition achieved comparable performance to the seated condition, while user experience, subjective impressions, and performance measurements showed no significant disparities between the two setups [21].

2.4. Identified Deficiencies

Based on the summarized experiments, several shortcomings could be identified that affected user experience with the prototype. These deficiencies had impact on user performance, and despite the haptic feedback enhancing immersiveness in this particular use case, performance was comparable between the seated and mobile VR conditions. This suggests that improving performance could potentially further enhance immersiveness and depth of presence. One significant deficiency is related to the comfort of the wearable harness, which received varied feedback from the participants. The weight of the system emerged as a primary concern, indicating sub optimal material choices. The wearable harness was only comfortable to wear for a limited duration, typically 15 min to around half an hour. Another deficiency lies in the haptic feedback, where some effects were reported to lack the necessary strength (force) to replicate real sensations accurately. This indicates a shortfall in delivering realistic haptic feedback to users. The combination of inputs proved to be usable but posed certain challenges as well. A rotational reset feature was implemented to address user avatar misalignment with the stylus after locomotion. However, the virtual stylus sometimes drifted from its real-world representation, making it unreliable as a rotational reset function. This limitation highlights the difficulty in maintaining consistent alignment and synchronization between the virtual and physical aspects of the system. While hand-tracking was generally considered reasonably accurate, transitioning between using the haptic stylus and hand gestures required users to temporarily hide their hand and bring it back into view to reload the hand-tracking mode. This process was deemed inconvenient, suggesting a deficiency in achieving smooth transitions between different interaction modes. These identified deficiencies indicated aspects that required improvement to enhance the user experience. Addressing these concerns involves improving comfort, enhancing the realism of touch sensations, refining input usability, ensuring alignment accuracy, and facilitating smoother transitions between different interaction modes in the system.

3. Resulting Enhancements

3.1. Weight Reduction

The prototype's comfort aspects and certain software-related shortcomings were the main foci for improvement, based on the identified deficits mentioned above. A key objective of this new edition was to achieve significant weight reduction by streamlining components attached to the harness. As part of this effort, the entire structure of the prototype underwent redesign, starting with the removal of the aluminum plating. The front shelf carrying 3D Systems Touch was also eliminated, making way for a more robust frame made of extruded aluminum that reduced flexing, particularly on the shelf in front of the user. The front cantilever was replaced with an extruded aluminum bar as well. To further reduce weight, the stylus device underwent additional modifications, stripping it down to essentials: servo motors, pulleys, and wiring. This resulted in less strain on the user due to reduced weight at the end of the cantilever. Moreover, the laptop computer formerly carried at the back of the user was removed, allowing the stylus control board to be relocated alongside the addition of a Raspberry Pi 3. While it would have been feasible to remove and replace these components without encountering difficulties, our specific implementation and use case demanded the preservation of untethered six-degrees-of-freedom movement within the virtual space. Consequently, we had to revector connectivity between Quest 2 and 3D Systems Touch, transitioning from a wired to a wireless setup to maintain the desired freedom of movement. The redesigned harness is shown in Figures 2 and 3.

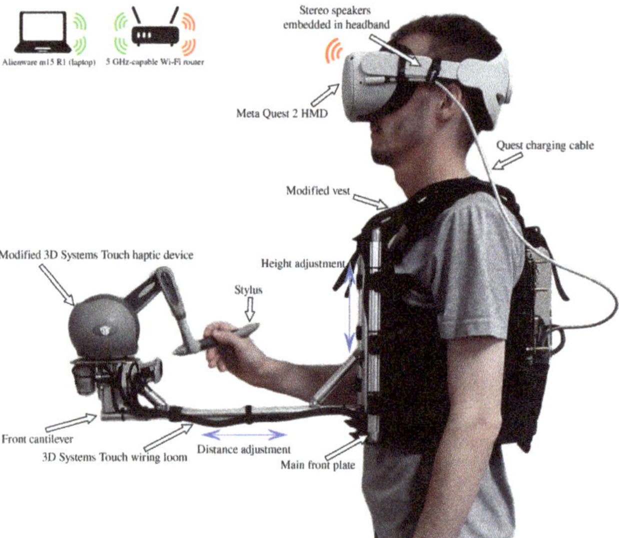

Figure 2. System assembly profile depicting the wearable harness and its adjustability, supporting the devices used for personal sensing. Head-mounted display (Meta Quest 2) and modified haptic force-feedback stylus (3D Systems Touch).

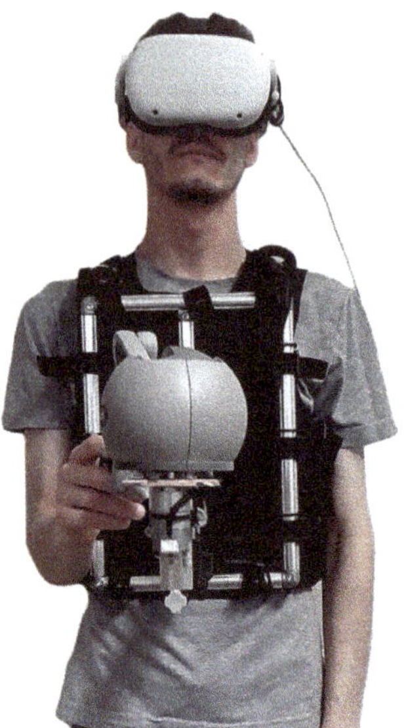

Figure 3. System assembly front view.

3.2. Wireless Connectivity

Switching from Quest Link to Air Link, which involves the HMD acting as a thin client of a PC, was a rather straightforward process. However, it required upgrading to newer versions of our development environment, Unity, and its Oculus XR libraries to ensure reliable functionality of Air Link. The conversion of the haptic device from wired to wireless connection posed significant challenges. We initially explored the possibility of hardware-level conversion from USB to Bluetooth, but ultimately opted for USB over network. To accomplish this, we employed a VirtualHere server running on Raspberry Pi 3, augmented with a Wi-Fi adapter capable of a 5 GHz Wi-Fi connection. VirtualHere allows the network to act as a pathway for transmitting USB signals, effectively enabling USB over IP [22]. This means that the USB device, such as the stylus apparatus, behaves as though it is directly connected to the client machine (e.g., a laptop computer), even though it is physically plugged into a remote server, specifically the Raspberry Pi 3. As a result, existing drivers and software function without requiring special modifications. However, there is a slight latency penalty that, to a small extent, negatively impacts our implementation under certain conditions. As described in Section 2.2.2, Unity utilizes the `OHToUnityBridge.dll` library, whereas vanilla applications from 3D Systems directly interface with the device via HD/HL APIs. When the stylus is driven directly using HD/HL libraries, the latency and reliability of haptic feedback are perceived as identical to when the device is physically connected to the target machine. However, when utilizing an application built in Unity, the additional overhead of the translation layer, which involves conversion of native API calls through middleware to accurately express virtual objects' physical properties in real life via the stylus, introduces occasional perceivable delays in force-feedback response and subtle stylus jitter. The occurrence of these issues depends on various factors, such as the environment and Wi-Fi signal quality, which may be beyond our control. Furthermore, even without perceptible issues, a slight degradation in the servoloop frequency, which facilitates bi-directional communication between the application and the device, can be observed when comparing wired and wireless communication. However, there are plans to address these concerns and improve the application in future updates by exploring ways to minimize overhead in the middleware.

3.3. Power Delivery

Wireless communication between the host laptop, client HMD, and the stylus allows for reduction in power delivery requirements for all the devices involved. Previously, we were constrained by the built-in battery of our host machine, a 90 Wh Li-ion battery. With several power-saving techniques in place, such as undervolting the CPU and underclocking the CPU and GPU, we achieved approximately 60 to 90 min of screen-on time. However, by eliminating the need for the user to carry the laptop, we can disable all power-saving measures, increase rendering performance output, and strive for higher texture quality and visuals while considering the limitations of the Air Link function of Meta Quest 2, rather than being restricted by the power consumption of the PC. The power delivery for the stylus device remains the same as described in Section 2.2.1, with an external power bank. However, the battery now needs to supply power to the Raspberry Pi 3 as well, which reduces the stylus's power-on time. Previously estimated at approximately 12 h, the stylus's system-on time can now be estimated to a maximum of about 10 h due to the added power requirements of Raspberry Pi 3. As the 10-h run-time of this system exceeds the presumed continuous session duration by a single user, we utilize this external power source to extend the battery life of Quest 2 as well. By taking into account the average power consumption of Quest 2 (between 4.7 and 7 W), Raspberry Pi 3 (between 1.3 and 3.7 W), and 3D Systems Touch (between 18 and 31.5 W), we can estimate that the entire system can be run on a single charge of a 20-ampere–hour battery for approximately 4.5 to 8 h, effectively quadrupling the minimum system up-time. Considering (as mentioned in Section 2.4) that the comfortable usability of this prototype is up to 30 min, the extended power management parameters significantly exceeded this use-case. Even if the weight reduction and ergonomic improvements described in Section 3.1 only quadruple usability time to 2 h, the power requirements would not become a significant limitation. All components of the wireless connectivity and power delivery are illustrated in Figure 4.

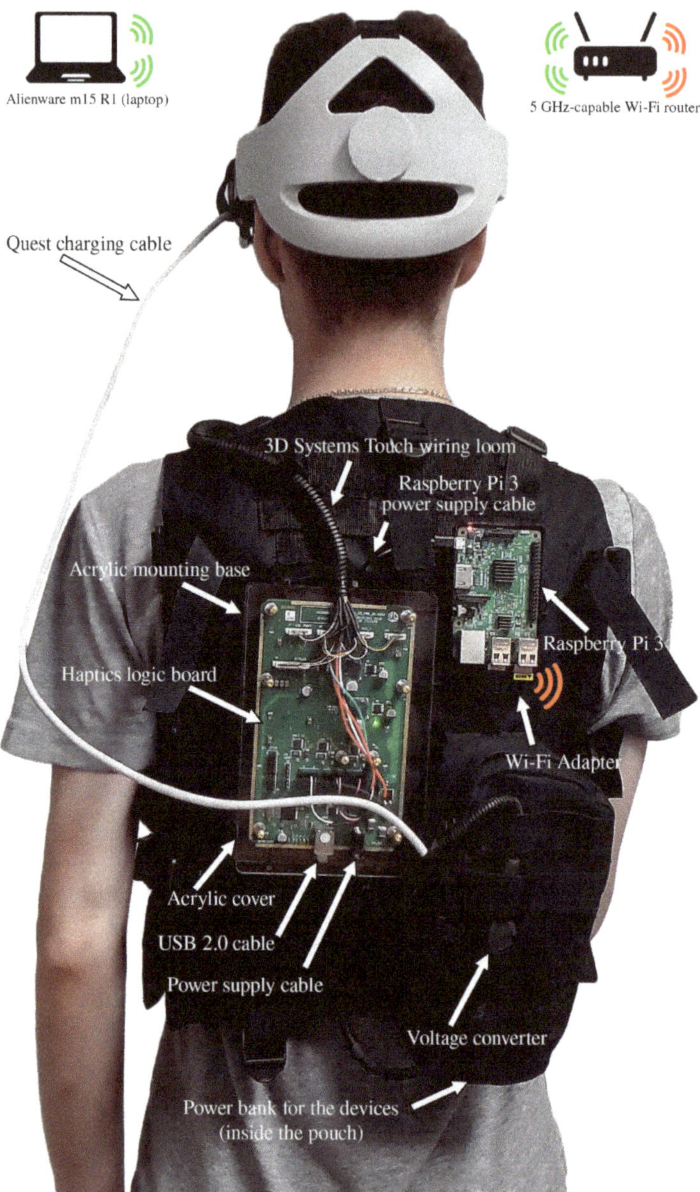

Figure 4. System assembly rear view showing the Haptics logic board, as well as the devices enabling wireless communication (Raspberry Pi 3 and a 5-GHz-capable Wi-Fi router), their power source, and the wired connections between Pi 3 and haptics logic board.

3.4. Haptic Feedback Intensity

In the previous section (Section 2.4), it was mentioned that the strength of certain haptic effects was perceived as inadequate. To address this, adjustments were made to enhance the modeling of virtual objects' physical properties. Unity's physics engine handles the simulation, which is then rendered into forces exerted through the stylus. Various properties of virtual objects were reviewed and modified to simulate realistic physical attributes and their corresponding responses when interacting with using the stylus. These properties include perceived weight, drag, bounciness, friction coefficient between different materials, their respective smoothness, stiffness, and the intensity of damping when the stylus interacts with a touched object. By fine-tuning these properties, we aimed to create a more immersive experience that closely resembles real-life interaction. The adjustments made allow users to perceive and interact with virtual objects in a manner that aligns with their expectations and provides a more satisfying haptic experience.

3.5. Stylus-to-Avatar Alignment and Reset Function

Previously, there were issues encountered in maintaining consistent alignment and synchronization between the avatar of the user and the stylus device. Whenever a user utilized the locomotion feature, which involved initiating teleportation through hand gestures, slight drift or rotational misalignment occurred between the virtual representation of the stylus and its real-life counterpart. In the implementation described in Section 2.2.2, the MRTK was utilized to enable features such as hand-tracking and gesture-controlled teleportation. However, the teleportation process resulted in the avatar being independently rotated around its local gravitational vertical axis (yaw), separate from the orientation of the stylus attached to the avatar. To address this discrepancy, a reset function was implemented, allowing users to request realignment by simultaneously pressing two buttons on the stylus. Due to the decision to switch to the latest version of the Oculus XR Plugin, departure from the outdated MRTK was necessary. This switch required reimplementation of the features previously provided by MRTK into the application. The newer XR Interaction Toolkit was utilized for this purpose and served as an aid in implementing the updated locomotion system. From the user's perspective, the locomotion function operates in an indistinguishable manner. However, the reimplementation resolves the issue of independently rotating the avatar and stylus after each teleportation. The updated system ensures that the user and their stylus face the same direction as they did before initiating the teleportation process.

3.6. Transition between Stylus Use and Hand-Tracking

The availability of newer libraries and prefabs within the Meta Quest Interaction SDK allows integration of enhanced hand-tracking and controller tracking into Unity applications [23]. By incorporating updated hand models and modifying relevant scripts to suit our specific use case, the problem of inconsistent transitions between using the stylus and hand-tracking with the user's dominant hand, based on their selected chirality (handedness), was resolved. The adjustments made to the hand-tracking functionality ensure that the transition between using the stylus and hand-tracking is now seamless. Users no longer experience a sense of disconnection or inconsistency between the perceived tracking performance and the movements of their real hands. This improvement significantly enhances the overall accuracy of tracking and reinforces the naturalness of the user's hand movements. This enhancement becomes particularly important in light of the introduction of a new feature to this prototype's software, a mixed reality pass-through, described below. All aforementioned software changes and the overall software architecture are illustrated in Figure 5.

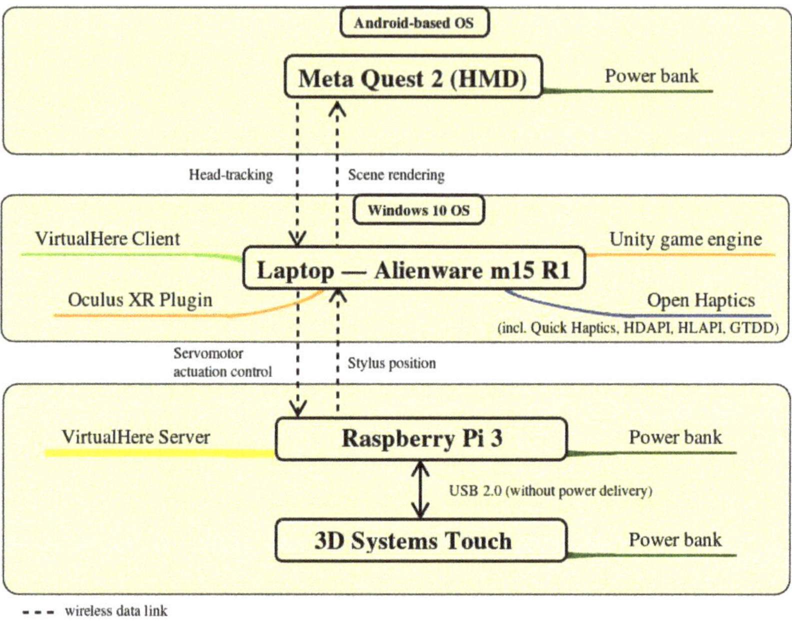

Figure 5. Wireless software architecture depicting the four devices used in this prototype, their operating systems, applications, and libraries utilized, as well as physical connections via cables and wireless data links, including the type of data being transmitted.

3.7. Mixed Reality Pass-Through

The mixed reality pass-through feature on Meta Quest 2 enhances the virtual reality (VR) experience by integrating real-world surroundings into the virtual environment. This advanced capability utilizes the built-in cameras of the headset to capture a live graphic stereo video feed of the user's physical environment, which is then blended with the virtual content. As a result, users can view and interact with their surroundings while wearing the headset, allowing virtual objects and characters to coexist with the real-world environment. This feature enables users to move around, navigate obstacles, and interact with physical objects while still being immersed in the virtual world. It enhances situational awareness, promoting user safety and preventing collisions with physical objects. The integration of real-world surroundings into virtual experiences creates an augmented reality-like effect, offering a more comprehensive and immersive encounter for users. Combining the mixed reality pass-through feature with haptic force-feedback devices further enriches the experience, providing users with heightened sensory stimulation and a seamless blend of virtual and physical realms.

4. Enhancements Validation

To assess the impact of the implemented enhancements, we conducted an experiment similar to the one outlined above in Section 2.3.2. However, the focus this time was on evaluating the effects of the enhancements on user performance and overall experience.

4.1. Conditions: Virtual Reality and Mixed Reality

In the baseline condition, each participant was equipped with our harness and guided through a set of tasks in virtual reality, including block-sorting and Jenga games. To precisely replicate the mixed reality environment, we re-implemented the virtual reality environment. The laboratory room where the experiment took place was scanned using a LiDAR sensor on an iPad Pro. The resulting mesh model was edited to remove non-manifold vertices, and its textures were adjusted to appear monochromatic with reduced resolution, while a grain effect was added to match the resolution and image quality of the Quest 2 pass-through function. Figures 6 and 7 illustrate the compositions of these scenes.

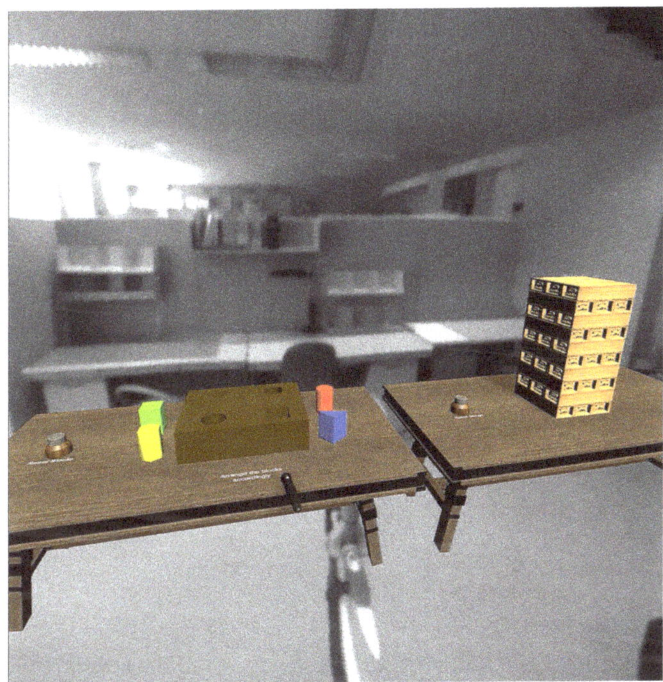

Figure 6. The mixed reality scene showcases virtual Jenga and shape-sorter desks created with computer graphics, overlaid onto photographic pass-through imagery rendered as a skybox within the Unity game engine. This scene incorporates three types of personal sensing: hand-tracking, head-tracking, and stylus-sensing. The user's hands are accurately represented by their corresponding 3D models. However, the virtual stylus is positioned slightly in front of the real electro-mechanical device to avoid the need for the user to be in close proximity to interactable objects.

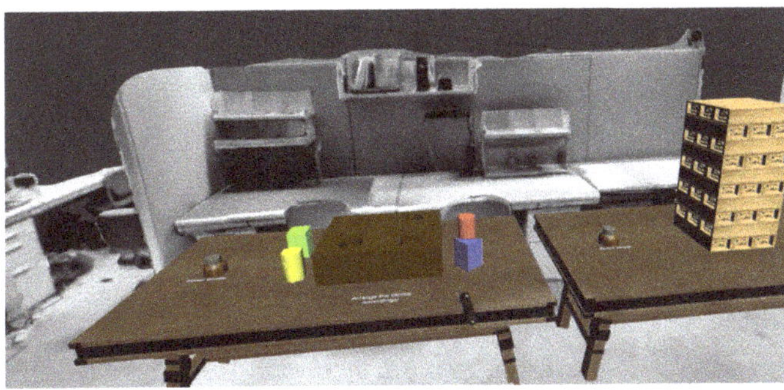

Figure 7. Similar to the mixed reality scene, the virtual reality scene showcases virtual Jenga and block-sorter desks. However, the seemingly real environment lacks the dynamicity that the MR provides as it is a static texture map.

4.1.1. Procedure and Controls

In the Jenga game, both VR and MR segments had a 4-min time limit for participants to remove as many blocks as possible without toppling the tower. If the tower toppled, the participants had to reset and start again. To reset, the users pushed a physics-reactive virtual button, placed next to the block sorter and Jenga tower, using their index finger (enabled via hand-tracking). The highest number of removed blocks from any number of runs was recorded, along with the number of resets. Similarly, in the block-sorting game, participants had 4 min to successfully sort the full range of shapes. The score was incremented only if all blocks were successfully sorted, preventing participants from sorting only easier shapes. To minimize the influence of learning on results, participants were divided into two groups: (1) those who first participated in the virtual reality segment and then in the mixed reality block, and (2) those who first participated in the mixed reality segment and then in the virtual reality block. Each experiment segment took approximately 10 min, excluding the introduction, warm-up sessions, and questionnaire answering. The warm-up session, lasting about 2 min for each game segment (8 min in total), primarily focused on harness adjustment to ensure comfort and familiarity with the interface. The questionnaire, administered after each segment, took up to 10 min, resulting in an overall experiment duration of 30 to 40 min per participant. Participants removed the harness and HMD while answering the questionnaire between the VR and MR segments. Furthermore, as described in Section 3.2, due to the varied latency introduced by the middleware in the wireless set-up and dependency on the quality of the network, the participants were tethered to a nearby laptop via 5 m long Quest Link and 5 m long USB cables for the HMD and haptic devices, respectively. This configuration ensured consistency between the experiment runs, as interference of the university network could impact consistency of the experiment. No hardware that would be otherwise present in the wireless set-up was removed or adjusted, and only the connectivity was changed from self-contained to external.

4.1.2. Participants

Ten adults were recruited to participate in this experiment, with 40% falling between the ages of 18 and 25, and the remaining 60% between 26 and 35. Among the participants, 60% were males and 40% were females. All subjects were right-handed and had a background in Computer Science or Software Engineering, making them familiar with standard human–computer interactions. However, some participants had no prior experience with VR or haptic devices. They were given a basic introduction to VR concepts and usage, followed by a brief explanation of interactions with force-feedback haptics. Participants were compensated JPY 1000 (approximately USD 8) for their involvement in this study.

4.1.3. Data Acquisition and Composition

The experimental supervisor collected the measured data using a stopwatch and Google Forms to record the scores. After each measured segment (VR or MR), each participant completed a User Experience Questionnaire (UEQ) consisting of 26 pairs of bipolar extremes, such as "complicated/easy" and "inventive/conventional," evaluated on a quantized Likert scale from 1 to 7 [24,25]. Additionally, participants were asked to rate their agreement with eight statements regarding the improvements, such as "The harness was heavy." or "The tethered connection bothered me." on a seven-point Likert scale, where 1 indicated "Strongly Disagree" and 7 "Strongly Agree." Participants were also asked to estimate the duration they could comfortably wear the harness and to rate their overall experience on a scale from 1 to 10.

4.1.4. Results

The User Experience Questionnaire (UEQ) was utilized to evaluate participants' experiences across six dimensions: Attractiveness, Dependability, Efficiency, Novelty, Perspicuity, and Stimulation. The questionnaire responses were collected on a zero-centered seven-point scale ($-3 \rightarrow +3$). Analysis of the UEQ data revealed that there were no statistically significant differences between the VR and MR experiences. Moreover, a comparison was made between the UEQ results obtained from the ambulatory condition in the previous experiment, which utilized the old hardware, and the VR condition in the current experiment, where the enhanced hardware was employed. The results of estimated marginal means showed no significant differences between these conditions. To calculate the estimated marginal means of the fitted linear model, we employed the "lme4" and "emmeans" libraries in the R programming language. This statistical analysis allowed us to assess the differences between treatment conditions while considering the influence of potential covariates and interaction effects, providing valuable insights into the user experience across the three conditions [26–28]. These results are presented in Figure 8. Furthermore, the UEQ also includes benchmarking of our results against data from 21,175 individuals who participated in 468 studies involving various products, such as business software, web pages, web shops, and social networks. Our study found that both conditions, VR and MR, performed similarly when compared to this extensive dataset.

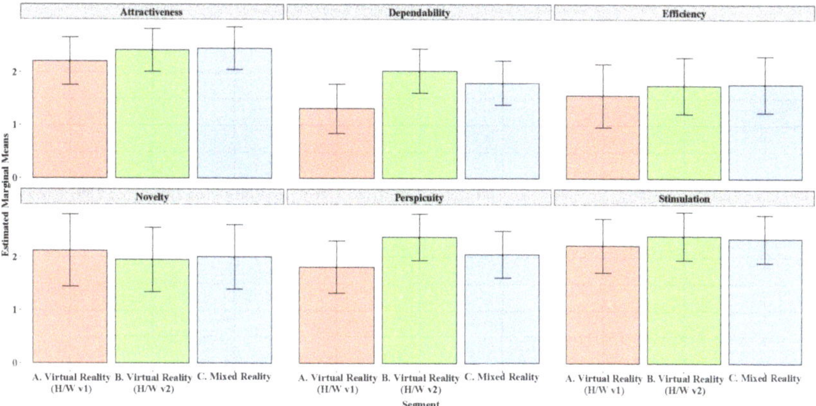

Figure 8. Comparison of User Experience Questionnaire Scores in (**A**) VR using the previous version ("v1") of hardware, (**B**) VR using the latest version ("v2") of hardware, and (**C**) MR using the latest version of hardware. Used confidence level: 95%.

Performance measurements were analyzed similarly to the UEQ, using estimated marginal means, but with a fitted generalized linear mixed-effects model [29]. Three dimensions were considered: the maximum number of removed and stacked Jenga blocks, the number of Jenga trials (resets + 1), and the number of filled block-sorter boards. Regarding comparison between the VR and MR segments, results showed that there was no significant performance difference. However, when comparing the VR segments using the previous version of the hardware with those using the latest one, results indicated a significant difference in performance across all three metrics, favoring the revised hardware implementation. These results are presented in Figure 9.

Additionally, the collected responses concerning the prototype enhancements indicate that participants perceived both the VR and MR segments as immersive, with a slight preference, albeit not statistically significant, toward the MR segment. The comfort of the harness was also perceived positively, although the standard deviation suggests high variability in the responses. Notably, participants strongly disagreed with the statement that the harness was heavy. As described in Section 4.1.1, users were tethered to a nearby PC to ensure system stability; while the key features of this prototype are freedom of movement and ambulatory aspects, in this experiment, every part of the virtual environment was reachable by walking without being limited by the wired connection. Nonetheless, participants were asked if the wired connection inconvenienced them in any way. The results indicate that the wired connection was generally not cumbersome. Furthermore, responses regarding the software enhancements, such as the strength of simulated forces, hand-tracking, stylus precision, and transitions between them, demonstrate that these improvements positively influenced the strength of force-feedback, precision, reliability of tracking, and seamlessness of transitions between the two types of control. These results are summarized in Table 1.

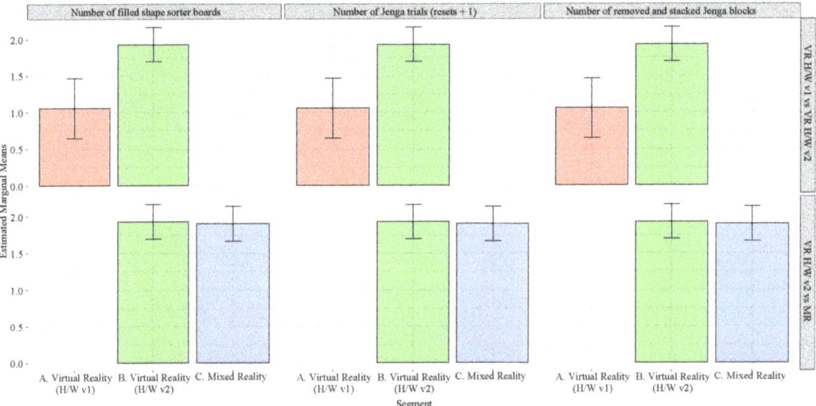

Figure 9. Comparison of performance results in (**A**) VR using the previous version ("v1") of hardware, (**B**) VR using the latest version ("v2") of hardware, and (**C**) MR using the latest version of hardware. Used confidence level: 95%.

Upon completion of both the VR and MR segments of this experiment, participants were surveyed to ascertain their estimated time limit for comfortable wear. The aggregate of all responses indicates that the wearability limit significantly improved to approximately 1.2 h (or 72 min). This represents a substantial enhancement, doubling the prototype's wearability time compared to its earlier version.

Upon evaluating the overall experience on a scale ranging from 1 to 10, participants provided remarkably positive feedback. The quality of our proof of concept received an average score of 9.1/10, affirming strong reception and high satisfaction with our refreshed implementation.

Table 1. User Perceptions and Responses to Prototype Enhancements in VR and MR Segments.

Statement	Mean ± SD
The VR mode was immersive.	5.5 ± 0.92
The MR mode was immersive.	6.1 ± 0.83
The wearable harness was comfortable.	5.9 ± 1.22
The harness was heavy.	2.5 ± 0.81
The tethered connection bothered me.	2.7 ± 1.27
The intensity of simulated force-feedback was adequate.	5.7 ± 1.27
The stylus and hand-tracking were adequately precise and reliable.	5.1 ± 1.51
The transition between stylus use and hand-tracking was seamless.	5.5 ± 1.12

5. Discussion

The results of our study, based on the analysis of the User Experience Questionnaire (UEQ) data, provide valuable insights into the user experiences in both virtual reality (VR) and mixed reality (MR) segments. Our statistical analysis revealed no significant differences between the VR and MR experiences, indicating that the environments offered comparable user experiences. This finding is particularly interesting, as it suggests that the advanced mixed reality pass-through feature on the HMD successfully achieved seamless integration of the virtual and physical realms, allowing users to interact with the real world while remaining immersed in the virtual environment.

However, it is worth noting that the understood definitions of terms such as "immersion" and "presence" lack uniformity. Some studies define immersion as a subjective property and presence as objective, while others assume the opposite [30]. Nonetheless, by considering immersion as an inherent characteristic of a VR system that assesses the extent to which it can faithfully replicate natural sensorimotor contingencies for perception and appropriately respond to perceptual actions [31], one can claim that a VR headset employing visual pass-through via cameras can impact one's sense of immersion and interaction between augmented virtuality and augmented reality.

Presence can be understood as a perceptual illusion of being present in a virtual environment while knowing that you are not physically there [31]. With this definition, we can explore the notion that the difference in levels of presence between virtual and mixed realities could be negligible, given that a virtual reality system can render the virtual representation of a real environment, excluding the augmented elements, to the same quality as the video pass-through.

To assess the impact of hardware enhancements, we compared the UEQ results obtained from the previous ambulatory condition, which utilized the old hardware, with the current VR condition employing the enhanced hardware, as well as an MR version of the current system implementation. Statistical analysis using the estimated marginal means did not show any significant differences among any of the conditions. This indicates that the hardware enhancements did not significantly alter the user experience when comparing VR with the previous ambulatory setup, nor with the MR extension.

Comparison of performance measurements between the VR and MR segments, using a fitted generalized linear mixed-effects model, demonstrated no significant performance difference. This suggests that both environments were equally effective in allowing participants to interact with virtual objects. However, when comparing the VR segments utilizing the previous version of the hardware with the latest one, a significant difference in performance was observed across all three metrics: the maximum number of removed and stacked Jenga blocks, the number of Jenga trials (resets + 1), and the number of filled shape-sorter boards. This result supports the notion that the hardware enhancements led to improved performance and precision in VR interactions.

The responses regarding the hardware enhancements indicated that participants perceived both the VR and MR segments as immersive, with a slight preference toward the MR segment, although this preference was not statistically significant. The comfort of the harness was generally perceived positively, though some variability was observed in the responses. Notably, participants strongly disagreed with the statement that the harness was heavy, indicating that the redesigned prototype effectively addressed previous weight-related concerns.

The software enhancements—including the strength of simulated forces, hand-tracking, stylus precision, and transitions between control modes—received positive feedback from subjects. This indicates that the software modifications contributed to enhancing the overall immersion in the virtual and virtually augmented environments.

Lastly, the improved wearability of the prototype represents an advancement, as participants reported an average anticipated wearability time of approximately 1.2 h (or 72 min), doubling the previous version's wearability time. This improvement is crucial for longer virtual experiences, such as immersive content-creation tasks, where user comfort and sustained engagement are essential.

Using the definitions stated above, these results suggest that there is no significant difference in the sense of presence within these three environments (based on the results of UEQ). However, there is a potential for differences in levels of immersion (based on the performance results), as the level of immersion depends on the properties of a system, such as clarity, responsiveness, smoothness, precision, endurance, stability, and comfort.

Overall, our findings suggest that the implemented hardware and software enhancements in the prototype led to a more comfortable and immersive user experience. The integration of VR and MR segments, combined with the refined hardware and software components, positively influenced user interactions and performance within the virtual environment. The positive feedback received from participants further validates the successful implementation of the proof of concept. These outcomes highlight the potential of our prototype in diverse applications, ranging from training simulations to content creation or computer-aided design, where users can benefit from free-range haptic and tactile interfaces, thereby bridging the gap between the virtual and physical worlds.

Future Work

Currently, the haptic stylus's base is not being explicitly tracked; rather, it is positioned based on the user's height and arm's length. The Unity scene takes into account the distance from the user's torso (X-axis) and chin (Y-axis), which determines the virtual representation of the stylus and its offset from the headset's anchor point. However, this setup has a limitation. If a user leans sideways without moving their hips, the virtual stylus follows such movement, while the real stylus remains stationary, causing a tracking disconnect.

To overcome this limitation in the future, we propose the use of an additional pair of cameras for image or object recognition. Deploying technologies like OpenCV or other image-processing frameworks would enable an analysis of the real space around the user, allowing accurate estimation of the true position of the haptic interface. This improvement would enhance tracking accuracy, providing a more realistic experience for users.

Additionally, the current prototype only provides force-feedback for one hand, while immersive experiences often involve both hands. Hence, enabling force-feedback for bimanual manipulation would be beneficial. One idea is to replace the stylus with a haptic glove attached to the gimbal part of the device at the wrist, allowing for the positional arrest of the user's arm and individual finger simulation, including touching and grabbing. This concept could be scaled to a bimanual setup where two Touch devices are used.

Furthermore, the current system only facilitates tactile perception with pre-made virtual elements within a scene. However, with future enhancements involving depth cameras or environment-scanning technologies, real-world objects could be rendered in real-time into simplified virtual representations. This extension would usher in the realm of augmented virtuality, seamlessly blending real and virtual objects.

The potential outcome of these developments is the creation of a "portable room" experience for users, regardless of their physical location. This advancement opens up exciting possibilities for various applications—such as remote collaboration, training simulations, and interactive experiences—that merge the virtual and physical worlds. The combination of improved tracking accuracy and augmented virtuality would provide users with immersive and engaging experiences beyond the confines of traditional virtual environments.

6. Conclusions

In conclusion, analysis of our ambulatory haptic system found no significant differences in user experiences between virtual and mixed reality segments, indicating seamless integration. The hardware enhancements positively impacted interactions, improving performance and precision. Participants reported enhanced wearability, doubling the previous version's duration of comfortable use. Positive feedback validated the prototype's success, showcasing its potential in various applications and bridging the virtual and physical worlds. Future developments in VR and MR technologies can build upon these findings, enriching immersive interactions via haptic feedback.

Author Contributions: The primary development of this project was carried out by the first author, P.K., under the guidance and support of the supervising professor, M.C. All authors have read and agreed to the published version of this manuscript.

Funding: The Spatial Media Group at the University of Aizu provided financial support for the development of the prototype of the wearable haptic device.

Institutional Review Board Statement: Subjective experiments complied with the requirements established by the University of Aizu Research Ethics Committee.

Informed Consent Statement: Prior to their participation, experimental participants provided written consent after being fully informed about the contents of this study. Additionally, explicit written consent was obtained from the subjects regarding publication of any potentially identifiable images or data.

Data Availability Statement: Publicly available datasets were analyzed in this study. These data can be found online: https://github.com/peterukud/WearableHapticsVRMR (accessed 23 July 2023).

Acknowledgments: We express our gratitude to Julián Alberto Villegas Orozco, Camilo Arévalo Arboleda, Wen Wen, James Pinkl, and experimental subjects for their valuable support during the implementation of this project. Additionally, we extend our appreciation to the anonymous referees for their useful and thoughtful suggestions on this submission.

Conflicts of Interest: The authors declare no conflict of interest.

Abbreviations

The following abbreviations, initializations, and acronyms are used in this manuscript:

3D	Three-dimensional
API	Application programming interface
AR	Augmented reality
CAD	Computer-aided design
DLL	Dynamic-link library
GTDD	Geomagic touch device drivers
HDAPI	Haptic device API
HLAPI	Haptic library API
LAN	Local area network
HMD	Head-mounted display
MRTK	Mixed reality toolkit
SDK	Software development kit
USB	Universal serial bus
VR	Virtual reality
XR	Extended reality
MR	Mixed reality

References

1. Skarbez, R.; Smith, M.; Whitton, M.C. Revisiting Milgram and Kishino's Reality-Virtuality Continuum. *Front. Virtual Real.* **2021**, *2*, 647997. [CrossRef]
2. Hillmann, C. Comparing the Gear VR, Oculus Go, and Oculus Quest. In *Unreal Mob. Standalone VR*; Apress: Berkeley, CA, USA, 2019; pp. 141–167. [CrossRef]
3. Azuma, R.; Baillot, Y.; Behringer, R.; Feiner, S.; Julier, S.; MacIntyre, B. Recent advances in augmented reality. *IEEE Comput. Graph. Appl.* **2001**, *21*, 34–47. [CrossRef]
4. Billinghurst, M.; Kato, H. Collaborative augmented reality. *Commun. ACM* **2002**, *45*, 64–70. [CrossRef]
5. Burdea, G. *Force and Touch Feedback for Virtual Reality*; John Wiley & Son, Inc.: Hoboken, NJ, USA, 1996.
6. Slater, M. Place illusion and plausibility can lead to realistic behaviour in immersive virtual environments. *Philos. Trans. R. Soc. B Biol. Sci.* **2009**, *364*, 3549–3557. [CrossRef]
7. Dangxiao, W.; Yuan, G.; Shiyi, L.; Yuru, Z.; Weiliang, X.; Jing, X. Haptic display for virtual reality: Progress and challenges. *Virtual Real. Intell. Hardw.* **2019**, *1*, 136. [CrossRef]
8. Perret, J.; Vander Poorten, E. Touching Virtual Reality: A Review of Haptic Gloves. In Proceedings of the ACTUATOR International Conference on New Actuators, Bremen, Germany, 25–27 June 2018; pp. 1–5.
9. Kang, D.; Lee, C.G.; Kwon, O. Pneumatic and acoustic suit: Multimodal haptic suit for enhanced virtual reality simulation. *Virtual Real.* **2023**, *1*, 1–23. [CrossRef]
10. Valentini, P.P.; Biancolini, M.E. Interactive Sculpting Using Augmented-Reality, Mesh Morphing, and Force Feedback: Force-Feedback Capabilities in an Augmented Reality Environment. *IEEE Consum. Electron. Mag.* **2018**, *7*, 83–90. [CrossRef]

11. Tzemanaki, A.; Al, G.A.; Melhuish, C.; Dogramadzi, S. Design of a wearable fingertip haptic device for remote palpation: Characterisation and interface with a virtual environment. *Front. Robot. AI* **2018**, *5*, 62. [CrossRef] [PubMed]
12. Kishishita, Y.; Das, S.; Ramirez, A.V.; Thakur, C.; Tadayon, R.; Kurita, Y. Muscleblazer: Force-feedback suit for immersive experience. In Proceedings of the IEEE Conference on Virtual Reality and 3D User Interfaces (VR), Osaka, Japan, 23–27 March 2019; pp. 1813–1818. [CrossRef]
13. Son, E.; Song, H.; Nam, S.; Kim, Y. Development of a Virtual Object Weight Recognition Algorithm Based on Pseudo-Haptics and the Development of Immersion Evaluation Technology. *Electronics* **2022**, *11*, 2274. [CrossRef]
14. Bermejo, C.; Hui, P. A Survey on Haptic Technologies for Mobile Augmented Reality. *ACM Comput. Surv.* **2021**, *54*, 184. [CrossRef]
15. Kudry, P.; Cohen, M. Prototype of a wearable force-feedback mechanism for free-range immersive experience. In Proceedings of the RACS International Conference on Research in Adaptive and Convergent Systems, Virtual, 3–6 October 2022; pp. 178–184. [CrossRef]
16. *Haptic Devices*; 3D Systems: Rock Hill, SC, USA, 2022.
17. About the Oculus XR Plugin | Oculus XR Plugin | 3.3.0. 2023. Available online: https://docs.unity3d.com/Packages/com.unity.xr.oculus@3.3/ (accessed on 16 June 2023).
18. MRTK2-Unity Developer Documentation—MRTK 2 | Microsoft Learn. Available online: https://learn.microsoft.com/en-us/windows/mixed-reality/mrtk-unity/mrtk2/?view=mrtkunity-2022-05 (accessed on 16 June 2023).
19. OpenHaptics® Toolkit Version 3.5.0 API Reference Guide Original Instructions. 2018. Available online: https://s3.amazonaws.com/dl.3dsystems.com/binaries/Sensable/OH/3.5/OpenHaptics_Toolkit_API_Reference_Guide.pdf (accessed on 16 June 2023).
20. OpenHaptics® Toolkit Version 3.5.0 Programmer's Guide. 2018. Available online: https://s3.amazonaws.com/dl.3dsystems.com/binaries/Sensable/OH/3.5/OpenHaptics_Toolkit_ProgrammersGuide.pdf (accessed on 16 June 2023).
21. Kudry, P.; Cohen, M. Development of a wearable force-feedback mechanism for free-range haptic immersive experience. *Front. Virtual Real.* **2022**, *3*. [CrossRef]
22. Home | VirtualHere. Available online: https://www.virtualhere.com/ (accessed on 16 June 2023).
23. Interaction SDK Overview | Oculus Developers. Available online: https://developer.oculus.com/ (accessed on 16 June 2023).
24. Schrepp, M.; Hinderks, A.; Thomaschewski, J. Construction of a benchmark for the User Experience Questionnaire (UEQ). *Int. J. Interact. Multimed. Artif. Intell.* **2017**, *4*, 40. [CrossRef]
25. Schrepp, M.; Hinderks, A.; Thomaschewski, J. Design and evaluation of a short version of the User Experience Questionnaire (UEQ-S). *Int. J. Interact. Multimed. Artif. Intell.* **2017**, *4*, 103. [CrossRef]
26. R Core Team. *R: A Language and Environment for Statistical Computing*; R Foundation for Statistical Computing: Vienna, Austria, 2018.
27. R Core Team. *lm: Fitting Linear Models*; R Foundation for Statistical Computing: Vienna, Austria, 2023.
28. Russell, V.; Lenth, E.A. *emmeans: Estimated Marginal Means, aka Least-Squares Means*, R Package Version 1.8.7. 2023. Available online: https://cran.r-project.org/web/packages/emmeans/index.html (accessed on 16 June 2023).
29. R Core Team. *glmer: Fitting Generalized Linear Mixed-Effects Models*; R Foundation for Statistical Computing: Vienna, Austria, 2023.
30. Wilkinson, M.; Brantley, S.; Feng, J. A Mini Review of Presence and Immersion in Virtual Reality. *Proc. Hum. Factors Ergon. Soc. Annu. Meet.* **2021**, *65*, 1099–1103. [CrossRef]
31. Slater, M. Immersion and the illusion of presence in virtual reality. *Br. J. Psychol.* **2018**, *109*, 431–433. [CrossRef] [PubMed]

Article

An Internet of Things-Based Home Telehealth System for Smart Healthcare by Monitoring Sleep and Water Usage: A Preliminary Study

Zunyi Tang [1,2,*], Linlin Jiang [2,3], Xin Zhu [4] and Ming Huang [5]

1   Center for Medical Education and Career Development, Fukushima Medical University, Fukushima 960-1295, Japan
2   Department of Electrical and Electronic Engineering, College of Engineering, Nihon University, Koriyama 963-8642, Japan
3   School of Nursing, Fukushima Medical University, Fukushima 960-1295, Japan
4   Division of Information Systems, The University of Aizu, Aizu-Wakamatsu 965-8580, Japan
5   School of Data Science, Nagoya City University, Nagoya 467-8501, Japan
*   Correspondence: tangzunyi@gmail.com

Abstract: Recently, the Internet of Things (IoT) has attracted wide attention from many fields, especially healthcare, because of its large capacities for information perception and collection. In this paper, we present an IoT-based home telehealth system for providing smart healthcare management for individuals, especially older people. Each client node of the system is mainly composed of an electronic water meter that records the user's daily water usage, in order to analyze their living patterns and lifestyle as well as ascertain their well-being, and an unobtrusive sleep sensor that monitors the user's physiological parameters during sleep, such as heart rate (HR), respiratory rate (RR), body movement (BM), and their states on the bed or outside the bed. The collected data can be transmitted to a remote centralized cloud service by a wireless home gateway for analyzing the living pattern and rhythm of users. Furthermore, the periodic feedback of results can be provided to users themselves, as well as their family and health advisers. In the present study, data was collected from a total of 18 older subjects for one year to evaluate the effectiveness of the proposed system. By analyzing living patterns and rhythm, preliminary results indicate the effectiveness of the telehealth system and suggest the potential of the system regarding improvement in the quality of life (QoL) of older people and promotion of their health.

Keywords: IoT; ubiquitous healthcare; telehealth system; daily water usage; sleep monitoring; solitary deaths; Node-RED

Citation: Tang, Z.; Jiang, L.; Zhu, X.; Huang, M. An Internet of Things-Based Home Telehealth System for Smart Healthcare by Monitoring Sleep and Water Usage: A Preliminary Study. *Electronics* **2023**, *12*, 3652. https://doi.org/10.3390/electronics12173652

Academic Editor: Ivan Ganchev

Received: 31 July 2023
Revised: 24 August 2023
Accepted: 27 August 2023
Published: 29 August 2023

1. Introduction

The world is experiencing rapid rates of aging population, which is manifested by an increasing median age in the population of a country or a region due to declining fertility rates and/or rising life expectancies. The rapidly aging global population has led to a significant increase in the prevalence of chronic diseases, such as asthma, chronic obstructive pulmonary disease (COPD), diabetes, heart failure, and hypertension, resulting in the rapidly increasing need for medical and welfare service resources. However, very low fertility rates in many countries and regions have resulted in serious shortages of doctors, nurses, and home caregivers. The situation has forced the medical community to resort to currently available advanced technologies to solve the shortage of medical and welfare service resources.

Telehealth (also called telemedicine) is one of the most recent forms of medical and healthcare services. Its purpose is to provide clinical health care from a distance by using information and communication/telecommunication technology (ICT). Telehealth refers to a broader spectrum of remote healthcare services. Therefore, telehealth is considered one of the ways to solve the problems mentioned above. As a result of technological progress with the goal of meeting the needs of older people, various health care/monitoring systems have been developed and applied over the past two decades. The solutions include telephone-based [1,2], video [3,4], web-based [5–8], and telemetry/remote monitoring [9–13]. These solutions and technologies have the potential to change the way healthcare is currently delivered [9,13–16], and many are in use in clinical settings, improving the diagnosis and treatment of diseases and establishing new healthcare management methods for individuals and healthcare providers and administrators. Especially during the COVID-19 pandemic, the use of telehealth has seen significant growth and has played a significant role in addressing the severe shortage of medical resources in many countries [17].

In the last few years, the "concept" of the Internet of Things (IoT) has attracted attention from various fields, including healthcare. Numerous researchers have explored the potential of IoT in the healthcare field, addressing diverse practical challenges [18–24]. For instance, Dziak et al. designed an approach to construct an IoT-based information system for healthcare applications [25]. Park et al. proposed an IoT system for remote patient monitoring at home [26]. Guan et al. developed a remote health monitoring system for collecting older individuals' electrocardiogram (ECG) and motion signals via smart home gateway equipment [27]. Similarly, Abdelgawad et al. implemented an IoT-based health monitoring system to support active and assisted living with the goal of improving older people's lifestyle [28]. Additionally, Kumar et al. designed and developed a wristwatch-based wireless sensor platform for IoT wearable health monitoring [29]. Stefanova-Pavlova et al. proposed a refined generalized net model capable of tracking the changes in the health status (diabetes) of adult patients [30]. Kalogeropoulos et al. introduced teleophthalmology's role in diabetic retinopathy screening-a form of telehealth-offering a means to meet the rising demand for eye care among diabetes patients [31]. Alongside the mentioned literature review, we scrutinized various representative articles related to at-home health monitoring systems, comparing their characteristics and limitations. This step was essential as the proposed method in this article is tailored for a household context. The detailed outcomes of this review are summarized in Table 1. Furthermore, beyond the fundamental functions of telehealth monitoring, its utility and practical applications could be enhanced with additional value and features. Considering the emerging concern over solitary deaths or deaths in isolation, respectively referred to as "kodokushi" or "koritsushi" in Japanese, have recently been recognized as a social issue of concern in Japan [32,33], more efforts are needed to reduce the number of solitary deaths, especially for older people who are living alone and have a lack of social contact and connections. IoT technology can play an important role in preventing or reducing the number of solitary deaths due to its attributes, such as continuous, cost-effective, and long-term monitoring. It's noteworthy that there have been few instances where IoT has been applied to address the social issue of solitary deaths.

In the present article, the authors propose a novel IoT-based telehealth monitoring system not only to monitor the healthcare of users but also to potentially reduce the number of solitary deaths. Recognizing that solitary deaths typically occur at home, we developed a telehealth monitoring system centered around an electronic water meter for tracking daily water usage and an unobtrusive sleep monitoring device for capturing users' vital signs during the night, including heart rate (HR), respiratory rate (RR), and body movement (BM). This approach enables continuous 24-h monitoring of users' lifestyles. To validate the effectiveness of the proposed system, we employed 18 households from the Oyamada area in the city of Koriyama, Fukushima, to participate in the empirical trial of the system. Through analysis of the gathered data and providing direct feedback to the participating households, we gained valuable insights into their health conditions. Whenever assistance was needed for improving their well-being, designated care administrators or physicians extended their support by offering advice, suggestions, and care plans, ultimately enhancing their overall quality of life (QoL).

Table 1. A brief review of several at-home telehealth/telecare systems.

Author, Year	Aims and Methodology	Hardware/Software (Technology)	Evaluation Metrics	Features	Limitations
S. Ohta, 2002 [9]	tracking movements of elderly living alone using infrared sensors for health monitoring	AMP2109, AMN1111-2	movements	inexpensive, unrestricted	difficult to detect an emergency
Cheng, 2010 [34]	tracking the location and movement of an in-home patient for facilitating early detection of Alzheimer's disease	Bluetooth AP, PC, database	locations, movements	inexpensive, unrestricted, unobtrusive	lack of real-time interaction and decision engine feedback
Tsukiyama, 2015 [35]	evaluating of the health status of solitary elderly by monitoring their daily water usage and motion	water flow sensor, infrared sensor, RFID, Raspberry Pi	water usage, living activities	unobtrusive low-cost	non-24-h monitoring
Guan, 2017 [27]	long-term and continuous home health monitoring for the elderly using wearable smart clothing	STM32F401, Bluetooth, accelerometer, Android APP	ECG, motion	wearable	obtrusive
Chatrati, 2020 [36]	predicting hypertension and type 2 diabetes using supervised classification machine learning approach	smartphone, PC	BP, glucose	user-friendly GUI	manually inputting
Dhruba, 2021 [37]	monitoring sleep apnea in real-time by measuring various physiological parameters with six kinds of sensors	AD8232, MAX30102, Bluetooth, GSR and sound sensor, Arduino UNO	ECG, HR, PR, SpO2, GSR	data processing in real-time	obtrusive, restrictive
Jeong, 2022 [38]	physiological monitoring of patients at risk of stroke using wearable wireless sensor	AD8233, Bluetooth, smpartphone, cloud server	ECG/EMG	wearable, high accuracy of HR	obtrusive, restrictive
Wu, 2023 [39]	remotely monitoring the locations and physical conditions of quarantined individuals in real-time	watch with thermistor MF5B, PPG MAX30102 sensor, Nano 33 BLE, cloud server	BT, HR, SpO2	total design, rich utility	high power consumption
Tang, 2023 (proposed)	evaluating the lifestyle of the living elderly alone and preventing solitary death by monitoring daily water usage and vital signs during sleep	electronic water meter, sleep sensor, Note-RED, cloud server	water usage, HR, RR, BM	unrestricted, unobtrusive	difficult to rescue critical patients

In this table, BM, BP, BT, HR, PR, and RR indicate body movement, blood pressure, body temperature, heart rate, pulse rate, and respiratory rate, respectively.

2. Materials and Methods

2.1. System Design

A centralized cloud service is the core of the system, and it is responsible for (i) collecting and storing the vital sign data from client devices and (ii) analyzing and visualizing the data to provide feedback to users. The system diagram of the proposed system is illustrated in Figure 1. Data collection and storage within this system were established on a CentOS server and were implemented using Node-RED, which is an open-source, low-code programming tool for IoT. Node-RED is a visual programming tool built on top of the Node.js server-side framework, which has recently gained popularity in the development of IoT applications due to its flexibility in rapidly prototyping software systems. Node-RED also serves as an event-processing engine that includes built-in function nodes for message transmission based on the Message Queuing Telemetry Transport (MQTT) protocol and PostgreSQL connections. It also allows writing ad hoc JavaScript functions to process incoming and outgoing data, thus greatly simplifying and expediting the development process.

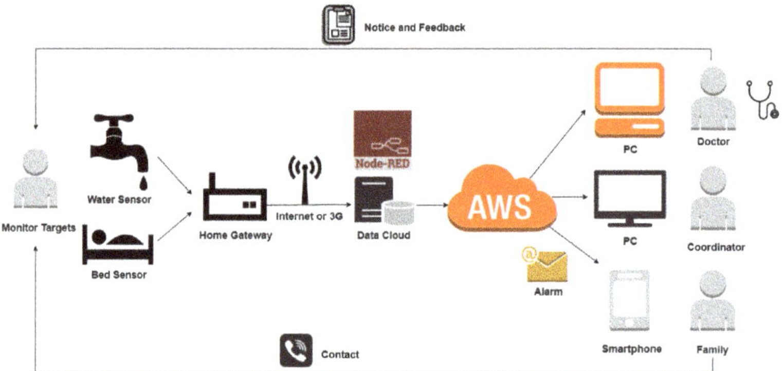

Figure 1. Systemic diagram of the proposed unobtrusive health monitoring system for home healthcare.

As shown in Figure 1, the developed cloud service acts as an MQTT broker for distributed sensors and devices. Upon receiving an MQTT measurements message from a client, the central service collects data for vital signs during sleep and water usage, which is then stored in the PostgreSQL database. By utilizing the timer trigger function of the database, various data analyses can be executed in a timely manner to generate further derived data. These lightweight data are then uploaded in real-time to Amazon Web Services (AWS) to provide prompt feedback to users, their families or caregivers, and/or their doctors. For instance, as part of the systemic functions, when an abnormal condition is detected and identified, the server will automatically trigger a response to notify the predefined contact of the user, enabling them to take timely or emergency action. Figure 2 illustrates the module flow built using Node-RED for collecting and storing bed sensor and electronic water meter data in the local PostgreSQL database and remote AWS in real time.

2.2. Distributed Sensors

As a typical IoT-based system, the distributed sensors constitute the edge nodes of the proposed telehealth monitoring system. As depicted in Figures 3 and 4, each edge node primarily consists of a sleep monitoring sensor and an electronic water meter. The sleep monitoring sensor (Anshin Hitsuji, NJI Corp., Koriyama, Japan) is primarily composed of 12 piezoelectric sensors and a control circuit with LAN communication functionality. The piezoelectric sensors are arranged in a matrix and enclosed within a soft and durable plastic container with dimensions of 800 mm × 100 mm × 5 mm. Placing the sleep monitoring sensor under the body or mattress during sleep allows for the measurement of HR, RR, and BM. Additionally, the sensor can determine whether a person is lying on the bed by sensing changes in pressure via the piezoelectric sensors. The sensor's accuracy for HR and RR measurements reaches 96.9 ± 0.1 % and 90.5 ± 0.7 %, respectively [40]. Consequently, the sleep monitoring sensor transmits the measured HR, RR, BM, the status whether the user is lying on the bed, and measurement time, to our remote cloud service.

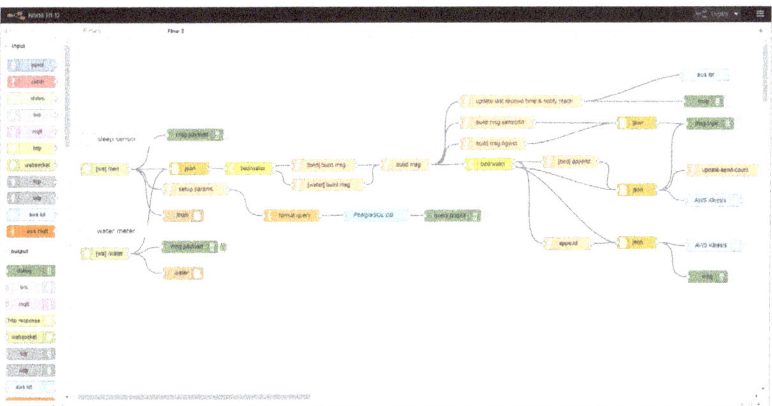

Figure 2. The module flow built using Node-RED for data collection from the bed sensor and electronic water meter in real time.

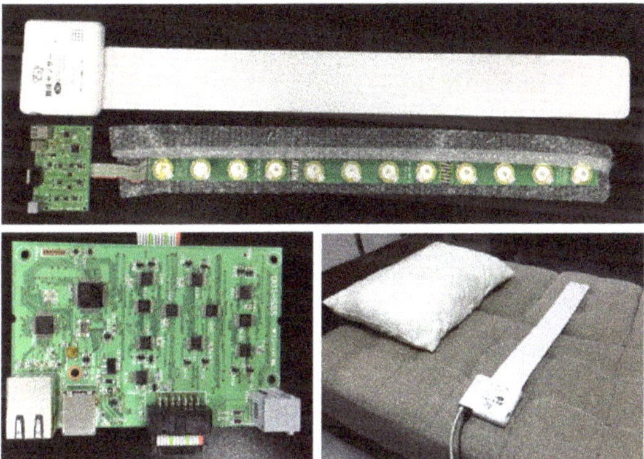

Figure 3. The sleep monitoring sensor used in the study is mainly composed of 12 piezoelectric sensors as the upper image and a control circuit with LAN communication functionality as the bottom left image. The scenario of sleep monitoring sensor being placed on a bed is shown in the bottom right image.

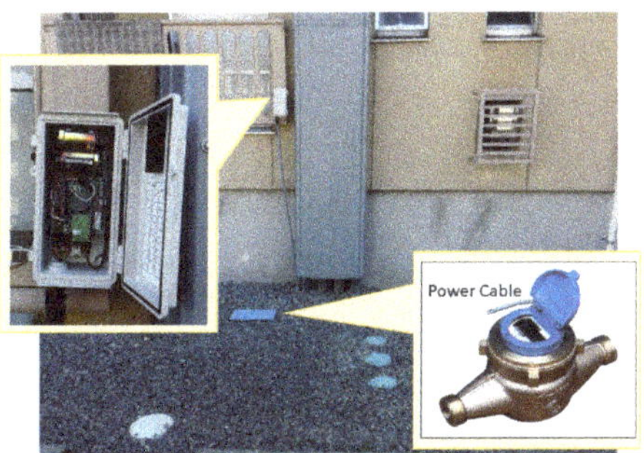

Figure 4. The interface circuit using Arduino Uno R3 (left) and water meter (right) used in the system.

The electronic water meters (EKDA13, Azbil Kimmon Corp., Tokyo, Japan) were employed to measure daily water usage. Each meter is comprised of 8-bit telegraphic communication output, LSI, LCD, and a lithium battery. The exterior resembles that of a conventional home-used water meter, but unlike the latter, it can display the cumulative water usage and wirelessly transmit the data to the telecommunication metering system, once at each predefined time interval, such as every hour. To enable our wireless home gateway to receive water usage data from the electronic water meters, we designed a transformation interface circuit using Arduino Uno R3. This interface circuit receives the data and subsequently transmits them to our cloud service. A 3G communication unit with a SIM card (SORACOM IoT SIM, SORACOM INC., Tokyo, Japan) is connected to the gateway and is responsible for communication with the remote cloud service. Finally, the collected data are stored in a PostgreSQL database on the cloud services. The electronic water meter adopted in this study, along with its transformation communication interface between the home gateway and the water meter, is depicted in Figure 4. The obtained water data includes consumed water volume (unit: liters), measured time, and an ID for distinguishing the water meters. For the purpose of transmitting the two types of raw data from the sleep monitoring sensor and water meter, the JavaScript Object Notation (JSON) is employed to facilitate data transmission between the client home gateway and Node-RED on remote servers.

2.3. Subjects and Sensor Settings

A total of 18 local older people (11 males and 7 females, 77.2 ± 6.6 years old) were recruited for this study, with 10 of them selected to have electronic water meters installed in their homes (refer to Figure 4). The characteristics of the subjects are summarized in Table 2. Sleep monitoring devices (refer to Figure 3) were set under either their mattress or a thick bed quilt. The electronic water meter was installed to connect with each subject's home's original water meter in series to measure daily water usage. Home gateways with 3G communication units using the IoT connectivity service known as "SORACOM Air for cellular" (SORACOM INC., Tokyo, Japan) were also set up at each subject's household. This study was approved by the Ethics Committee of Southern TOHOKU Hospital, Fukushima, Japan. Following a detailed explanation of the investigation's objective, informed consent was obtained from each subject or their household before data collection.

Table 2. A summary of subjects.

Subject ID	Age	Sex	Living Alone	Water Meter
1	68	Male	No	No
2	81	Male	Yes	Yes
3	77	Male	No	Yes
4	75	Male	No	Yes
5	77	Male	No	Yes
6	79	Female	Yes	Yes
7	86	Male	Yes	Yes
8	84	Female	Yes	Yes
9	86	Female	Yes	Yes
10	67	Female	Yes	Yes
11	65	Male	No	No
12	81	Female	Yes	No
13	67	Male	-	No
14	83	Female	Yes	No
15	81	Male	No	No
16	77	Male	-	Yes
17	79	Male	-	No
18	77	Female	-	No

2.4. Data Collection and Evaluation

Based on the raw data, four aspects of data analysis were performed to assess the proposed system.

2.4.1. Long-Term Measurement of HR and RR

HR was recorded in seconds on a beat-by-beat basis, and RR was recorded on a breath-by-breath basis during sleep. The averaged HR and RR of each subject every night were calculated to track long-term changes and corresponding trends.

2.4.2. Measurement of BM

The accumulative time and number of BM durations during sleep, changing with sleep duration time, were measured. In the present study, we used the average BM rate (BMR) to assess the frequency of BM during sleep. The BMR was defined as the number of BM occurrences every hour. The BMR is considered an index related to the quality of sleep [41]. The average BMR was calculated for each subject every night.

2.4.3. Measurement of Daily Water Usage

The daily water usage data of each subject or their household was collected every hour. To understand their daily water usage habits, the variation of daily water usage was first analyzed. Through analysis, we can observe the time distributions of daily living activities, including cooking, washing, bathing, and using the toilet. Furthermore, we plotted the time distributions of averaged water usage over a 24-h period based on the one-year data to analyze the variation of water usage and outline the regular pattern of water usage.

2.5. Feedback on the Analyzed Results

By analyzing the collected data and utilizing the opportunity to provide feedback to the households, we regularly organized gatherings where designated care administrators or physicians would provide face-to-face feedback of these results to the residents involved. Through this feedback process, we gained insight into their health conditions. When they required assistance about how to improve their health, designated care administrators or physicians offered help, suggestions, and care plans, thereby enhancing their quality of life. These gatherings also provided a platform for mutual communication, strengthening their connections, and preventing isolation, thus reducing the risk of solitary deaths.

3. Results

As a result, the measurements of the three kinds of vital signs during sleep for all subjects, as well as their daily water usage data, were consistently taken for one year from April 2016. In total, about 400,000,000 records were gathered. The typical data obtained from two subjects for a single day are shown in Figure 5, where it can be seen that the HR, RR, and BM profiles of each subject during sleep and water usage distributions over a 24-h period. The short bars on the vertical axes indicate the time and duration of BM. By comparing the results, the different trends of HR in the two subjects during sleep can be observed: the top one is relatively stable and the bottom one seems to be more jittery. Different sleep habits can be seen, such as the top lasting about 9.5 h while leaving the bed one time and the bottom showed about 7.5 h of sleep while leaving 3 times, maybe to go to the toilet. The top one did not only record sleep at night, but it also recorded the nap activity of the subject. Through the two examples above, it can be seen that the different living habits of users can be recorded using the proposed system.

Figure 6 shows the daily water usage of two subjects over one year. The Y-axis denotes the volume of the daily consumed water in liters (L). The local area experiences summer from June to August every year, and therefore, it can be seen that the water usage during this period is relatively higher than other times. The blank periods denote no record or data loss.

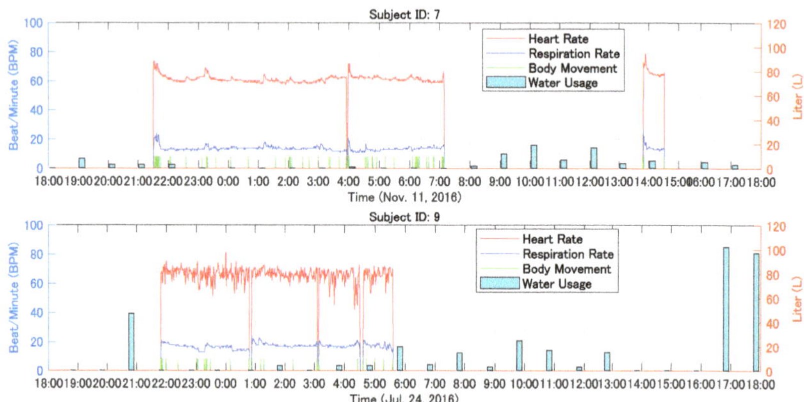

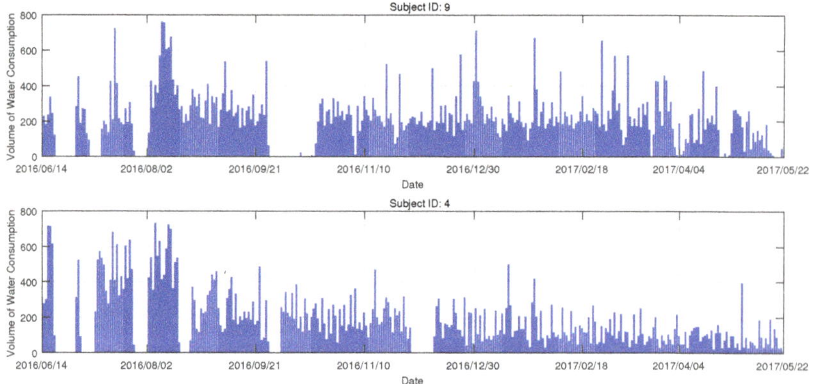

Figure 5. Typical trends of HR, RR, BM, and water usage from two different subjects with time for one sleep cycle. Note that the measured time of data sent from water meters are not the same.

Figure 6. Trends of the daily water usage from two subjects, respectively. X axis shows the period of about one year.

In order to analyze subjects' daily habits and living patterns using water, we used the water usage data for about a year to generate histograms of the averaged water usage each hour for the 10 subjects whose homes were equipped with electronic water meters. These results are shown in Figure 7.

Figure 8 shows histograms of sleep duration, latency, averaged HR, averaged RR, accumulated time of body movements, and leaving times of the subject with ID 13 while sleeping at night for one year. To track the long-term changes in the physical conditions of the subjects, we plotted the nightly averaged HR and RR as well as their standard deviations for the subjects. A typical example is shown in Figure 9. In the figure, the vertical axis denotes HR (beats/min) and RR (breaths/min). The curves and vertical bars with short horizontal lines at their upper and lower ends indicate the averaged values and standard deviations of detected HR and RR for the corresponding day.

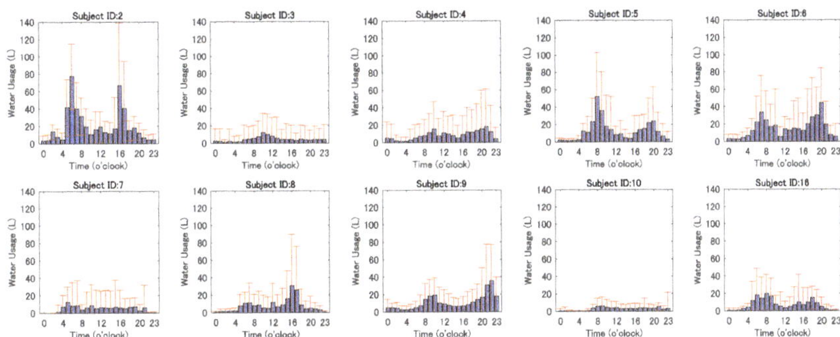

Figure 7. The histograms display the means and deviations of water usage per hour over one year for ten subjects whose homes were equipped with electronic water meters. The subfigures for subjects with IDs 2, 4, 5, 6, 8, 9, and 16, respectively, exhibit the characteristic pattern of typical double-peak water usage.

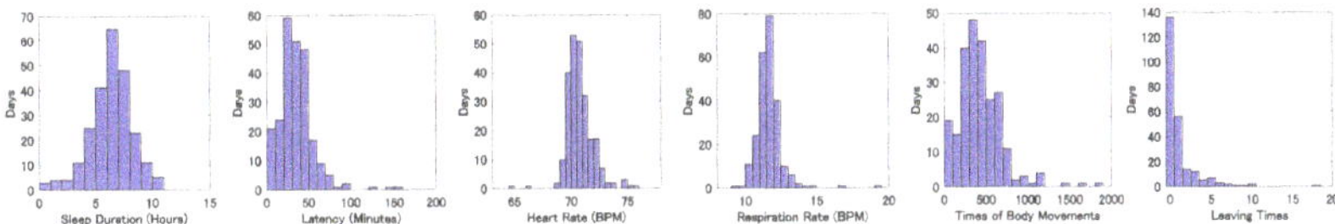

Figure 8. The histograms of sleep duration, latency, average HR, average RR, the number of body movements, and leaving times of the subject with ID 13 each night for one year.

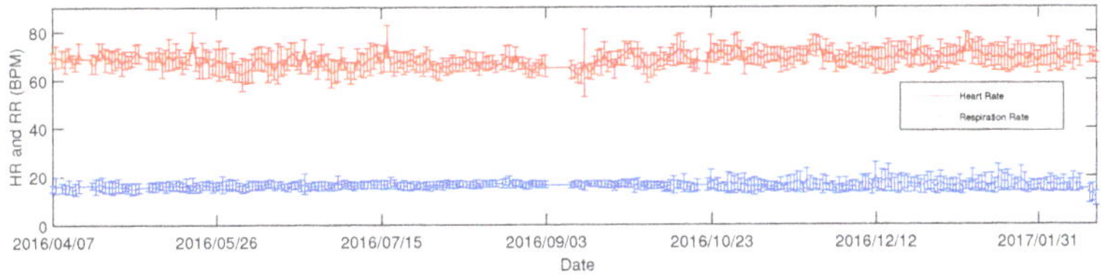

Figure 9. The complete profiles of HR and RR from the subject with ID 13 for about one year. In this figure, the Y-axis denotes the HR (beats/min) and RR (breaths/min). The curves and vertical bars with short horizontal lines at their upper and lower ends indicate the mean values and standard deviations of the detected HR and RR for the corresponding day. Note that the vertical lines stand for the loss of data.

Figure 10 shows the trends of HR, RR, and BM from the subject with ID 13 during sleep for four days using a web-based GUI.

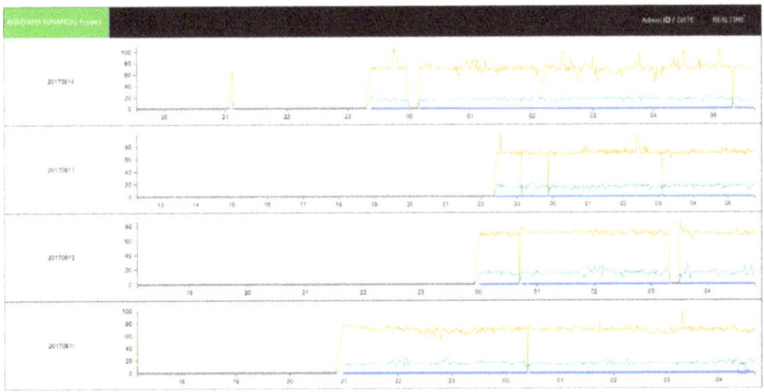

Figure 10. The trends of HR (orange lines), RR (green lines), and BM (blue lines) from the subject with ID 13 during sleep for four days are displayed in a web-based GUI.

4. Discussion

The goal of this study was to provide a solution that would allow older people to maintain their autonomy and to increase their QoL, as well as being able to notify and contact the designated people in the case of an emergency to avoid solitary death as much as possible. Whether the system is effective depends on several major factors, including accurate, long-term measurements of vital signs, timely data analysis and feedback, and rapid response when an emergency is detected. For the telehealth monitoring system, data measurements and collections can be achieved unconstrainedly, unlike wearable devices that need users to wear, like Fitbits and smartwatches. Wearable devices are designed to collect users' personal health data, but their inconveniences are easy to forget, especially for older people. Therefore, the continuity of the long-term monitoring approach proposed in the present article is ensured because it doesn't require cumbersome operations and is less likely to be forgotten. Other types of measurement devices for vital signs, for example, blood pressure monitors with Bluetooth communication [42] and wearable watches [43], are compatible with the proposed system, but they were not adopted in the current system due to their obtrusiveness and intrusiveness of usage.

Simple water usage data may not directly reflect the health conditions of users, but long-term information can be used to analyze and judge whether the daily living pattern of the users is regular or not, as a regular life is generally important for everyone, especially older people. For example, when there is no water usage over a certain time interval, e.g., 12 h, for older people living alone and rarely leaving their home, it may indicate that they have suddenly suffered a critical disease onset or a serious accident that makes self-aid impossible. Therefore, the data can be used to detect some emergencies. Certainly, there are some limitations to this method because a relatively long time, for example, several hours, is usually required to detect an emergency, meaning that the method may be helpless for rescuing some critical patients due to serious diseases, such as heart attacks and strokes. In addition, family activities such as cooking, laundry, bathing, and cleaning all require a certain level of physical effort. By analyzing changes and patterns in water usage, it is possible to assess whether users, especially those living alone, require support and care in their daily lives. This approach can serve as an indicator of an individual's ability to carry out daily activities.

Further data analysis and comparison based on the telehealth monitoring system can detect normal/abnormal states, living patterns, and rhythms. For example, in Figure 7, it can be seen that the averaged water usage each hour between the subjects was significantly different, but most results clearly show water usage peaking twice in one day, i.e., two intervals of 7 to 12 o'clock and 16 to 20 o'clock. This normal state reflects one aspect of subjects' regular life, which is significant for their health. On the contrary, a relatively 'abnormal state' indicates a lack of habitual routines, irregular cooking with frequent reliance on takeout, and infrequent laundering or bathing. These abnormal states could be attributed to changes in physical or health conditions, as well as poor lifestyle habits. During the validation phase, the detection of these states was achieved through offline comparison of histogram graphs. It is important to note that the concepts of normal and abnormal states are relative and lack absolute significance. On the other hand, when an emergency is detected, a fast reaction from family members, relatives, or assigned healthcare caregivers is necessary. This work was supported by the local city government to develop and evaluate a solution to maintain the autonomy of the local older people living alone or the older family needing support, e.g., regular visiting and cooking, to improve their quality of life. Rapid aging is a serious social problem in Japan, which is caused by the current rapid aging and sub-replacement fertility. Based on the proposed telehealth monitoring system, older people can be monitored unobtrusively 24 h a day, while their privacy can also be protected as much as possible compared to other kinds of approaches, e.g., video monitoring systems [3,4]. The framework of the system is flexible and compatible with many other types of methods for collecting vital signs, such as infrared and Doppler radar. Thus, various telehealth monitoring systems can be rapidly derived and constructed to meet specific needs.

The proposed telehealth monitoring system is also helpful in preventing and quickly identifying solitary death (dying alone), which is an increasingly serious problem in Japan. According to the statistics [44] from the government of Japan, thousands of older people nationwide die alone every year. Solitary death results in high costs with regard to dealing with the remains and personal belongings of the deceased, which has a very negative impact on their family and neighbors. The phenomenon of solitary death is mainly attributed to a decreasing proportion of older Japanese people living in multi-generational housing and instead living alone [32,45]. Older people who live alone are more likely to lack social contacts with family and neighbors, and are therefore more likely to die alone and remain undiscovered. By introducing the proposed telehealth monitoring system, the life rhythms and living patterns of older people living alone can be recorded and analyzed. When remarkable changes are detected, the early warning and decision support functions of this system can promptly notify their families and health advisors for acquiring further responses. Further work related to data analysis is still in progress, and the results will be presented in the near future.

As mentioned above in Figures 6 and 9, the entire procedure of data collection lasting about one year suffered from data loss (several blank intervals) due to several reasons below. First, the home gateway routers responsible for data communication were accidentally switched off by users. Second, data transmission units encountered troubles due to long-time communication overload. Additionally, traveling away from home or hospitalization were also reasons for some data loss. Despite several cases of data loss or blank periods, the time trend of daily water usage is clear, and changes in at-home activities can be identified using water usage. Because the continuity of the data is crucial for predicting the risk of potential diseases or detecting emergency states, we are currently addressing the mentioned issues to ensure data continuity and maintain the system's reliability, such as detecting accidental interruptions in power supply and strengthening the load capacity of data transmission units.

In the present study, we reported an IoT-based telehealth monitoring system that integrates sleep monitoring sensors and electronic water meters based on Node-RED, an open-source integration framework for IoT-based application development. There are currently several alternative technologies or frameworks similar to Node-RED, such as Eclipse Kura and Flogo. Here, we provided an example of how to construct an IoT-based health monitoring system using current popular IoT technologies. We believe that more health monitoring systems related to IoT will be constantly developed not only for older people but also for various other groups, contributing to improving people's health and addressing the shortage of doctors, nurses, and home caregivers. Moreover, it may be beneficial to construct a next-generation telemedicine system that can predict users' health levels and the risks of lifestyle-related diseases. Finally, the data collected from the sleep sensor also have the potential to be used for detecting various diseases such as cardiac arrhythmia, sleep apnea syndrome, and sleep disorders. Our research in this area is still in progress, and the results will be reported in the near future.

5. Conclusions

The increasing demand for home health care has led to the development of various home telehealth monitoring systems based on IoT and ICT. By integrating an unconstrained sleep monitoring sensor and an electronic water meter, and connecting them with a centralized cloud service, we successfully constructed an IoT-based home telehealth monitoring system aimed at monitoring users' daily activities over the long term. Compared to existing healthcare monitoring solutions, this system conducts entirely unobtrusive measurements and provides continuous 24-h monitoring. Our results demonstrate the validity of the proposed system for daily, unconstrained sleep monitoring and living activities. Therefore, we consider it an ideal solution for long-term home healthcare monitoring. Finally, while users are living in their homes, sleep monitoring and water usage can be automatically and unconsciously recorded. This approach demonstrates its convenience compared to monitoring methods based on traditional video/audio communication and/or wearable devices.It also suggests that the proposed unobtrusive telehealth monitoring system has great potential for supporting long-term home healthcare monitoring.

Author Contributions: Conceptualization, Z.T. and L.J.; methodology, Z.T. and L.J.; software, Z.T.; validation, Z.T. and L.J.; data curation, L.J.; writing—original draft preparation, Z.T.; writing—review and editing, X.Z., L.J., and M.H.; visualization, L.J.; supervision, X.Z. and M.H.; project administration, Z.T. All authors have read and agreed to the published version of the manuscript.

Funding: This research was partially funded by grants-in-aid from the Koriyama Demonstration Experiment Project of Health Monitoring System 2016–2019, Koriyama City, Fukushima, Japan.

Institutional Review Board Statement: The study was conducted in accordance with the Declaration of Helsinki, and approved by the Ethics Committee of Southern TOHOKU Hospital, Fukushima, Japan.

Informed Consent Statement: Informed consent was obtained from all subjects involved in the study.

Data Availability Statement: Not applicable.

Acknowledgments: We thank Kaoru Sakatani, Hiroyuki Hashimoto, Kaori Matsumoto for providing great support and Tadao Munakata, Katsuhiko Nishimaki, and Satoshi Maeda for their help in implementing Node-RED based data collection and AWS-based data analysis. We would like to thank Lizhen Hu, Harumi Igarashi, Yuta Murayama, and Yutaka Sato for their advice and assistance during the course of this project. We also thank Koji Otani for providing many valuable suggestions on this paper.

Conflicts of Interest: The authors declare no conflict of interest.

References

1. Grant, J.S.; Elliott, T.R.; Weaver, M.; Bartolucci, A.A.; Giger, J.N. Telephone Intervention with Family Caregivers of Stroke Survivors After Rehabilitation. *Stroke* **2002**, *33*, 2060–2065. [CrossRef] [PubMed]
2. Badger, T.A.; Segrin, C.; Hepworth, J.T.; Pasvogel, A.; Weihs, K.; Lopez, A.M. Telephone-delivered health education and interpersonal counseling improve quality of life for Latinas with breast cancer and their supportive partners. *Psycho-Oncology* **2013**, *22*, 1035–1042. [CrossRef] [PubMed]
3. Dick, P.T.; Filler, R.; Pavan, A. Participant satisfaction and comfort with multidisciplinary pediatric telemedicine consultations. *J. Pediatr. Surg.* **1999**, *34*, 137–142. [CrossRef] [PubMed]
4. Chan, D.S.; Callahan, C.W.; Sheets, S.J.; Moreno, C.N.; Malone, F.J. An Internet-based store-and-forward video home telehealth system for improving asthma outcomes in children. *Am. J. Health-Syst. Pharm.* **2003**, *60*, 1976–1981. [CrossRef]
5. Artinian, N.T.; Harden, J.K.; Kronenberg, M.W.; Vander Wal, J.S.; Daher, E.; Stephens, Q.; Bazzi, R.I. Pilot study of a Web-based compliance monitoring device for patients with congestive heart failure. *Heart Lung J. Acute Crit. Care* **2003**, *32*, 226–233. [CrossRef]
6. Wantland, D.J.; Portillo, C.J.; Holzemer, W.L.; Slaughter, R.; McGhee, E.M. The effectiveness of Web-based vs. non-Web-based interventions: A meta-analysis of behavioral change outcomes. *J. Med. Internet Res.* **2004**, *6*, e40. [CrossRef]
7. Lazakidou, A.A. *Web-Based Applications in Healthcare and Biomedicine*; Springer: New York City, NY, USA, 2010.
8. Dhillon, J.S.; Ramos, C.; Wünsche, B.C.; Lutteroth, C. Designing a web-based telehealth system for elderly people: An interview study in New Zealand. In Proceedings of the 2011 24th International Symposium on Computer-Based Medical Systems (CBMS), Bristol, UK, 27–30 June 2011.
9. Ohta, S.; Nakamoto, H.; Shinagawa, Y.; Tanikawa, T. A health monitoring system for elderly people living alone. *J. Telemed. Telecare* **2002**, *8*, 151–156. [CrossRef]
10. DelliFraine, J.L.; Dansky, K. Home-based telehealth: A review and meta-analysis. *J. Telemed. Telecare* **2008**, *14*, 62–66. [CrossRef]
11. Koch, S. Home telehealth—Current state and future trends. *Int. J. Med. Inform.* **2006**, *75*, 565–576. [CrossRef]
12. Peng, Y.T.; Lin, C.Y.; Sun, M.T.; Landis, C.A. Multimodality Sensor System for Long-Term Sleep Quality Monitoring. *Telemed. e-Health* **2010**, *16*, 244–253. [CrossRef]
13. Jiang, L.; Tang, Z.; Liu, Z.; Chen, W.; Kitamura, K.I.; Nemoto, T. Automatic sleep monitoring system for home healthcare. In Proceedings of the Proceedings of 2012 IEEE-EMBS International Conference on Biomedical and Health Informatics, Hong Kong, China, 5–7 January 2012; pp. 894–897.
14. Jaana, M.; Paré, G. Home telemonitoring of patients with diabetes: A systematic assessment of observed effects. *J. Eval. Clin. Pract.* **2007**, *13*, 242–253. [CrossRef]

15. Tamura, T.; Sekine, M.; Tang, Z.; Yoshida, M.; Takeuchi, Y.; Imai, M. Preliminary study of a new home healthcare monitoring to prevent the recurrence of stroke. In Proceedings of the 2015 37th Annual International Conference of the IEEE Engineering in Medicine and Biology Society (EMBC), Milan, Italy, 25–29 August 2015; pp. 5489–5492.
16. Weiner, M.; Callahan, C.M.; Tierney, W.M.; Overhage, J.M.; Mamlin, B.; Dexter, P.R.; McDonald, C.J. USing information technology to improve the health care of older adults. *Ann. Intern. Med.* **2003**, *139*, 430–436. [CrossRef] [PubMed]
17. Bouabida, K.; Lebouché, B.; Pomey, M.P. Telehealth and COVID-19 Pandemic: An Overview of the Telehealth Use, Advantages, Challenges, and Opportunities during COVID-19 Pandemic. *Healthcare* **2022**, *10*, 2293. [CrossRef]
18. Hassanalieragh, M.; Page, A.; Soyata, T.; Sharma, G.; Aktas, M.; Mateos, G.; Kantarci, B.; Andreescu, S. Health monitoring and management using Internet-of-Things (IoT) sensing with cloud-based processing: Opportunities and challenges. In Proceedings of the 2015 IEEE international conference on services computing, New York, NY, USA, 27 June–2 July 2015; pp. 285–292.
19. Islam, S.R.; Kwak, D.; Kabir, M.H.; Hossain, M.; Kwak, K.S. The Internet of Things for Health Care: A Comprehensive Survey. *IEEE Access* **2015**, *3*, 678–708. [CrossRef]
20. Yuehong, Y.I.N.; Zeng, Y.; Chen, X.; Fan, Y. The internet of things in healthcare: An overview. *J. Ind. Inf. Integr.* **2016**, *1*, 3–13.
21. Scuro, C.; Sciammarella, P.F.; Lamonaca, F.; Olivito, R.S.; Carnì, D.L. IoT for Structural Health Monitoring. *IEEE Instrum. Meas. Mag.* **2018**, *1094*, 4–14. [CrossRef]
22. Ruman, M.R.; Barua, A.; Rahman, W.; Jahan, K.R.; Roni, M.J.; Rahman, M.F. IoT based emergency health monitoring system. In Proceedings of the 2020 International Conference on Industry 4.0 Technology (I4Tech), IEEE, Pune, India, 13–15 February 2020; pp. 159–162.
23. Siam, A.I.; Almaiah, M.A.; Al-Zahrani, A.; Elazm, A.A.; El Banby, G.M.; El-Shafai, W.; El-Samie, F.E.A.; El-Bahnasawy, N.A. Secure Health Monitoring Communication Systems Based on IoT and Cloud Computing for Medical Emergency Applications. *Comput. Intell. Neurosci.* **2021**, *2021*, 8016525. [CrossRef] [PubMed]
24. Cao, S.; Lin, X.; Hu, K.; Wang, L.; Li, W.; Wang, M.; Le, Y. Cloud computing-based medical health monitoring IoT system design. *Mob. Inf. Syst.* **2021**, *2021*, 8278612. [CrossRef]
25. Dziak, D.; Jachimczyk, B.; Kulesza, W.J. IoT-Based Information System for Healthcare Application: Design Methodology Approach. *Appl. Sci.* **2017**, *7*, 596. [CrossRef]
26. Park, K.; Park, J.; Lee, J. An IoT System for Remote Monitoring of Patients at Home. *Appl. Sci.* **2017**, *7*, 260. [CrossRef]
27. Guan, K.; Shao, M.; Wu, S. A Remote Health Monitoring System for the Elderly Based on Smart Home Gateway. *J. Healthc. Eng.* **2017**, *2017*, 5843504. [CrossRef]
28. Abdelgawad, A.; Yelamarthi, K.; Khattab, A. IoT-Based Health Monitoring System for Active and Assisted Living. In Proceedings of the Smart Objects and Technologies for Social Good: Second International Conference, GOODTECHS 2016, Venice, Italy, 30 November–1 December 2017; pp. 11–20.
29. Kumar, S.; Buckley, J.L.; Barton, J.; Pigeon, M.; Newberry, R.; Rodencal, M.; Hajzeraj, A.; Hannon, T.; Rogers, K.; Casey, D.; et al. A wristwatch-based wireless sensor platform for IoT health monitoring applications. *Sensors* **2020**, *20*, 1675. [CrossRef]
30. Stefanova-Pavlova, M.; Andonov, V.; Stoyanov, T.; Angelova, M.; Cook, G.; Klein, B.; Vassilev, P.; Stefanova, E. Modeling telehealth services with generalized nets. *Recent Contrib. Intell. Syst.* **2017**, *657*, 279–290.
31. Kalogeropoulos, D.; Kalogeropoulos, C.; Stefaniotou, M.; Neofytou, M. The role of tele-ophthalmology in diabetic retinopathy screening. *J. Optom.* **2020**, *13*, 262–268. [CrossRef]
32. Nomura, M.; McLean, S.; Miyamori, D.; Kakiuchi, Y.; Ikegaya, H. Isolation and unnatural death of elderly people in the aging Japanese society. *Sci. Justice* **2016**, *56*, 80–83. [CrossRef]
33. Nakazawa, E.; Yamamoto, K.; London, A.J.; Akabayashi, A. Solitary death and new lifestyles during and after COVID-19: Wearable devices and public health ethics. *BMC Med. Ethics* **2021**, *22*, 89. [CrossRef]
34. Cheng, H.T.; Zhuang, W. Bluetooth-enabled in-home patient monitoring system: Early detection of Alzheimer's disease. *IEEE Wirel. Commun.* **2010**, *17*, 74–79. [CrossRef]
35. Tsukiyama, T. In-home health monitoring system for solitary elderly. *Procedia Comput. Sci.* **2015**, *63*, 229–235. [CrossRef]
36. Chatrati, S.P.; Hossain, G.; Goyal, A.; Bhan, A.; Bhattacharya, S.; Gaurav, D.; Tiwari, S.M. Smart home health monitoring system for predicting type 2 diabetes and hypertension. *J. King Saud-Univ.-Comput. Inf. Sci.* **2022**, *34*, 862–870. [CrossRef]
37. Dhruba, A.R.; Alam, K.N.; Khan, M.S.; Bourouis, S.; Khan, M.M. Development of an IoT-Based Sleep Apnea Monitoring System for Healthcare Applications. *Comput. Math. Methods Med.* **2021**, *2021*, 7152576. [CrossRef]
38. Jeong, J.W.; Lee, W.; Kim, Y.J. A real-time wearable physiological monitoring system for home-based healthcare applications. *Sensors* **2022**, *22*, 104. [CrossRef]
39. Wu, J.Y.; Wang, Y.; Ching, C.T.S.; Wang, H.M.D.; Liao, L.D. IoT-based wearable health monitoring device and its validation for potential critical and emergency applications. *Front. Public Health* **2023**, *11*, 1188304. [CrossRef] [PubMed]
40. Tang, Z.; Jiang, L.; Hu, L.; Sato, Y.; Komuro, Y.; Sakatani, K. Preliminary Study of Sleep Quality Assessment for Elderly People Based on Unobtrusive Sleep Monitoring System. *Trans. Jpn. Soc. Med. Biol. Eng.* **2017**, *55*, 552–553.
41. Lu, L.; Tamura, T.; Togawa, T. Detection of body movements during sleep by monitoring of bed temperature. *Physiol. Meas.* **1999**, *20*, 137. [CrossRef] [PubMed]
42. AND Digital Blood Pressure Monitor Model UA-767. Available online: https://www.aandd.co.jp/adhome/products/me/ua767pbt-c.html (accessed on 17 October 2022).

43. Zheng, Y.L.; Ding, X.R.; Poon, C.C.Y.; Lo, B.P.L.; Zhang, H.; Zhou, X.L.; Yang, G.Z.; Zhao, N.; Zhang, Y.T. Unobtrusive sensing and wearable devices for health informatics. *IEEE Trans. Biomed. Eng.* **2014**, *61*, 1538–1554. [CrossRef] [PubMed]
44. Summary of the 2014 White Paper on Aging Society of Japan. Available online: http://www8.cao.go.jp/kourei/whitepaper/w-2014/gaiyou/s1_2_6.html (accessed on 17 October 2022).
45. Morita, S.; Nishi, K.; Furukawa, S.; Hitosugi, M. A Survey of Solitary Deaths in Japan for Shortening Postmortem Interval Until Discover. *Prilozi* **2015**, *36*, 47–51. [CrossRef]

 *electronics*

MDPI

Article

A Viscoelastic Model to Evidence Reduced Upper-Limb-Swing Capabilities during Gait for Parkinson's Disease-Affected Subjects

Luca Pietrosanti [1], Cristiano Maria Verrelli [1], Franco Giannini [1], Antonio Suppa [2,3], Francesco Fattapposta [2], Alessandro Zampogna [2], Martina Patera [2], Viviana Rosati [4] and Giovanni Saggio [1,*]

[1] Department of Electronic Engineering, University of Rome Tor Vergata, 00133 Rome, Italy
[2] Department of Human Neurosciences, Sapienza University of Rome, 00185 Rome, Italy
[3] IRCCS Neuromed, 86077 Pozzilli, Italy
[4] A.O.U. Policlinico Umberto I, 00161 Rome, Italy
[*] Correspondence: saggio@uniroma2.it

Abstract: Parkinson's disease (PD) is a chronic neurodegenerative disorder with high worldwide prevalence that manifests with muscle rigidity, tremor, postural instability, and slowness of movement. These motor symptoms are mainly evaluated by clinicians via direct observations of patients and, as such, can potentially be influenced by personal biases and inter- and intra-rater differences. In order to provide more objective assessments, researchers have been developing technology-based systems aimed at objective measurements of motor symptoms, among which are the reduced and/or trembling swings of the lower limbs during gait tests, resulting in data that are potentially prone to more objective evaluations. Within this frame, although the swings of the upper limbs during walking are likewise important, no efforts have been made to reveal their support significance. To fill this lack, this work concerns a technology-based assessment of the forearm-swing capabilities of PD patients with respect to their healthy counterparts. This was obtained by adopting a viscoelastic model validated via measurements during gait tests tackled as an inverse dynamic problem aimed at determining the torque forces acting on the forearms. The obtained results evidence differences in the forearm movements during gait tests of healthy subjects and PD patients with different pathology levels, and, in particular, we evidenced how the worsening of the disease can cause the worsening of the mechanical support offered by the forearm's swing to the walking process.

Keywords: Parkinson's disease; motor impairment; TUG test; upper-limb swings

Citation: Pietrosanti, L.; Verrelli, C.M.; Giannini, F.; Suppa, A.; Fattapposta, F.; Zampogna, A.; Patera, M.; Rosati, V.; Saggio, G. A Viscoelastic Model to Evidence Reduced Upper-Limb-Swing Capabilities during Gait for Parkinson's Disease-Affected Subjects. *Electronics* **2023**, *12*, 3347. https://doi.org/10.3390/electronics12153347

Academic Editor: Lei Jing

Received: 30 June 2023
Revised: 28 July 2023
Accepted: 2 August 2023
Published: 4 August 2023

1. Introduction

Parkinson's disease (PD) is a chronic and progressive neurodegenerative disorder that affects the central nervous system. The disease manifests when there is a significant loss of dopaminergic cells in a region of the brainstem called the *substantia nigra*. The dopaminergic system plays a crucial role in motor control, and the decrement in dopamine levels in the brain is responsible for the genesis of its motor symptoms.

The cardinal motor symptoms of PD include tremor, bradykinesia (slowness of movement), rigidity (stiffness of muscles), dyskinesia (movements similar to tics or chorea) [1], and postural instability (impaired balance and coordination) [2,3]. These symptoms progressively worsen over time and can significantly impact a patient's quality of life.

In addition to motor symptoms, PD may also present non-motor symptoms [4], such as cognitive difficulties, speech impairments [5,6], mood changes, sleep disturbances, and autonomic dysfunction. These non-motor symptoms can vary widely among individuals and may appear before or after the onset of motor symptoms.

Among the most adopted protocols for PD assessment is the MDS-UPDRS scale [7], consisting of a 0–4-interval rating assessment of motor and non-motor symptoms, mostly based on the qualitative rater's considerations [8] and therefore prone to intra- and inter-rater variability and experience [9], potentially mitigated by technology-based objective measurements. In such a frame, Lonini et al. [10] adopted inertial measurement units (IMUs) placed on the hand and classified the motor tremors and bradykinesia by means of a convolutional neural network (CNN) during the patient's daily activities. In a similar test, Di Biase et al. [11] explored the optimal configuration of IMUs for symptom detection in upper-arm tasks, demonstrating that, to discriminate between healthy and pathological subjects and between the ON and OFF medicine conditions, a single inertial unit placed on the distal location of the upper limb is sufficient. Cesarelli et al. [12] focused on motor tasks performed by the upper limbs to evidence kinematic features useful for classifying PD vs. healthy conditions through the so-called Knime Analytics Platform. Ricci et al. [13–15] employed a network of wearable IMUs and analyzed the collected data by means of the k-nearest neighbor (kNN) and Support Vector Machine (SVM) algorithms for assessing motor impairments in a PD de novo subject, and they found that the early stages of PD are characterized by a set of features that are not visible to the naked eye.

While researchers have been devoting efforts to objectively evaluate a number of different motor impairments, the particular (but crucial) aspect of the resistance to motion, which is rigidity [16], remains insufficiently analyzed. Rigidity in PD is characterized by limb stiffness, which is generally assessed by clinicians via the passive movements of the limbs.

This cardinal motor symptom greatly affects the daily lives of PD patients, contributing to their global motor slowness, gait impairment, and loss of independence. The objective evaluation of rigidity, better if performed remotely in an ecological setting, could improve clinical management with targeted interventions such as tailored therapies or focused physiotherapy.

We have recently shown that rigidity in PD can be objectively estimated by examining the frequency content of arm-swing movements during gait [17].

Arm-swing reduction is frequently observed in PD patients from the early stages of the disease, so it has been proposed as a prodromic marker of the disease. As an example, Mirelman et al. [18] adopted wearable sensors on the participant's trunk, ankles, and wrists during a gait test, and they found that 60% of the discriminative motor characteristics come from the wrists' motion only.

The pathophysiological mechanisms underlying reduced arm swing during gait in PD is still a matter of debate, but it has been demonstrated to be strongly associated with the severity of rigidity [17,19]. However, as far as we know, there are no validated methods to quantitatively assess rigidity in pathological subjects.

The present study aims to propose an easy-to-interpret model to analyze upper-limb mobility and impairments and quantitatively estimate rigidity in PD subjects from gait measurements.

To achieve this, here, we introduce an innovative descriptive model of the forearm swings during gait to underline the possible differences in the acting torque on the elbows of PD patients with respect to healthy subjects, and we quantify the reduced motor capabilities of the forearms by means of a finite set of coefficients representing biomechanical aspects of the forearm. In particular, the model refers to the Lagrangian formalism to describe the upper arms' swing and a double-pendulum system, taking a cue from works on healthy individuals [20–22], with elastic and viscosity properties embedded into the physical constraints.

This model is fed with data gathered from inertial sensors and is capable of computing information about the elastic and viscous properties of forearm muscles. Moreover, we aim to quantify the reduced arm swing in PD patients in order to examine possible clinical–behavioral correlations. We hypothesize that our model could be a new analytic tool for the instrumental assessment of patients with PD.

2. Materials and Methods

2.1. Subjects

A total of 37 volunteer participants were recruited at the Movement Disorders Outpatient Service of Sapienza University of Rome in Italy. Each participant was instructed on the procedure and gave informed consent according to the standard protocol of Helsinki. The participants were divided into two groups: one consisting of 18 PD patients and another consisting of 22 age-matched healthy subjects (HC, hereafter).

The clinical assessment included MDS-UPDRS scores, based on main motor symptoms' evaluation, including cardinal signs of PD (bradykinesia, rigidity, and tremor) and axial disorders (gait and balance issues), and H&Y scale, according to symptom distribution (symmetric/asymmetric involvement) or axial impairment. Cognitive functions were examined through MMSE, a multi-domain cognitive scale for evaluating several abilities such as orientation, memory, fluency, attention, visuospatial abilities, etc. These assessments provided valuable data for the analysis and comparison of the PD patients versus the control group, as well as for patient classification based on disease severity and clinical–behavioral correlations.

Inclusion criteria for PD patients were as follows: diagnosis of idiopathic PD based on current consensus criteria [23]; Hoehn and Yahr scale (H&Y) $\leq$ 2.5; ability to maintain an upright stance and walk independently. The exclusion criteria were as follows: diagnosis of atypical parkinsonism; severe PD (H&Y > 2.5); presence of L-Dopa-induced dyskinesia, dementia (as reflected by Mini-Mental State Examination—MMSE < 24) and comorbidities affecting gait or arm movements (including orthopedic or rheumatologic issues, polyneuropathies, and other neurological diseases).

PD patients were divided according to the severity of their upper limb rigidity as defined by item 3.3 of MDS-UPDRS-III. This item identifies five levels of severity: (0) no rigidity; (1) slight rigidity (only detected asking the patient to activate the contralateral limb); (2) mild rigidity (when full range of passive motion is easily achieved); (3) moderate rigidity (when full range of passive motion is achieved with effort); (4) severe rigidity (when full range of passive motion is not achieved at all). No patients with moderate or severe rigidity were enrolled, so the participants were divided into 3 groups (PD0, PD1, PD2) according to their 3.3 item of the MDS-UPDRS-III score (0, 1, 2, respectively).

PD patients were under chronic dopaminergic treatment. Arm swing during gait, as rigidity, bradykinesia, or tremor, can benefit from levodopa intake [19,24]. To observe rigidity without the influence of medication, we examined our patients when in the OFF-state of therapy, as intended after 12 h of L-Dopa withdrawal according to standardized procedures [25,26].

Concerning healthy subjects, only individuals not suffering from significant clinical disorders affecting gait and arm swing were enrolled.

Participants were matched for age and demographic characteristics (Table 1). Anthropometric data (mass, length, center of mass, and turning radius for both the upper arm and forearm) were computed from height and weight of each participant, as reported in [27], to feed the later detailed model (as in the following Section 2.4).

Table 1. Clinical and anthropometric features of patients with Parkinson's disease (PD) and healthy controls (HC).

	PD	HC
Age (years)	65 ± 8	69 ± 11.2
Gender	12 M, 6 F	10 M, 22 F
Height (cm)	170.5 ± 11.5	163.1 ± 10.7
Weight (Kg)	73 ± 16.0	67.5 ± 12.4
MDS-UPDRS III	29.8 ± 10.7	

MDS-UPDRS III: Movement disorder society—Unified Parkinson's disease rating scale part III.

2.2. Wearable Electronics

A number of different technologies have been developed and applied to gather movement data from subjects performing gait. These technologies are classified according to the locations of sources and sensors (if attached to the body being measured or elsewhere in the world, respectively [3]), or according to the type of adopted technology (basically: optical, electromagnetic, and physical ones). Within technologies, the camera-based systems are the most reliable and represent the gold-standard, but are expensive and require dedicated rooms and experienced technicians, whilst the wearable-based systems are less accurate but are low-cost, potentially ubiquitous, and offer an easy-to-use solution. In particular, the wearable based-systems refer to those electronics that can be unobtrusively embedded into clothes within supporting fabrics, such as bend of stretch sensors [28,29] or inertial measurement units (IMUs).

For our work, the motor capabilities of upper limbs were gathered by means of a network of five lightweight (<20 g for each) IMUs, termed Movit® (model G1, by Captiks Srl, Rome, Italy), previously successfully adopted [30,31] and validated against an optical-based gold-standard gait measuring system, and thus proven to perform more than sufficiently for our purposes [32]. Each IMU was placed on a specific district of the subject's body, that is, the upper back (thoracic vertebrae T7) and arms and forearms, by means of Velcro straps, as evidenced in Figure 1.

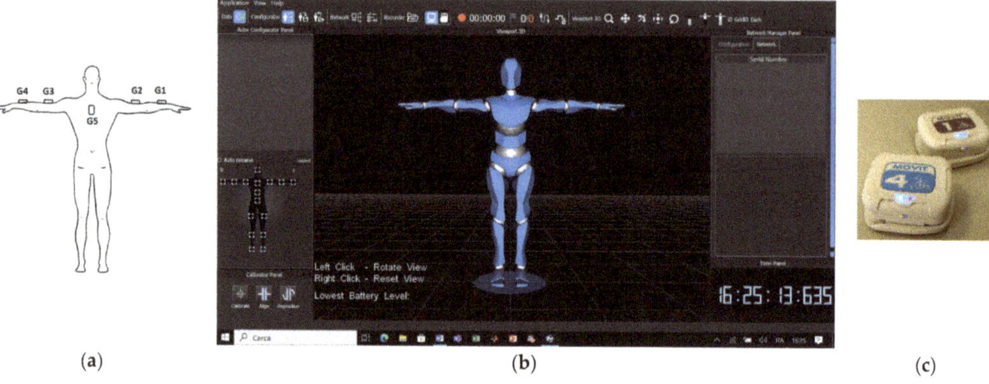

Figure 1. (**a**) The IMU sensors, termed Movit, labeled from G1 to G5, as seen on the back of the subject's body; (**b**) a screen of Motion Studio Software version 2.4 (capable to manage 16 IMUs at once) used for data acquisition; and (**c**) two Movit G1 units.

Each IMU (two examples are shown in Figure 1c) is equipped with a tri-axis accelerometer capable of measuring up to ±16 g with 16.384 LSB/g and a tri-axis gyroscope with a measurement range up to ±2000°/s and a resolution of 32.8 LSB/°/s [33,34]. The recording process is managed, at 200 Hz sampling frequency, by the Motion Studio software version 2.4 (a proprietary one, Figure 1b) running on a personal computer that receives data in wireless mode via ZigBee protocol [35]. The Motion Analyzer software version 1.2.98 (from the same manufacturer) furnishes the elbow angles (in degrees) from the gathered raw data.

2.3. Test Protocol

This study focuses on analyzing the forearm swings during gait performances gathered from the standard Timed Up and Go (TUG) test, a widely used protocol for evaluating gait impairments associated with various pathologies. The TUG test comprises multiple sub-phases (Figure 2a):

1. Sit to Walk: Starting from sitting, the participant gets up from the chair without the aid of the arms;
2. First linear walking: The subject walks (away from the chair) at a comfortable speed along a straight line for six meters;
3. Turning: The participant turns 180° around a pin to go back;
4. Second linear walking: Same as point 2, but in the opposite direction (toward the chair);
5. Stand To Sit: The participant turns themself around 180° and sits down.

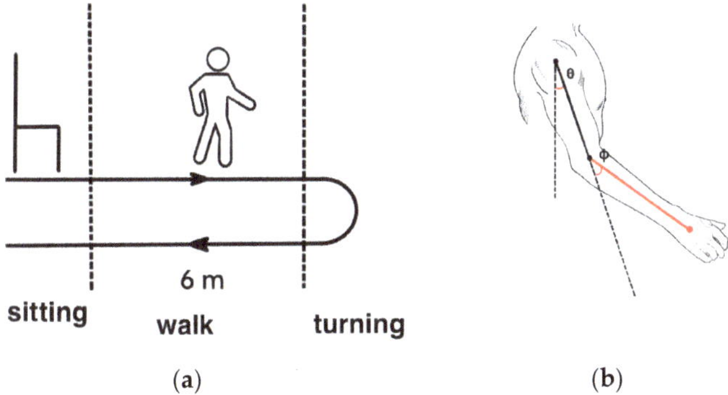

Figure 2. (**a**) Scheme of the Timed Up and Go (TUG) test and its sub-phases; (**b**) arm and forearm as a double pendulum system.

We intentionally asked each participant to follow the entire TUG protocol for completeness, being a standard test; however, our interest was focused on the linear walking phases only, and in particular on the upper arms' pendular movements in supporting the gait.

2.4. Double Pendulum Model

Arms' swings during gait are complex and not sufficiently explained tasks. Some studies [36,37] have shown their dependence on the activation and synergies of different muscles. By the way, for the purposes of the present study focused on the torque forces acting on the upper arms, we adopted a simple model based on the 2 link arm inverse kinematics. In particular, we modelled the arm as a double pendulum system [38] (Figure 2b). This model takes into account only the movement of the upper limb in the sagittal plane and does not consider wrist degree of freedom. Despite these simplifications, this model shows good agreement capabilities with the real data [39]. Accordingly, here, we express the dynamic behavior of the upper limb through the Lagrangian formulation, L being the Lagrangian function as the difference between kinetic and potential energies of the double pendulum, T and V, respectively:

$$L = T - V, \tag{1}$$

In particular, we model the two segments (arm and forearm) of the double pendulum adopting the anthropometric data previously computed (Section 2.1), with each segment characterized by mass m_i, length l_i, center of mass (with respect to the proximal joint) r_i, and moment of inertia I_i.

Considering the upper arm, the kinetic energy T_1, and the potential energy V_1 results in:

$$T_1 = \frac{1}{2} I_1 \theta^2, \tag{2}$$

$$V_1 = m_1 g r_1 (1 - \sin \theta), \tag{3}$$

while for the forearm, we obtain the following expression:

$$T_2 = \frac{1}{2}m_2 v_2^2 + \frac{1}{2}I_2(\dot{\theta} + \dot{\varphi})^2, \tag{4}$$

$$V_2 = m_2 g\left[l_1(1 - \sin\theta) + r_2(1 - \sin(\theta + \varphi))\right], \tag{5}$$

with v_2 being the center of mass velocity, defined by the following expression:

$$v_2^2 = (l_1\dot{\theta})^2 + 2l_1 r_2\dot{\theta}(\dot{\theta} + \dot{\varphi})\cos\varphi + (r_2\dot{\varphi})^2, \tag{6}$$

with θ and φ being the angles formed by the arm and the forearm, respectively (Figure 2b), and g the gravity acceleration.

The torque acting on shoulder and elbow joints can thus be obtained by differentiating the Lagrangian function as follows:

$$\frac{d}{dt}\frac{\partial L}{\partial \dot{\theta}} - \frac{dL}{d\theta} = \tau_\theta, \tag{7}$$

$$\frac{d}{dt}\frac{\partial L}{\partial \dot{\varphi}} - \frac{dL}{d\varphi} = \tau_\varphi, \tag{8}$$

where τ_θ and τ_φ are the torques acting on the upper arm and the forearm, respectively. Specifically, and according to the MDS-UPDRS protocol, we focus on the elbow's dynamics and, consecutively, on forearm movements.

2.5. Dynamic of the Forearm

Mechanical properties of muscles can be represented as a system of springs and damping elements, as in Figure 3a, known as the Kelvin–Voight model [40,41]. According to this model, in this study, we express the torque acting on the elbow joint as a function of elastic and viscous elements, as depicted in Figure 3b:

$$\tau_\varphi = \beta_0 + \beta_1(\varphi_m - \varphi) + \beta_2\varphi + \beta_3\dot{\varphi}, \tag{9}$$

where the constant β_0 represents the forcing constant (due to muscle's activity); β_1 and β_2 represent the flexion and the extension elastic constants, respectively; β_3 represents the damping factor (due to the "viscosity" aspect of the muscle's movements); and φ_m represents the maximum flexion angle of the elbow joint.

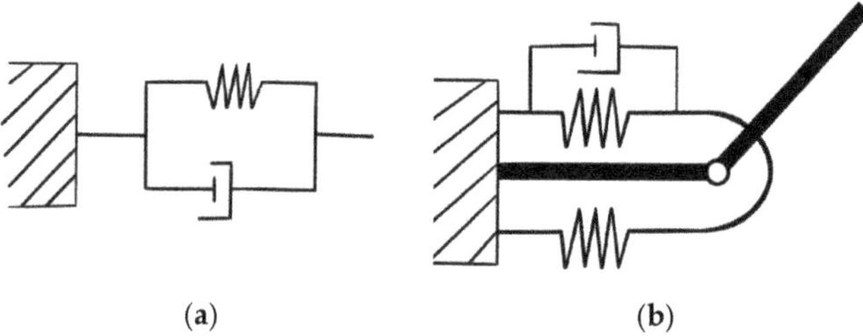

(**a**)   (**b**)

Figure 3. Representation of (**a**) single muscle with the Kelvin–Voight model and (**b**) biceps and triceps muscles as system of spring-damping elements.

2.6. Data Processing

In order to increase the signal-to-noise ratio of the analog signals gathered from the elbows' angles data during the TUG test, we approximated the measured data with an 8-term Fourier series, similarly to [1]. This step allows maintaining the significant information of the signal, removing unwanted frequencies and noise. Fourier series can be applied to approximate periodic signals; however, if the signal loses its periodicity, Fourier series can occur in a loss of information. Due to the adopted measurement protocol and task, the signal recorded from arm oscillations are quasi-periodic, meaning that Fourier series can be applied with a negligible information loss. Smoothed data were inserted in the double pendulum model to obtain τ_φ. The β coefficients were then computed by fitting Equation (8) with measured τ_φ using the least-squares method. The optimization of β coefficients was performed by means of a Matlab routine (by Mathworks Inc., Massachusetts, US) based on the trust-region-reflective algorithm. This procedure was first applied to healthy subjects to determine the mean reference values for β features that characterize average values related to their swing capabilities. We then repeated the same procedure for each of the PD patients to evidence their differences with respect to the average values of their healthy counterparts, so as to underline the impact (if any) of the disease on the range of motion of the elbows' swings and on the elastic and damping aspects of the forearm movements for both upper limbs.

In particular, to highlight differences (if any) between the HC and PD groups of the β coefficients, specifically related to elastic and dumping aspects of muscles acting on the forearm, we employed a Mann–Whitney U test (significant p-value set to 0.05).

Moreover, to evaluate any correlation between obtained data and clinical scores, we computed the Spearman Rank correlation coefficient (significant p-value set to 0.05) between the aforementioned β_0, β_1, β_2, β_3 coefficients and the score from item 3.3 of the MDS-UPDRS scale (particularly related to rigidity aspects of upper limbs). Spearman correlation was preferred instead of classical Pearson correlation because of the ordinal nature of UPDRS values.

3. Results

Figure 4a,b show the mean torque τ_φ values as obtained during a single complete oscillation of the forearm (starting when in maximum flexion) with the method previously detailed (Equation (8)), for both HC and PD groups, respectively.

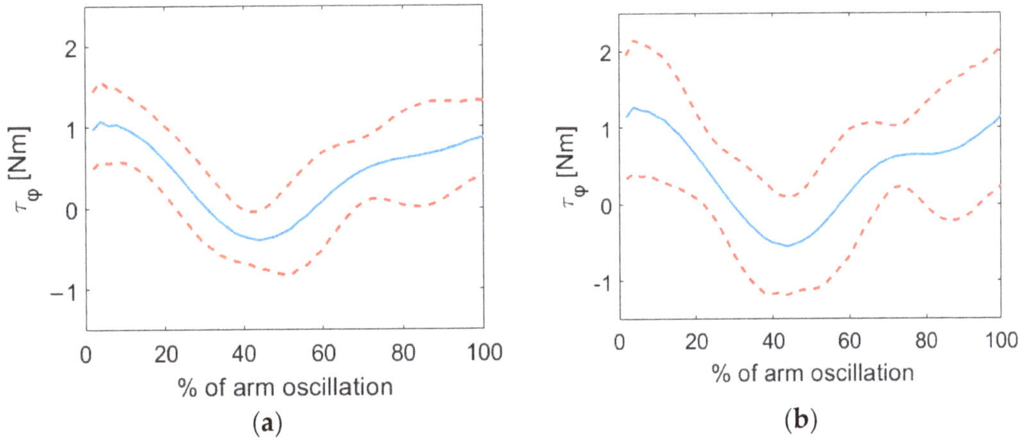

Figure 4. Average elbow torque (blue, solid lines) and related standard deviation (red, dashed lines) in a complete gait cycle for (**a**) HC group and (**b**) PD patients.

Figure 5 shows the RMSEs of the τ_φ values as determined by Equation (8) using the estimated four β coefficients, with respect to the τ_φ value as obtained directly from the measurements, for the HC and PD group, respectively.

Statistics of the estimated coefficients are reported in Table 2 for both HC and PD groups, together with the results from the Mann–Whitney U test.

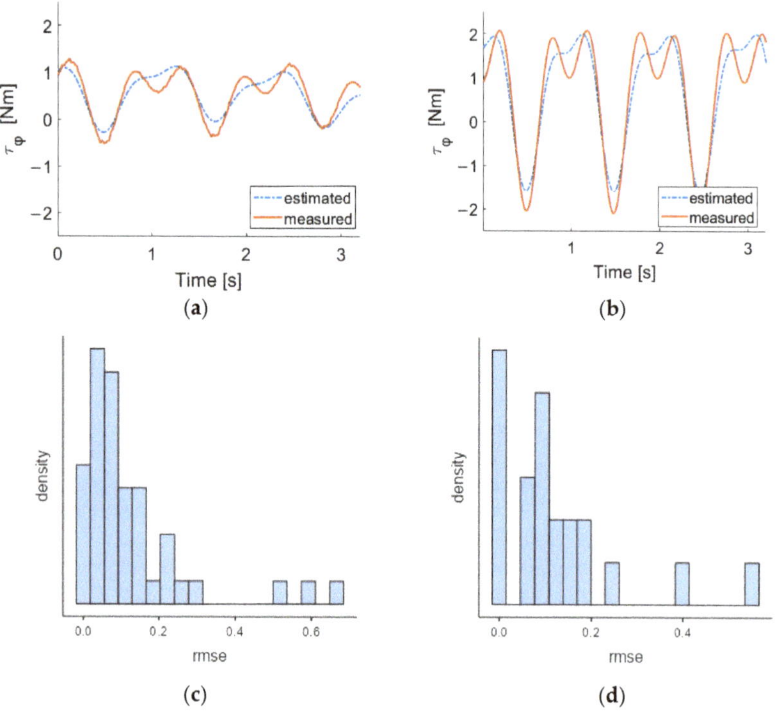

Figure 5. Torques measured (orange solid line) and estimated with β coefficients (blue dashed line) for (**a**) HC and (**b**) PD subjects. Distribution of RMSE in (**c**) HC and (**d**) PD groups.

Table 2. Median and standard deviation (SD) values of beta coefficients for different PD subgroups, and results of the Mann–Whitney test. The meaningful p-values are evidenced in bold.

	PD (MDS-UPDRS Item 3.3 Score)			HC (Median $\pm$ Sd)	p-Value		
	PD (0) (Median $\pm$ Sd)	PD (1) (Median $\pm$ Sd)	PD (2) (Median $\pm$ Sd)		PD 0 vs. HC	PD 1 vs. HC	PD 2 vs. HC
β_0	3.53 ± 1.81	2.56 ± 1.71	2.27 ± 0.70	2.69 ± 1.02	0.419	0.386	**0.047**
β_1	-6.09 ± 2.01	-6.49 ± 4.89	-6.74 ± 7.21	-6.23 ± 2.30	0.760	0.792	0.348
β_2	-15.0 ± 5.31	-12.3 ± 6.85	-13.6 ± 9.63	-11.6 ± 3.88	0.075	0.522	0.402
β_3	-0.56 ± 0.44	-0.47 ± 0.63	-0.83 ± 0.63	-0.26 ± 0.35	0.166	**0.049**	**0.040**

Table 3 shows the results for Spearman's rank correlation between the β_0, β_1, β_2, β_3 coefficients of PD subjects and item 3.3 "limb rigidity" score.

Table 3. Spearman's rank correlations between β coefficients and MDS-UPDRS item 3.3 score. The meaningful p-value is reported in bold.

Feature	Rigidity (Sub-Item 3.3)	
	r	p-Value
β_0	-0.378	**0.041**
β_1	-0.155	0.245
β_2	0.116	0.697
β_3	-0.262	0.119

Figure 6 reports the distribution of beta coefficients for different MDS-UPDRS item 3.3 scores.

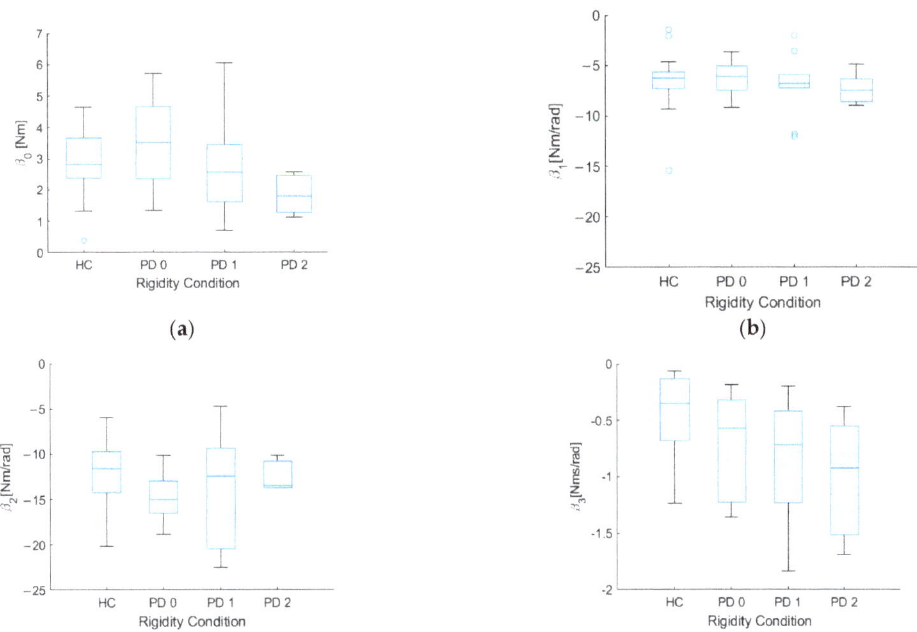

Figure 6. Distribution of (**a**) the forcing constant (because of the muscles' activities) β_0, (**b**) the flexion elastic constant β_1, (**c**) the extension elastic constant β_2, and (**d**) the dumping factor β_3 for different rigidity conditions. Circles outside the boxes represent outliers.

4. Discussion

In this work, we introduce and adopt a novel mathematical model to analyze the arm swing during linear walking to estimate the forces and rigidity acting on the elbow joints on the basis of measurements gathered during gait tests, with numerical coefficients representing the viscoelastic properties of the forearm.

Figure 4 illustrates the average curve depicting the relationship between the measured τ_φ and the percentage of arm oscillation. It is observed that when starting from a position of maximum flexion, the torque initially exhibits a positive value during the initiation of the forward oscillation phase. The point of maximum elbow flexion corresponds to roughly 50% of the oscillation, wherein the torque reaches its minimum value. Notably, in PD subjects, the τ_φ reaches lower values, opposing the forward movement, acting like a spring at its maximum extension trying to return to equilibrium position. In the second half of the arm oscillation, from maximum flexion to maximum extension, the torque returns to positive values. This can be attributed to the elbow blocking the backward oscillation of the forearm.

Figure 5a,b illustrate the torques extracted from the measurement values inserted into the double pendulum model (orange solid line), and the others as derived from β coefficients (blue dashed line) for, by way of example, but not limited to, one healthy subject (Figure 5a) and one PD patient (Figure 5b). In such a case, the extracted and derived coefficients are mostly coincident for negative values, which corresponds to the flexion phase, as explained for Figure 4, while for positive values, the model has a poor fit with the measured torque. This discrepancy can be addressed to the fact that we did not insert into our linear model (Equation (9)) a term to represent elbow constraint, which is characterized by strong nonlinearity. We are aware that such an omission has an impact on the results; however, in this study, we intended to focus only on the viscoelastic behavior of the limb in flexion–extension. It is worth noting that for these two subjects, the negative values associated with the force counteracting elbow flexion are greater (considering absolute values) for the PD subject (Figure 5a) with respect to the healthy one (Figure 5b).

Figure 5c,d show the distribution of the RMSE between the double pendulum model and the β coefficient model for the HC (Figure 5c) and for PD (Figure 5d) groups, respectively, with the possibility of appreciating the good fit since most of the subjects' RMSEs are limited to low values (<0.2).

Figure 6 presents boxplots illustrating the β values under various rigidity conditions. Here, we define as "positive" the torque that induces a counterclockwise rotation around the shoulder's mediolateral axis, i.e., a forward oscillation of the arm. Consequently, we observe that β_0, representing the constant force associated with muscle activity, is positive. On the other hand, β_1, β_2, and β_3 exhibit negative values as they contribute to the generation of restoring forces, which act in opposition to changes in joint angles.

Figure 6a shows a trend for the forcing constant β_0 that mostly decrease (but not as marked) inversely with the MDS-UPDRS 3.3 item score. Indeed, PD groups with higher rigidity scores show smaller β_0 values, with statistical significance when discerning PD2 from HC ($p < 0.05$). As already stated, the β_0 coefficient represents the forcing constant due to muscle's activity. The progressive reduction in this coefficient, which proceeds according to disease severity, could be explained by the progressive loss of upper limb synkinesis of gait that occurs in PD patients starting in the early stages of the disease [42]. We obtained statistical significance comparing HC with PD02, that is, patients with more severe rigidity. These results suggest that our model could be more reliable in advanced disease stages.

Figure 6b evidences how the flexion elastic constant values β_1 do not have statistical significance since they do not result in particular differences among different groups of participants.

Figure 6c shows a decrease in the extension elastic constant β_2 with increasing PD severity. It is worth mentioning the differences between the flexion and extension elastic constants, β_1 and β_2, respectively, probably due to asymmetric effect of disease on flexors and extensors muscles.

Figure 6d reports differences in the dumping factor β_3 (associated with viscous behavior of the upper limbs) among groups. The differences are not so pronounced, but a downward trend with PD severity can be observed. This result suggests a possible impact of viscous-elastic factors in the limitation of upper limb movements in PD patients. Given that the viscous-elastic coefficient not associated with muscle contraction, we can hypothesize that β_3 could include the restraints on movement induced by ligaments, muscle sclerosis, and articular deformations. This interpretation could possibly explain the statistical significance obtained in the comparison between patients with higher rigidity scores (PD1, PD2) and HC. In fact, osteo-articular deformities are common comorbidities associated with more advanced stages of PD (i.e., antero/latero-collis, stooped posture, Pisa's syndrome) [24]. The progressive increase in viscoelastic forces could result in more abnormal and deformed postures that can eventually be associated with the origin of osteo-articular comorbidities.

Regarding the clinical–behavioral correlations, the results reported in Table 3 show a mild correlation between β_0 and the UPDRS-III rigidity score. This correlation supports the hypothesis that rigidity is partly responsible for reduced upper limb swing during gait in PD. The limited strength of this correlation, however, would suggest that other factors, such as bradykinesia, may also contribute to reduced arm swing during gait in PD. Arm swing reduction during gait has been recently associated with higher risk of falls in PD patients [43]. Relying on this model, objectively identifying PD patients with reduced arm swing and, therefore, patients prone to higher risk of falls could represent a valid instrument for fall prevention and treatment. For instance, applying our algorithm to closed-loop systems, possibly with sensory cues to improve arm swing during gait, could represent a promising clinical application of this model.

When considering the present study, a number of possible limitations should be taken into account. First, the cohort of participants is rather limited, thus requiring the replication of findings in wider samples of patients, including those with more severe rigidity (i.e., UPDRS-III 3.3 item $\geq$ 3). Second, we assessed patients when in the OFF state of therapy, but arm swing and rigidity in PD can be significantly influenced by dopaminergic therapy [19,24]. Accordingly, future studies should address the effects of L-Dopa on the proposed model to clarify the impact of the pharmacological therapy on its accuracy.

Another limitation comes from condensing all the muscles acting on the limb in three coefficients only. The results suggest that this is a good approximation, since we are able to fit real data with a reduced error, but the model can be improved by differentiating the contribution of each muscle. Moreover, this model limits the analysis to planar movements, neglecting the arm's motion in three-dimensional space during oscillation. Conversely, the increase in complexity needs a more detailed analysis, so that it can be convenient to determine the best accuracy–complexity balance for a viable solution.

5. Conclusions

By considering that the swings of the upper limbs play a key, but often underrated, role during gait, here, we report a method for evidencing differences in the rigidity of forearm swings during gait for a group of PD patients with respect to a group of heathy subjects, for a total of 37 volunteer participants, and its relationship with the related MDS-UPDRS score.

To this aim, we model the arm swing as a double pendulum, focusing on the forearm movements as a key aspect, as the item 3.3 of the MDS-UPDRS rating scale evidences.

In particular, we use the double pendulum equations to derive the total torque acting on the forearm from data gathered using wearable sensors. The obtained torque is then described by means of a reduced set of coefficients representing the viscoelastic behaviour of the forearm. Accordingly, we differentiate swing aspects of the upper limbs during gait, evidencing a different viscoelasticity of PD patients with respect to their healthy counterparts.

Our results indicate a trend suggesting that the worse the pathology, the worse the capability to support walking with forearm swinging.

Moreover, the coefficients estimated through the model exhibit a correlation with the MDS-UPDRS item 3.3 scores, related to upper limb's rigidity.

In summary, our analysis demonstrates a correlation between UPDRS scores with respect to rigidity aspects of forearm swings. Nevertheless, we are aware that the number of involved participants must be improved to gain statistical effectiveness, that the model can be improved to take into account other second-order aspects in movements, and that other types of measurements, such as electromyographic ones, can add significance to the completeness of the study.

The model we propose, once validated in a more complex fashion too, could be associated with real-time evaluation of rigidity in PD patients in ecological settings. As the adoption of sensors and wearable devices increases in everyday clinical practice, our model could provide an easy tool for motor fluctuation detection and, therefore, for possible improvement in therapeutic strategies. Objective methods to assess tremor, bradykinesia, or other motor signs such as gait disorder (i.e., freezing of gait) are already largely available to use also in domestic environments. A new approach to quantitatively assess rigidity would help improve telemedicine strategies by implementing the remote examination of cardinal motor signs in PD through quantitative approaches.

Author Contributions: Conceptualization, L.P., F.G., C.M.V. and G.S.; methodology, A.S. and G.S.; software, L.P.; validation, G.S.; formal analysis, L.P. and G.S.; investigation, L.P., A.S., F.F., A.Z., M.P., V.R. and G.S.; resources, G.S. and A.S.; data curation, L.P.; writing—original draft preparation, L.P. and G.S.; writing—review and editing, G.S.; visualization, L.P.; supervision, A.S., F.G. and G.S.; project administration, G.S.; funding acquisition, G.S. All authors have read and agreed to the published version of the manuscript.

Funding: This research received no external funding.

Data Availability Statement: Data are available upon reasonable request.

Acknowledgments: We want to thank Captiks Srl for providing the wearable sensors.

Conflicts of Interest: One of the authors, G.S., declares involvement in Captiks Srl as a founder, but also declares to have not received money, goods, commodities, or any other advantages from this involvement other than scientific interest.

References

1. Ricci, M.; Giannini, F.; Saggio, G.; Cenci, C.; Di Lazzaro, G.; Pisani, A. A Novel Analytical Approach to Assess Dyskinesia in Patients with Parkinson Disease. In Proceedings of the 2018 IEEE International Symposium on Medical Measurements and Applications (MeMeA), Rome, Italy, 11–13 June 2018; pp. 975–979. [CrossRef]
2. Zampogna, A.; Cavallieri, F.; Bove, F.; Suppa, A.; Castrioto, A.; Meoni, S.; Pélissier, P.; Schmitt, E.; Bichon, A.; Lhommée, E.; et al. Axial Impairment and Falls in Parkinson's Disease: 15 Years of Subthalamic Deep Brain Stimulation. *NPJ Park. Dis.* **2022**, *8*, 121. [CrossRef] [PubMed]
3. Saggio, G.; Sbernini, L. New Scenarios in Human Trunk Posture Measurements for Clinical Applications. In Proceedings of the 2011 IEEE International Symposium on Medical Measurements and Applications, Bari, Italy, 30–31 May 2011; pp. 13–17. [CrossRef]
4. Poewe, W. Non-Motor Symptoms in Parkinson's Disease. *Eur. J. Neurol.* **2008**, *15* (Suppl. 1), 14–20. [CrossRef] [PubMed]
5. Suppa, A.; Asci, F.; Saggio, G.; Marsili, L.; Casali, D.; Zarezadeh, Z.; Ruoppolo, G.; Berardelli, A.; Costantini, G. Voice Analysis in Adductor Spasmodic Dysphonia: Objective Diagnosis and Response to Botulinum Toxin. *Park. Relat. Disord.* **2020**, *73*, 23–30. [CrossRef] [PubMed]
6. Suppa, A.; Costantini, G.; Asci, F.; Di Leo, P.; Al-Wardat, M.S.; Di Lazzaro, G.; Scalise, S.; Pisani, A.; Saggio, G. Voice in Parkinson's Disease: A Machine Learning Study. *Front. Neurol.* **2022**, *13*, 831428. [CrossRef]
7. Goetz, C.G.; Tilley, B.C.; Shaftman, S.R.; Stebbins, G.T.; Fahn, S.; Martinez-Martin, P.; Poewe, W.; Sampaio, C.; Stern, M.B.; Dodel, R.; et al. Movement Disorder Society-Sponsored Revision of the Unified Parkinson's Disease Rating Scale (MDS-UPDRS): Scale Presentation and Clinimetric Testing Results. *Mov. Disord.* **2008**, *23*, 2129–2170. [CrossRef]
8. Fahn, S.; Marsden, C.; Goldstein, M.; Calne, D. State of the Art Review—The Unified Parkinson's Disease Rating Scale (UP-DRS):Status and Recommendations. *Mov. Disord.* **2003**, *18*, 738–750.
9. Siderowf, A.; McDermott, M.; Kieburtz, K.; Blindauer, K.; Plumb, S.; Shoulson, I. Test-Retest Reliability of the Unified Parkinson's Disease Rating Scale in Patients with Early Parkinson's Disease: Results from a Multicenter Clinical Trial. *Mov. Disord.* **2002**, *17*, 758–763. [CrossRef]
10. Lonini, L.; Dai, A.; Shawen, N.; Simuni, T.; Poon, C.; Shimanovich, L.; Daeschler, M.; Ghaffari, R.; Rogers, J.A.; Jayaraman, A. Wearable Sensors for Parkinson's Disease: Which Data Are Worth Collecting for Training Symptom Detection Models. *NPJ Digit. Med.* **2018**, *1*, 64. [CrossRef]
11. Di Biase, L.; Summa, S.; Tosi, J.; Taffoni, F.; Marano, M.; Rizzo, A.C.; Vecchio, F.; Formica, D.; Di Lazzaro, V.; Di Pino, G.; et al. Quantitative Analysis of Bradykinesia and Rigidity in Parkinson's Disease. *Front. Neurol.* **2018**, *9*, 121. [CrossRef]
12. Cesarelli, G.; Donisi, L.; Amato, F.; Romano, M.; Cesarelli, M.; D'Addio, G.; Ponsiglione, A.M.; Ricciardi, C. Using Features Extracted from Upper Limb Reaching Tasks to Detect Parkinson's Disease by Means of Machine Learning Models. *IEEE Trans. Neural Syst. Rehabil. Eng.* **2023**, *31*, 1056–1063. [CrossRef]
13. Ricci, M.; Di Lazzaro, G.; Pisani, A.; Mercuri, N.B.; Giannini, F.; Saggio, G. Assessment of Motor Impairments in Early Untreated Parkinson's Disease Patients: The Wearable Electronics Impact. *IEEE J. Biomed. Health Inform.* **2020**, *24*, 120–130. [CrossRef]
14. Salarian, A.; Russmann, H.; Vingerhoets, F.J.G.; Dehollain, C.; Blanc, Y.; Burkhard, P.R.; Aminian, K. Gait Assessment in Parkinson's Disease: Toward an Ambulatory System for Long-Term Monitoring. *IEEE Trans. Biomed. Eng.* **2004**, *51*, 1434–1443. [CrossRef]
15. Di Lazzaro, G.; Ricci, M.; Al-Wardat, M.; Schirinzi, T.; Scalise, S.; Giannini, F.; Mercuri, N.B.; Saggio, G.; Pisani, A. Technology-Based Objective Measures Detect Subclinical Axial Signs in Untreated, de Novo Parkinson's Disease. *J. Park. Dis.* **2020**, *10*, 113–122. [CrossRef]
16. Asci, F.; Falletti, M.; Zampogna, A.; Patera, M.; Hallett, M.; Rothwell, J.; Suppa, A. Rigidity in Parkinson's Disease: Evidence from Biomechanical and Neurophysiological Measures. *Brain* **2023**, awad114. [CrossRef]

17. Pietrosanti, L.; Calado, A.; Verrelli, C.M.; Pisani, A.; Suppa, A.; Fattapposta, F.; Zampogna, A.; Patera, M.; Rosati, V.; Giannini, F.; et al. Harmonic Distortion Aspects in Upper Limb Swings during Gait in Parkinson's Disease. *Electronics* **2023**, *12*, 625. [CrossRef]

18. Mirelman, A.; Ben Or Frank, M.; Melamed, M.; Granovsky, L.; Nieuwboer, A.; Rochester, L.; Del Din, S.; Avanzino, L.; Pelosin, E.; Bloem, B.R.; et al. Detecting Sensitive Mobility Features for Parkinson's Disease Stages Via Machine Learning. *Mov. Disord.* **2021**, *36*, 2144–2155. [CrossRef]

19. Navarro-López, V.; Fernández-Vázquez, D.; Molina-Rueda, F.; Cuesta-Gómez, A.; García-Prados, P.; del-Valle-Gratacós, M.; Carratalá-Tejada, M. Arm-Swing Kinematics in Parkinson's Disease: A Systematic Review and Meta-Analysis. *Gait Posture* **2022**, *98*, 85–95. [CrossRef]

20. Mackay, W.A.; Crammond, D.J.; Kwan, H.C.; Murphy, J.T. Measurements of Human Forearm Viscoelasticity. *J. Biomech.* **1986**, *19*, 231–238. [CrossRef]

21. Lacquaniti, F.; Carrozzo, M.; Borghese, N.A. Time-Varying Mechanical Behavior of Multijointed Arm in Man. *J. Neurophysiol.* **1993**, *69*, 1443–1464. [CrossRef]

22. Katayama, M.; Kawato, M. *Virtual Trajectory and Stiffness Ellipse during Multijoint Arm Movement Predicted by Neural Inverse Models*; Springer: Berlin/Heidelberg, Germany, 1993; Volume 69.

23. Postuma, R.B.; Berg, D.; Stern, M.; Poewe, W.; Olanow, C.W.; Oertel, W.; Obeso, J.; Marek, K.; Litvan, I.; Lang, A.E.; et al. MDS Clinical Diagnostic Criteria for Parkinson's Disease. *Mov. Disord.* **2015**, *30*, 1591–1601. [CrossRef]

24. Kalia, L.V.; Lang, A.E. Parkinson's Disease. *Lancet* **2015**, *386*, 896–912. [CrossRef] [PubMed]

25. Xia, R.; Sun, J.; Threlkeld, A.J. Analysis of Interactive Effect of Stretch Reflex and Shortening Reaction on Rigidity in Parkinson's Disease. *Clin. Neurophysiol.* **2009**, *120*, 1400–1407. [CrossRef] [PubMed]

26. Lewek, M.D.; Poole, R.; Johnson, J.; Halawa, O.; Huang, X. Arm Swing Magnitude and Asymmetry during Gait in the Early Stages of Parkinson's Disease. *Gait Posture* **2010**, *31*, 256–260. [CrossRef] [PubMed]

27. Plagenhoef, S.; Gaynor Evans, F.; Abdelnour, T. Anatomical Data for Analyzing Human Motion. *Res. Q. Exerc. Sport* **1983**, *54*, 169–178. [CrossRef]

28. Saggio, G.; Quitadamo, L.R.; Albero, L. Development and Evaluation of a Novel Low-Cost Sensor-Based Knee Flexion Angle Measurement System. *Knee* **2014**, *21*, 896–901. [CrossRef]

29. Saggio, G.; Bocchetti, S.; Pinto, C.A.; Orengo, G.; Giannini, F. A Novel Application Method for Wearable Bend Sensors. In Proceedings of the 2nd International Symposium on Applied Sciences in Biomedical and Communication Technologies, Bratislava, Slovakia, 24–27 November 2009.

30. Ricci, M.; Terribili, M.; Giannini, F.; Errico, V.; Pallotti, A.; Galasso, C.; Tomasello, L.; Sias, S.; Saggio, G. Wearable-Based Electronics to Objectively Support Diagnosis of Motor Impairments in School-Aged Children. *J. Biomech.* **2019**, *83*, 243–252. [CrossRef]

31. Alessandrini, M.; Micarelli, A.; Viziano, A.; Pavone, I.; Costantini, G.; Casali, D.; Paolizzo, F.; Saggio, G. Impiego Degli Accelerometri Triassiali Nel Deficit Vestibolare Unilaterale: Affidabilità Rispetto Alla Posturografia Statica. *Acta Otorhinolaryngol. Ital.* **2017**, *37*, 231–236. [CrossRef]

32. Saggio, G.; Tombolini, F.; Ruggiero, A. Technology-Based Complex Motor Tasks Assessment: A 6-DOF Inertial-Based System Versus a Gold-Standard Optoelectronic-Based One. *IEEE Sens. J.* **2021**, *21*, 1616–1624. [CrossRef]

33. Ricci, M.; Di Lazzaro, G.; Pisani, A.; Scalise, S.; Alwardat, M.; Salimei, C.; Giannini, F.; Saggio, G. Wearable Electronics Assess the Effectiveness of Transcranial Direct Current Stimulation on Balance and Gait in Parkinson's Disease Patients. *Sensors* **2019**, *19*, 5465. [CrossRef]

34. Saggio, G.; Cavallo, P.; Ricci, M.; Errico, V.; Zea, J.; Benalcázar, M.E. Sign Language Recognition Using Wearable Electronics: Implementing K-Nearest Neighbors with Dynamic Time Warping and Convolutional Neural Network Algorithms. *Sensors* **2020**, *20*, 3879. [CrossRef]

35. Ricci, M.; Lazzaro, G.D.; Errico, V.; Pisani, A.; Giannini, F.; Saggio, G. The Impact of Wearable Electronics in Assessing the Effectiveness of Levodopa Treatment in Parkinson's Disease. *IEEE J. Biomed. Health Inform.* **2022**, *26*, 2920–2928. [CrossRef]

36. Kuhtz-Buschbeck, J.P.; Jing, B. Activity of Upper Limb Muscles during Human Walking. *J. Electromyogr. Kinesiol.* **2012**, *22*, 199–206. [CrossRef]

37. Cappellini, G.; Ivanenko, Y.P.; Poppele, R.E.; Lacquaniti, F. Motor Patterns in Human Walking and Running. *J. Neurophysiol.* **2006**, *95*, 3426–3437. [CrossRef] [PubMed]

38. Myers, A.D.; Tempelman, J.R.; Petrushenko, D.; Khasawneh, F.A. Low-Cost Double Pendulum for High-Quality Data Collection with Open-Source Video Tracking and Analysis. *HardwareX* **2020**, *8*, e00138. [CrossRef] [PubMed]

39. Jackson, K.M.; Joseph, J.; Wyard, S.J. A Mathematical Model of Arm Swing during Human Locomotion. *J. Biomech.* **1978**, *11*, 277–289. [CrossRef] [PubMed]

40. Özkaya, N.; Goldsheyder, D.; Nordin, M.; Leger, D. Fundamentals of Biomechanics: Equilibrium, Motion, and Deformation, Fourth Edition. In *Fundamentals of Biomechanics*, 4th ed.; Springer: Cham, Switzerland, 2016; pp. 1–454. [CrossRef]

41. Takada, Y.; Nakamura, S.; Morioka, S.; Iwase, M. Development of Myoelectric Prostheses with Elbow Joint. In Proceedings of the International Conference on Advanced Mechatronic Systems, ICAMechS, Hanoi, Vietnam, 10–13 December 2020; pp. 60–64.

42. Mirelman, A.; Bernad-Elazari, H.; Thaler, A.; Giladi-Yacobi, E.; Gurevich, T.; Gana-Weisz, M.; Saunders-Pullman, R.; Raymond, D.; Doan, N.; Bressman, S.B.; et al. Arm Swing as a Potential New Prodromal Marker of Parkinson's Disease. *Mov. Disord.* **2016**, *31*, 1527–1534. [CrossRef]
43. Cole, M.H.; Silburn, P.A.; Wood, J.M.; Worringham, C.J.; Kerr, G.K. Falls in Parkinson's Disease: Kinematic Evidence for Impaired Head and Trunk Control. *Mov. Disord.* **2010**, *25*, 2369–2378. [CrossRef]

Article

Dynamic Fall Detection Using Graph-Based Spatial Temporal Convolution and Attention Network

Rei Egawa, Abu Saleh Musa Miah, Koki Hirooka, Yoichi Tomioka and Jungpil Shin *

School of Computer Science and Engineering, The University of Aizu, Aizuwakamatsu 965-8580, Fukushima, Japan
* Correspondence: jpshin@u-aizu.ac.jp

Abstract: The prevention of falls has become crucial in the modern healthcare domain and in society for improving ageing and supporting the daily activities of older people. Falling is mainly related to age and health problems such as muscle, cardiovascular, and locomotive syndrome weakness, etc. Among elderly people, the number of falls is increasing every year, and they can become life-threatening if detected too late. Most of the time, ageing people consume prescription medication after a fall and, in the Japanese community, the prevention of suicide attempts due to taking an overdose is urgent. Many researchers have been working to develop fall detection systems to observe and notify about falls in real-time using handcrafted features and machine learning approaches. Existing methods may face difficulties in achieving a satisfactory performance, such as limited robustness and generality, high computational complexity, light illuminations, data orientation, and camera view issues. We proposed a graph-based spatial-temporal convolutional and attention neural network (GSTCAN) with an attention model to overcome the current challenges and develop an advanced medical technology system. The spatial-temporal convolutional system has recently proven the power of its efficiency and effectiveness in various fields such as human activity recognition and text recognition tasks. In the procedure, we first calculated the motion along the consecutive frame, then constructed a graph and applied a graph-based spatial and temporal convolutional neural network to extract spatial and temporal contextual relationships among the joints. Then, an attention module selected channel-wise effective features. In the same procedure, we repeat it six times as a GSTCAN and then fed the spatial-temporal features to the network. Finally, we applied a softmax function as a classifier and achieved high accuracies of 99.93%, 99.74%, and 99.12% for ImViA, UR-Fall, and FDD datasets, respectively. The high-performance accuracy with three datasets proved the proposed system's superiority, efficiency, and generality.

Keywords: fall detection (FD); graph convolutional network (GCN); human activity recognition (HAR); computer vision; body pose detection; AlphaPose; channel attention; ageing people

Citation: Egawa, R.; Miah, A.S.M.; Hirooka, K.; Tomioka, Y.; Shin, J. Dynamic Fall Detection Using Graph-Based Spatial Temporal Convolution and Attention Network. *Electronics* **2023**, *12*, 3234. https://doi.org/10.3390/electronics12153234

Academic Editor: George A. Tsihrintzis

Received: 21 June 2023
Revised: 20 July 2023
Accepted: 24 July 2023
Published: 26 July 2023

1. Introduction

The ageing of the population has become a global phenomenon, and the number of elderly people in the world is projected to more than double over the next 30 years. Approximately 16.0% of the population is expected to be elderly by 2050 [1]. According to the World Health Organization (WHO) [2], falls are the second leading cause of unintentional death after traffic accidents, and adults over the age of 60 suffer the most fatal falls. Japan is one of the most ageing countries in the world, with a total of 29% of their people projected to be over 65 years old in the future; there is the possibility of increasing this ratio. When a person is unable to respond to stimuli and unable to maintain awareness of his surroundings, he becomes unconscious. As a result, they seem to be asleep and fall asleep. Falling is a significant issue for senior citizens in Japan and causes injuries and death [3]. Urgent treatment and intervention are crucial on losing consciousness, otherwise, there is a high risk to patients. It is very important to develop an automatic fall detection method to protect life in this situation. Some of these falls are serious enough to require medical attention. It has been shown that medical attention immediately after a fall effectively reduces the likelihood of death by 80% and the need for long-term hospitalization by 26% [4]. The response time to rescue a seriously injured person from a fall is critical to the survival of the elderly. Therefore, it is very important to develop an automatic fall detection method to prevent the risk of serious injury and death because of falls. Our main goal was to develop an indoor fall detection system that will be subject- and environment-independent. There are two types of existing fall detection methods—wearable sensor-based and vision-based methods [5]. The wearable sensor-based methods have the elderly person wear a sensor, which detects the sudden acceleration changes caused by a fall [6]. However, this is inconvenient because many elderly people are often forgetful and unwilling. In addition, the method is susceptible to noise, and everyday activities such as lying down or sitting up can lead to false detection [7]. Recently, cameras have become popular in many public and private spaces, such as train stations, bus stops, and office buildings. Also, the rapid development of computer vision under the influence of deep learning [8] has led to the development of vision-based methods [9–14]. Although vision-based methods eliminate the inconvenience of wearing a device, they may be subject to false detection due to lighting or complex backgrounds. Most previous vision-based fall detection systems were developed using threshold-based methods by comparing the settings reference with input data [15]. The main problem of the threshold-based system is that there is the possibility of missing a prediction of a fall event because of the high or low threshold. In addition, it may create some sensitive issues, such as sudden changes in the human body positions, like picking up anything from the floor, or Muslim prayer, which could be considered as a fall. Currently, researchers use various machine learning algorithms for fall detection, such as random forest, support vector machine (SVM), and k-nearest neighbors (KNN) [16]. Some researchers have demonstrated the comparison between machine learning and threshold-based systems to prove the effectiveness of the machine learning algorithms [17,18].

The author collected data from various sources such as a gyroscope, an accelerometer, and magneto meters located on the subject's wrist. They compared the threshold-based and machine learning methods and reported that machine learning performs well. The main problem of the machine learning-based system is the handcrafted feature because this feature must be closely related to and connected with human activities and similar actions. The main drawback of the handcrafted feature is that it is not guaranteed to find a good description. In addition, the system's robustness is not good because it needs strong domain background knowledge to select the handcrafted feature. The choice of features is not straightforward, and finding the best feature is very important to reflect the essence of a fall [19]. The main problem occurs when multitask classification is needed, when features are closely related. Researchers have also used depth sensors, infrared sensors, optical sensors, and RGB cameras [10–12,14,20,21]. To solve the handcrafted feature problem, researchers employed a deep learning-based method to explore data and extract effective features for the specific classification task using RGB images [20,22]. RGB image data-based deep learning for fall detection still faces problems in achieving a high performance because of the redundant background light illuminations and computational complexity. To solve the problems, many researchers have employed deep learning on the redundant backgrounds of images, such as geometric multi-grid (GMG), fuzzy methods, Gaussian mixture models (GMM), and RPCA methods [23–26] Also, many researchers have used deep learning-based background removal methods such as ANN [27–29], Faster R-CNN, and Yolov3 [30,31] but these have some computational issues because of the two-time deep learning for background reduction and class action. Recently, the skeleton data points of the human body, instead of the RGB image, has been used by many researchers to solve efficiency and accuracy-related issues. Chen et al. proposed a skeleton data point-based fall detection method where they calculated different geometrical and static features from the skeleton data. Then, using the machine learning method, they achieved 97.00% accuracy [32,33].

The main problem with these features is that the skeleton data are different from images and videos because they form a graph instead of 2D or 3D grids. Consequently, conventional feature extraction methods are not able to extract exact information and handle this data structure in its native form, which yields the preprocessing steps. The many existing approaches merge the joint points in a digestible type of data structure, such as metrics vectors. This transformation can lead to the loss of relevant, effective information and especially different joint relationships. To solve the problem, Yan et al. applied a new deep learning method, the spatial-temporal graph convolutional network (ST-GCN) [34]. It mainly extracts the various node relationships, specifically the spatial and temporal contextual relationships among the joints [32]. They constructed a graph instead of 2D grids and achieved a satisfactory performance in hand gesture and activity recognition. Keskes et al. applied the ST-GCN fall detection method to solve various challenges in the domain [35]. The main drawback of their method is that they used ten units of ST-GCN sequentially, which increases the high computational complexity. In addition, they did not consider the role of non-connected skeleton points in fall events during the spatiotemporal feature extraction. However, the positional relationship between some non-real connected points is very helpful for partially identifying events. To overcome the problems, we proposed a graph-based spatial-temporal convolution and attention network (GSTCAN) model to overcome the current challenges and developed an advanced medical technology system. The major contributions of this work are detailed below:

- We proposed a graph-based GSTCAN model in which we first calculated the motion among the consecutive frame. Then, we constructed a graph and applied a graph-based spatial-temporal convolutional neural network to extract intra-frame joints and an inter-frame joints relationship by considering the spatial and temporal domains.
- Secondly, we fed the spatial-temporal convolution feature into an attention module to select channel-wise effective features. The main purpose of the attention model is to improve the role of non-connected skeleton points in certain events during the spatial-temporal feature; we applied the attention model to GSTCAN, aiming to extract global and local features bound to impact model optimization. In the same procedure, we sequentially applied GSTCAN six times in a series, producing effective features that carry the skeleton joint's internal relationship structure.
- Finally, we applied a softmax function as a classifier and achieved high accuracies of 99.93%, 99.74%, and 99.12% for ImViA, UR-Fall, and FDD datasets, respectively. The high-performance accuracy with three datasets proved the superiority and efficiency of the proposed system.

The remainder of this paper is organized as follows: Section 2 summarizes the existing research work and related problems. Section 3 describes the three fall detection benchmark datasets, and Section 4 describes the architecture of the proposed system. Section 5 details the evaluation performed, including a comparison with a state-of-the-art approach. In Section 6, our conclusions and directions for future work are discussed.

2. Related Work

Many researchers have been working to develop fall detection systems with various feature extraction and classification approaches [18,36–40]. All the algorithms used in this domain can be divided into the following categories: (I) sensor-based systems for monitoring the person [41,42]; (II) radio frequency (RF) sensor-based systems, and (III) camera-based vision-related systems. Many researchers record various signals with various sensors such as gyroscopes, accelerometers, EMGs, and EEGs to collect information from many people, not just the elderly [43–52]. Then they extract various kinds of features, including angle, distance, the sum of X and Y with various directions and their derivatives, and geometrical, statistical, and mathematical formulas [39]. Wang et al. collected data on fall events using an accelerometer sensor and calculated the SVMA for the patients [53].

They first assigned a threshold value as an assumption and, if the SVMA value surpassed the threshold, they then calculated some features of the trunk angle and pressor pulse sensors. Moreover, if the two values were higher than the normal value, it can produce an emergency alarm and achieve 97.5% accuracy. Desai et al. used multiple sensors in combination, including an accelerometer, a gyroscope, a GSM module microcontroller, a battery, and an IMU sensor [54]. They used a logistic regression classifier and, if a fall event happened, GSM produced an emergency alarm for the helpline number. The drawback is that they only used the human activity dataset, not including any specific dataset for the fall events. Xu et al. reviewed the wearable accelerometers-based work, proving some advantages of the wearable sensor, such as low cost, portability, and efficiency at detecting falls with high-performance accuracy [55].

The main drawbacks of this type of work is that the patient needs to wear a sensor all day, as well as there being high noise, which leads to difficulty for ageing people, which badly affects their daily lives. To solve the wearable sensor problems, the second category of radar technologies and Wi-Fi was proposed to solve the mentioned problems. Tian et al. collected the RF reactions from the environment using frequency-modulated continuous-wave radio (FMCW) equipment [56]. They generated two heat maps from the reflection and applied a deep learning model for the classification, which achieved 92.00% precision and 94.00% sensitivity. RF is a non-intrusive sensor-based method that achieves good performance accuracy. It can solve the noise problem, but collecting data from each cell based on an antenna with interference is challenging. Researchers proposed a camera-based data collection system to solve the portability and high-cost of data collection problems. In recent years, camera-based fall detection approaches have been acceptable to researchers and consumers because of their low cost and portability properties.

Zerrouki et al. extracted curvelet transforms and area ratios to identify human posture in images, used SVM to identify posture, and used a hidden Markov model (HMM) [57] for activity recognition [58]. Chua et al. proposed an RGB image-based fall events detection method using human shape variation. After extracting the foreground information, they calculated three points with which to calculate the fall event-related features, reporting 90.5% accuracy [59]. Cai et al. applied the hourglass convolutional auto-encoder (HCAE) approach by combining with hourglass residual units (HRU) to extract the intermediate features from the RGB video dataset [60]. They extracted the features for the fall classification and then reconstructed the image to enhance the representation of the intermediate fall event-related features. After evaluating their model with the UR fall detection dataset, they achieved 96.20% accuracy. Chen et al. applied mask R-CNN to extract a feature, aiming to detect a fall event from the RGB image based on the CNN model [20]. Later, they applied bi-directional LSTM for the classification and achieved 96.7% accuracy for the UR fall detection dataset. Harrou et al. proposed a multi-step procedure for detecting fall events, including data processing and segmentation, splitting the foreground image into five regions based on the relevant features, extracting features from each region, and then calculating the generalized likelihood ratio (GLR) [61]. Finally, they evaluated their model with the FDD and URFD datasets, which produced 96.84% and 96.66% accuracies, respectively. Han et al. applied the MobileVGG network, which extracts the motion features from the RGB video to detect fall events, and they achieved 98.25% accuracy with their dataset [62]. Standard camera and image-based systems are sometimes not robust and their performance may be limited because of the complexity of distinguishing between foreground and background.

In addition, they still face problems of light illumination, partial occlusion, and redundant background complexity problems. To overcome the problems, many computer vision researchers have used skeleton datasets to detect fall events and human activity instead of the RGB pixel-based image to solve the mentioned problems. The skeleton-based dataset's main advantage is its robust scene variation, light illumination, and partial occlusion [34,63–65]. Yao et al. extracted the features from the skeleton joint, then applied the SVM, and achieved 93.56% accuracy with the TSTv2 dataset [63]. The main concept behind their task is that they divide the skeleton data into five parts based on the organs such as the head, neck, spine base, and spine centre. Tsai et al. extracted features from the selected potential joints of the skeleton dataset and then applied a 1DCNN for the classification [64]. After evaluating the NTU-RGBD dataset, they achieved high-performance accuracy compared to the previous system. The main drawback of the dataset is that the NTU-RGBD dataset does not include all types of fall events. Tran et al. proposed a handcrafted feature-based fall detection method where they first calculated the plane based on the floor of the room [66]. After that, they calculated the velocity distance of the head and the spine associated with the floor. After applying the SVM method, they achieved better accuracy than the previous method. Most of the existing work on fall event detection was developed with hand-crafted features, which faces difficulties in handling large datasets. In addition, effective feature extraction and potential feature selection approaches still face many challenges. The deep learning-based approach is the most powerful classification approach, and can extract the effective features and outperforms hand-crafted features because it can obtain many more features during training; however, it needs a large dataset. In this study, we proposed a skeleton-based GSTCAN model to recognize fall events through the skeleton data provided by the AlphaPose. Our main goal was to develop a robust fall detection system with high-performance accuracy, efficiency, and generality. We tested the proposed model with three datasets to prove its high-performance accuracy, efficiency, and generality according to the standard generality system.

3. Datasets

There are few dynamic fall detection benchmark datasets available online. For this study, we selected three benchmark dynamic fall detection datasets, namely: the UR Fall Dataset [36], ImViA Datasets(le2i) [38], and FDD [67]. Table 1 provides a summary of those datasets and their specifications, including features, people, and actions, etc.

Table 1. Summary of the datasets used in this study.

Dataset Name	Type	Classes	Sample
UR Fall Detection [36]	This consists of raw video and does not contain bounding box information.	2 Class Fall/ Non-Fall	3 K images in total
ImViA Datasets (le2i) [38]	This consists of raw video and does not contain bounding box information.	2 Class Fall/ Non-Fall	40 K images in total
FDD [67]	This consists of the image	Five classes	22 K images in total

3.1. UR Fall Detection Datasets

The videos in the UR Fall Detection Dataset [68] are short and correspond to fall and non-fall sequences. This dataset contains videos of 30 falls. We also used the UR Fall detection dataset [36]. This dataset is a fall detection dataset provided by the University of Rochester's Rehabilitation Medicine Research Group. The dataset includes video data captured from multiple cameras and corresponding annotation data for fall events. The video data were captured by RGB and Depth cameras with a resolution of 640 × 480. The annotation data include information such as the time of the fall event and the posture of the person before and after the fall. It has been a useful resource for fall detection research and has been widely used in various studies. It is also used as a benchmark for fall detection tasks. Table 1 demonstrates the two most usable fall detection datasets. The videos in the UR Fall detection dataset were recorded by two different cameras, with 70 activated cameras and 3000 images. Among them, 30 activities are considered to be falls and 40 activities are normal daily living activities.

3.2. ImViA Datasets(le2i)

The ImViA dataset [38] is a dataset including videos from a single camera in a realistic video surveillance setting. It includes daily activities such as going from a chair to a sofa, exercising, and falling. Only one person is displayed at a time, the frame rate is 25 frames/s, and the resolution is 320 × 240 pixels. The background of the video is fixed and simple, while the texture of the images is complex.

3.3. FallDetection DATASET (FDD)

This dataset was recorded with a single uncalibrated Kinect sensor and resized at 320 × 240—the original size was 640 × 480. They collected 21,499 images in total and divided them into training and testing. The total number of images in the training dataset is 16,794, the validation dataset includes 3299 images, and the testing dataset includes 2543 images. The dataset was recorded in five different rooms and from eight different angles. They collected the dataset from five different participants, among them two male participants aged between 32 and 50, and three females aged between 10 and 40. The dataset includes five other classes: sitting, standing, bending, crawling, and lying [67].

4. Proposed Methodology

In this study, we proposed a graph-based spatial-temporal convolution and attention network (GSTCAN), mainly inspired by [34,35,65]. The main goal was to capture the pattern in the spatial domain from the motion version of the skeleton dataset. Our method was mainly a neural network designed with graphs and their structural information. We did not need to use the dimension reduction approach for the proposed model, but it works with the native form. It will overcome the limitation by extracting the complex internal pattern using contextual temporal information. Moreover, we needed a good benchmark dataset with sufficient samples representative of the action's diverse variability and a camera view to prove the power of the model. Many skeleton-based fall detection datasets still need to contain an efficient number of samples, yielding a lack of training. Many existing deep learning-based methods can adapt to the fall detection model [69,70]. These methods are not practical because of the limited size of the publicly available fall and human activity action-related datasets. This data inefficiency problem can be solved by transfer learning, a pre-trained model with a related dataset that can be used as an initial trained model for the novel task. It is highly effective at solving the existing data inefficiency problems. This method was mainly developed from the data reuse concept of learning new things. The main problem is that deep learning mainly works for specific data and domains. There is a need to newly train the model from scratch when needing to apply it to a new task or new domain. In this study, we proposed to develop an attention-based ST-GCN model to recognize fall detection by extracting complex spatial and temporal internal patterns. The working flow architecture of the proposed model is demonstrated in Figure 1, and the pseudocode of the proposed method is described in Algorithm 1.

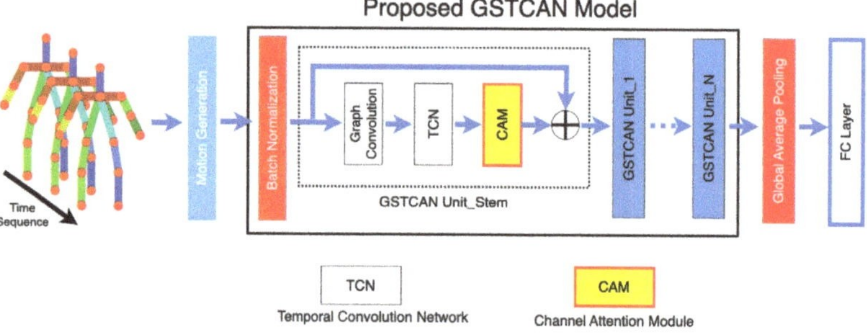

Figure 1. Proposed working flow diagram.

Algorithm 1 Pseudocode of the proposed system.

Input: Set of Input Dataset $P_i \in P(n)$
Number of Samples: N, 70% for Training and 30% for Test
Output: Set of vector s_i
define GSTCAN Model(input=InputLayer, outputs=ClassificationLayer):
$\quad Motion_{Feature} \leftarrow MotionModule(D)$
$\quad Normalize_{Feature} \leftarrow BatchNormalization_{Layer}(Motion_{Feature})$
$\quad GCN_{input} \leftarrow Normalize_{Feature}$
$\quad$**while** $i \neq 6$ **do**
$\quad\quad GCN \leftarrow GraphConvolutionalNetwork(GCN_{input})$
$\quad\quad TCN \leftarrow TemporalConvolutionalNetwork(GCN)$
$\quad\quad CAM \leftarrow ChannelAttention(TCN)$
$\quad\quad Features \leftarrow Concatenation(CAM, GCN_{input})$
$\quad\quad GCN_{input} \leftarrow Features$
$\quad GAP \leftarrow GlobalAveragePooling_{Layer}(Features)$
$\quad PredictedClass \leftarrow Classification_{Layer}(GAP)$
$\quad returnPredictedClass$
while $i \neq NumEpochs$ **do**
$\quad$// For Training
$\quad$**while** $Batch \neq NumberBatchTraining$ **do**
$\quad\quad PredictedClass \leftarrow Model(Batch)$
$\quad\quad Loss \leftarrow Criterion(PredictedClass, Train_Class)$
$\quad\quad Updatetheloss \leftarrow Loss.backward(), Optimizer.Step()$
$\quad$// For Testing
$\quad$**while** $Batch \neq NumberBatchTesting$ **do**
$\quad\quad PredictedClass \leftarrow Model(Batch)$
$\quad\quad Output \leftarrow CPerformanceMatrix(PredictedClass, TestClass)$

4.1. AlphaPose Estimation

We used AlphaPose to extract skeleton joints from the fall detection dataset, an open-source library for visual image processing tools. It was developed with a deep learning-based pre-trained model which can be perceptible in real-time for various applications such as face detection, object detection, pose estimation, computer vision, and hand gesture recognition. It was built with custom overflow integration with various Python libraries, such as OpenCV and tensor flow, and we used it here to extract body landmarks for fall detection. There are two main methods for joint point detection in posture estimation: bottom-up and top-down. The bottom-up method estimates all joint points in an image and summarizes the joint points that constitute each person. The bottom-up method is vulnerable because it estimates the pose from local areas. AlphaPose [71,72] uses a top-down framework to detect human bounding boxes and then individually estimates the posture within each box. The detection of each person's joint points is accurate [73]. The Algorithm of AlphaPose we used ResNet101-based Faster R-CNN . Table 2 shows the number of frames for which AlphaPose was able to obtain skeletons for each dataset. In the study, we used those frames' skeleton points which were successfully extracted by AlphaPose and discarded the rest of the frames.

Table 2. Number of frames for which AlphaPose was able to obtain skeletons for each dataset.

Dataset Name	Number of All Frames	Number of Frames for Which the Skeleton Could Be Acquired	Number of Frames for Which the Skeleton Could Not Be Acquired
ImViA	42,066	40,631	1435
UR fall dataset	2995	2142	853
FDD fall dataset	26,911	18,704	8207

This system can read the real-time camera video or recorded video and produce the corresponding skeleton points. In the study, we provided video from the UR fall dataset and generated the skeleton points for consecutive frames. It mainly generates 18 points for each frame, including nose, mouth, ear, shoulder, elbow, wrist, finger index, hip, knee, ankle, and foot, and these points are for both the left and right sides; details of the media pipe skeleton are visualized Figure 2 and Table 3. Although this system collects 18 key points, we selected 13 by excluding eyes and ears.

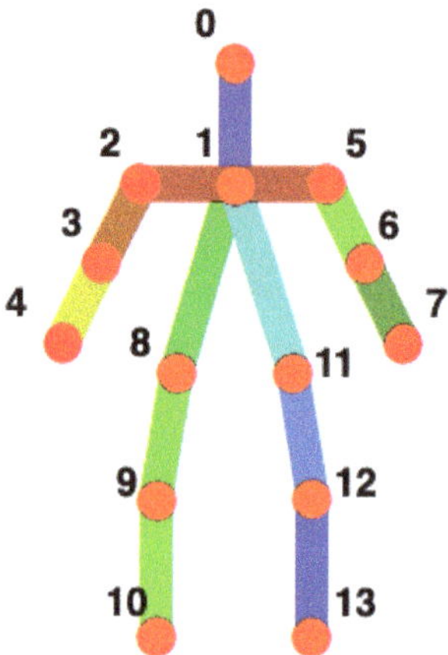

Figure 2. Body skeleton joint visualization generated with AlphaPose.

Table 3. AlphaPose landmarks name with index.

No.	Pose Name	No.	Pose Name	No.	Pose Name
0	Nose	7	Right-wrist	12	Right Knee
1	Neck	8	Left hip	14	Left eye
2	Left shoulder	11	Right hip	15	Left ear
3	Left elbow	9	Left Knee	16	Right eye
4	Left-wrist	12	Right Knee	17	Left ear
5	Right shoulder	10	Left ankle		
6	Right elbow	13	Right ankle		

4.2. Motion Calculation and Graph Construction

We mainly considered the dynamic fall detection dataset in this study; motion is one of the most effective features for the dynamic fall detection approach in terms of movement, alignment, and overall data structure effectiveness. This also directly affects the movement of the fall data. We calculated the motion using all the landmarks for X and Y as a two-dimensional vector. We mainly generated the difference between consecutive frame joint positions to calculate the motion. We calculated the motion for a specific joint by subtracting the consecutive frame joints, which are visualized in Figure 3. To calculate the motion M for a joint j, we used the formula shown in Equation (1).

$$M(j) = \begin{cases} ActivityMotion_X = X_t - X_{t-1} \\ ActivityMotion_Y = Y_t - Y_{t-1}. \end{cases} \tag{1}$$

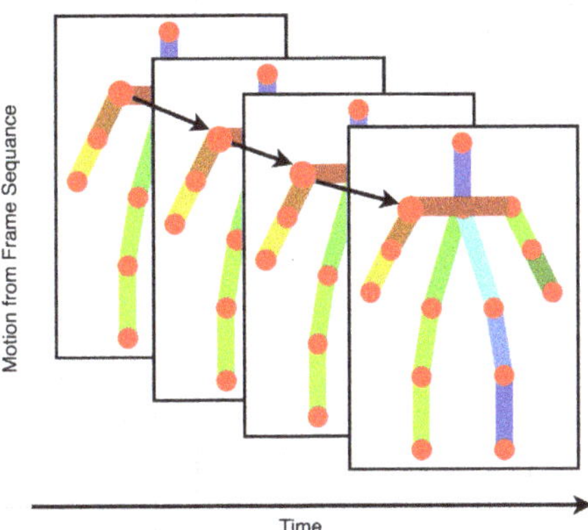

Figure 3. Example visualization of the motion calculation procedure.

Motion skeleton information represents the 2D coordinates of the human joint. In addition, full-body fall and non-fall events use multiple frames based on the sequence of relative structure and samples. The graph was mainly constructed based on the spatial and temporal domains by considering natural bone or connections among the joints. The underreacted graph was constructed using the following Equation (2).

$$G = (V, E). \tag{2}$$

Here, V and E denoted the set of nodes and edges where the graph node can be defined as $V = v_{(i,t)} \mid i = 1, \ldots, N, t = 1, \ldots, T$, which is mainly composed of the whole-body skeleton. After that, we constructed an adjacent matrix based on the graph using the following formulas in Equation (3):

$$f(x) = \begin{cases} 1 & \text{if the nodes are adjacent} \\ 0 & \text{if they are not adjacent.} \end{cases} \tag{3}$$

4.3. Graph Convolutional Network

The study extracted the potential embedded with the whole-body skeleton based on the spatial-temporal graph convolution network. We construed the graph using the below formulas [21,34]:

$$G_{out} = D^{-(1/2)}(A + I)D^{-(1/2)} \times W, \tag{4}$$

where D, I, and A represent the diagonal, identity matrix or self-connection, and inter-body connection, respectively. Where the diagonal degree can be expressed as $(A + I)$, the weight matrix is denoted by W. For implementing the graph-based convolution, we focused on the 2D convolution, and for the spatial graph convolution, we multiplied it with $D^{-(1/2)}(A + I)D^{-(1/2)}$. In the same way for the graph-based temporal convolution, we multiplied it with a $k_t \times 1$ kernel size.

4.4. GSTCAN Algorithm

The graph-based spatial temporal convolutional and attention network (GSTCAN) model was proposed here to enhance the work of [34,35,65], and is a GCN [74]-based motion recognition method that automatically learns spatial and temporal patterns from skeleton data. The main advantage of the proposed system is that the data can be treated in its original form [34]. The data based on convolutional networks (CNN) and the data based on the skeleton were obtained in Step 2. The sequence of body joints in two-dimensional form constructs a spatiotemporal graph, with the joints as the nodes of the graph and the natural connections between the structure and time of the human body as the edges of the graph. The inputs to GSTCAN are the joint coordinate vectors on the graph nodes, and a multi-layered spatiotemporal graph convolution operation is applied to generate higher-order feature maps. Then, whether it is falling or not is classified by the Softmax classifier. Figure 1 shows the overall flow of GSTCAN. Our proposed approach is mainly composed of a series of GSTCN + channel attention module [75] units, a pooling layer, and a fully connected layer. Each unit of the GSTCAN included the spatial and temporal convolutional neural network. Figure 2 demonstrates the node and joints where skeleton joints are considered the graph's node. As we considered dynamic fall detection, there is a sequence of frames that creates the intra-body and inter-body relationships. The intra-body connection comes from the natural connection of the human body joints, and the inter-body connection comes from the relationship between the consecutive frames established by the temporal convolution. We considered the input dimension of the tensor as (N,2, T,33, S). The batch is represented by N, the 2D joint coordinates are represented by 2 (x, y) and can be denoted as channel C, the number of frames are represented by T, the number of the skeletons from the media pipe is 33 and can be denoted by vertex V, and the total number of videos comes from the subject represented by S. After that, we modified $S \times N, C, T,$ V. After calculating the motion of the raw skeleton, we fed it into a spatial convolutional layer, aiming to extract the spatial information for each joint. This process is a little bit different from that of image convolution. Around the specific pixel location, the weight coefficient is multiplied in a spatial order for the image convolution, whereas the labeling process is followed in the GSTCAN with joint location and spatial configuration partitioning approaches. The labels considered here include root node, centripetal nodes near central gravity compared to the root node, and centrifugal nodes. After extracting the spatial features, we fed them into the temporal convolutional network (TCN) to extract temporal contextual information. The main concept of TCN is to calculate the relationship among the same joints in consecutive frames. We repeated the same process 6 times (except stem) consecutively and then applied the pooling layer to enhance the features. Finally, the output layer of the proposed model produced a vector p, which has the same size as the classes. This mainly represents the probability that is the same as the specified corresponding class. The motion of the skeletal points in the graph-based GSTCAN produced a better representation of the fall activity based on the exploitation of the spatial and temporal relationships between intra- and inter-body frame joints.

4.5. Attention Module

According to the skeleton landmark concept, there are some border skeleton points or leaf points, known as the non-connected skeleton. We can solve the problems with Graph CNN because in the graph, all key points are connected with each other through the undirected graph. In addition, we calculated the motion before feeding it into the spatial-temporal architecture. The joint motion and bone motion were calculated between the consecutive frames for each non-connected skeleton point. These motion vectors represent the motion of the non-connected points over time and can capture temporal dynamics. Moreover, our spatiotemporal and attention model can calculate and learn hierarchical representations with temporal dependencies in the long term or short term based on the sequence of non-connected skeleton points of both spatial and temporal features directly from the non-connected skeleton point sequences. We applied attention mechanisms here after the spatial, temporal feature, which can be beneficial for capturing both global and local features in a spatiotemporal model for the non-connected skeleton point, and they can significantly impact model optimization. Our study also included a channel attention model to handle the role of non-connected skeleton issues.

We added an attention mechanism at the end of each GSTCN unit. The added attention mechanism is shown in Figure 4. The layer was used here sequentially and we can define it as (1) GlobalAveragePooling, (2) Dense (N/4), (3) BatchNorm, (4) Dense (N), and (5) Sigmoid. A value between 0 and 1 was output for each channel using the Sigmoid function. Important features had an output of 1 or a value close to 1, and unimportant features output 0 or a value close to 0. Then, a strong feature graph could be created by multiplying the previously learned feature graph because important features remained.

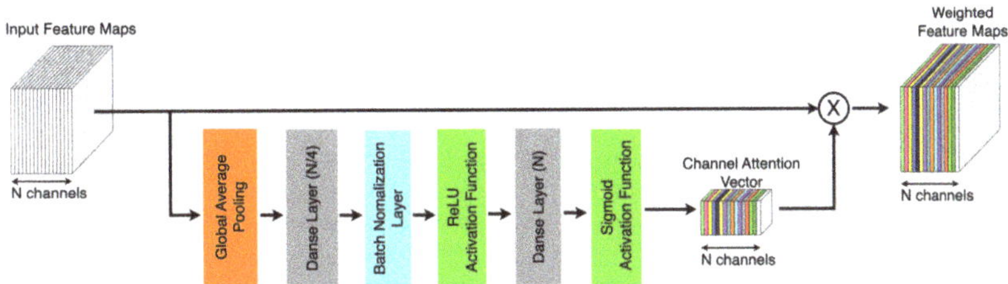

Figure 4. Channel attention mechanism.

4.6. Fully-Coupled Layer

Finally, we considered the Softmax function or Softmax activation layer as a classification or outputs layer to predict the value for each label. The loss function for classification tasks uses a cross-entropy loss function.

4.7. Network Architecture

Figure 1 demonstrates the proposed method, showing that we first calculated the motion and then fed it into a series of N. GSTCAN units in our study, $N = 6$, which leads to reducing the computational complexity. There are 64 output layers for the first two layers, 128 channels for two layers, and 256 output channels for the last two layers. The kernel size for each layer was set as 9, a residual or skip connection, and a dropout rate of 0.5 was used here to overcome the overfitting issues. We refined the feature with an attention module. Finally, we employed the Softmax activation function as a classifier. We employed an RMSprop [76] optimizer to learn the model, with 0.001 as the learning rate value.

5. Experimental Evaluation

To prove the system's superiority and effectiveness, we conducted various experiments with three benchmark datasets. We first demonstrated the training setting and evaluation matrix, then the performance of the proposed model with multiple datasets and, finally, we visualized the state-of-the-art comparison table.

5.1. Training Setting

To divide the training and testing, we followed the three-fold cross-validation approaches. In the training process, we used a learning rate of 0.001 and a batch size of 32. To implement the system, we used a GPU machine that has CUDA version 11.7, NVIDIA driver version 515, and GPU Geforce RTX 3090 24GB, with RAM 32 GB. Models were run for 100 epochs with the optimizer RMSprop [76] with the RTX3090. We also used Pytorch (version–1.13.1) [77], which has a low computational cost for deep learning, attention, transformer, OpenCV (version–4.7.0.72), pickle, and csv packages for the initial processing [78,79].

5.2. Evaluation Matrices

We used three benchmark datasets to evaluate the proposed model, mainly seen as a binary class classification problem. The evaluation metrics which we used here are included below [10]:

- Time: Perform a *t*-test from the mean and variance of processing speeds;
- Accuracy: most researchers used this, which denotes the percentage of total items classified correctly = (TP + TN)/(TP + TN + FP + FN);
- Recall/sensitivity: mainly denotes the true positive rate TP/(TP + FN);
- F1-score: denotes the harmonic mean between precision and recall = TP/(TP + 1/2 (FP + FN));

where TP comes from the true positives in our cases–the activity labeled is fall, and the system predicted it as a fall. FP denotes the false positives—the actual class is non-fall, but the system predicted a fall. TN denotes the true negatives—the actual class label is non-fall, and the system predicted non-fall. FN denotes the false negatives; the actual class label is fall, but the system predicted non-fall.

5.3. Evaluation Matrices

The processing speed of the ST-GCN model and the proposed system were compared. We evaluated the proposed system with three datasets, and the tables below visualize the proposed model's performance accuracy. Using the ImViA dataset, our proposed model achieved 99.57%, 99.68%, 99.63% and 99.93% for precision, sensitivity, F1-score, and accuracy, respectively. The UR fall detection dataset achieved 99.87%, 97.36%, 98.56%, and 99.75% for precision, sensitivity, f-score, and accuracy, respectively. In the same way, for the FDD dataset, our model achieved 97.98%, 97.21%, 97.55%, and 99.12% with precision, sensitivity, F-score, and accuracy, respectively.

5.3.1. Processing Speed Comparison

The mean and variance of the processing speeds of the ST-GCN and the proposed model are shown in Table 4. *T*-test results reject the hypothesis that the processing speeds are equal, indicating that the proposed model is more efficient.

Table 4. The mean and variance of the processing speeds.

Algorithm	Mean [sec]	Variance
ST-GCN	7.511	0.0001876
Proposed GSTCAN Model	6.787	0.0003385

5.3.2. Performance Result and State-of-the-Art Comparison for the UR Fall Dataset

The class-wise evaluation matrix table of the proposed model with the UR fall detection dataset is demonstrated in Table 5. We can see that the fall class reported 100%, 94.72%, and 97.25% for precision, sensitivity, and F1-score, respectively. In the same way for the non-fall class label, it achieved 99.73%, 100%, and 99.86% scores for precision, sensitivity, and F1-score, respectively. It also showed the performance accuracy for the all-class label simultaneously, which is 99.74%, 99.86%,97.36%, and 98.55% scores for accuracy precision, sensitivity, and F1-score, respectively.

Table 5. Class wise precision, sensitivity, and F1-score for UR fall dataset.

Label Name	Accuracy [%]	Precision [%]	Sensitivity [%]	F1-Score [%]
Fall	–	100	94.72	97.25
NonFall	–	99.73	100	99.86
Average	99.74	99.86	97.36	98.55

The state-of-the-art comparison for the proposed model is shown in Table 6 with the UR fall detection system. In this state-of-the-art comparisons table, we included the accuracy, precision, sensitivity, specificity, and F-score for a fair comparison with the previous model. We included the performances of seven previous state-of-the-art methods for the UR fall dataset. The authors of [36,37] extracted the hand-crafted features from the skeleton and depth information and, using the SVM method, they achieved 94.28% and 96.55% accuracies, respectively. The author of [60] employed a CNN-based encoder and decoder system and reported 90.50% accuracy. The author of [20] used mask-RCNN to segment and extract the features from the fall event video dataset and applied bi-directional LSTM and achieved 96.70% accuracy with the UR fall dataset. Zheng et al. extracted the skeleton points using AlphaPose, then employed ST-GCN and achieved 97.28%, 97.15%, 97.43%, 97.30%, and 97.29% scores for accuracy [65], precision, sensitivity, specificity, and F1-score, respectively. Wang et al. [80] extracted OpenPose key points and then applied MLP (multilayer perceptron) and random forest for the classification and achieved a high-performance accuracy of 97.33%. In the same way, the author of [61] applied the GLR scheme to design the system and achieved 96.66% accuracy.

Table 6. State -of-the-art comparison for UR fall dataset.

Algorithm	Dataset	Accuracy [%]	Precision [%]	Sensitivity [%]	Specificity [%]	F-Score [%]
Depth +SVM [36]	UR Fall	94.28	n/a	n/a	n/a	n/a
Skeleton +SVM [37]	UR Fall	96.55	n/a	n/a	n/a	n/a
HCAE [60]	UR Fall	90.5%	n/a	n/a	n/a	n/a
Bi-Directional LSTM [20]	UR Fall	96.70%	n/a	n/a	n/a	n/a
Hontago [65]	UR Fall	97.28	97.15	97.43	97.30	97.29
Wang [80]	UR Fall	97.33	97.78	97.78	96.67	97.78
Harrou [61]	UR Fall	96.66	94	100	94.93	96.91
Proposed GSTCAN Model	UR Fall	99.74	99.86	97.36	n/a	98.55

5.3.3. Performance Result and State-of-the-Art Comparison of the ImViA Dataset

In this section, we compared the performance of the proposed model with the state-of-the-art model. Table 7 demonstrates the state-of-the-art comparison for the ImViA dataset, where the proposed model achieved 99.93% accuracy whereas the previous model reported 96.86% accuracy, proving that our model has high effectiveness and efficiency.

Table 7. Class wise precision, sensitivity, and F1-score for ImViA fall dataset.

Label Name	Accuracy [%]	Precision [%]	Sensitivity [%]	F1-Score [%]
Fall	n/a	99.17	99.39	99.28
NonFal	n/a	99.96	99.95	99.96
Average	99.93	99.57	99.67	99.92

The state-of-the-art comparison of the proposed model using the ImViA dataset is shown in Table 8. Wang et al. [80] extracted OpenPose key points and then applied MLP (multilayer perceptron) and random forest for the classification and achieved a high-performance accuracy of 96.91%. Chalme et al. [81] demonstrated a performance of 79.31%, 79.41%, 83.47%, 73.07% and 81.39% accuracies. This accuracy proved that we could evaluate the system.

Table 8. State-of-the-art comparison for ImViA fall dataset.

Algorithm	Dataset	Accuracy [%]	Precision [%]	Sensitivity [%]	Specificity [%]	F-Score [%]
Hontago [65]	ImViA	96.86	97.01	96.71	96.81	96.77
Wang [80]	ImViA	96.91	97.65	96.51	97.37	97.08
Chamle [81]	ImViA	79.31	79.41	83.47	73.07	81.39
Proposed GSTCAN Model	ImViA	99.93	99.57	99.67	n/a	99.92

5.3.4. Performance Result and State-of-Art Comparison for the FDD Fall Dataset

In this section, we compared the performance of the proposed model with the state-of-the-art model. Table 9 demonstrates the state-of-the-art comparison performance for the ImViA dataset. The table shows that the FDD dataset has five rows including accuracy, precision, sensitivity, and F1-score.

Table 9. Class wise precision, sensitivity, and F1-score for FDD fall dataset.

Label Name	Accuracy [%]	Precision [%]	Sensitivity [%]	F1-Score [%]
Sitting	n/a	99.54	99.49	99.52
Standing	n/a	98.96	99.04	98.99
Bending	n/a	98.01	99.00	98.50
Crawling	n/a	96.64	89.92	93.09
Lying	n/a	96.75	98.60	97.63
Total	99.12	97.98	97.21	97.55

A state-of-the-art comparison for the FDD dataset is demonstrated in Table 10. The author of [61] applied the GLR scheme to design the system and achieved 96.6% accuracy for the FDD dataset, whereas our proposed method achieved 99.22% accuracy using the FDD dataset.

Table 10. State-of-the-art comparison for FDD fall dataset.

Algorithm	Dataset	Accuracy [%]	Precision [%]	Sensitivity [%]	F-Score[%]
Haroo [61]	FDD	96.84			
Proposed GSTCAN Model	FDD	99.12	97.98	97.21	97.55

5.4. Deliberation

In this study, we proposed using AlphaPose to extract and select the skeleton data points instead of the RGB image. Then, we constructed an undirected graph and applied a graph-based CNN like GSTCAN. The positional relationships between some of the non-real connected points are very helpful for partially identifying events. We proposed a graph-based spatial-temporal convolution and attention network (GSTCAN) model to overcome the current challenges and developed an advanced medical technology system. In the procedure, we first calculated the motion among the consecutive frames, then constructed a graph and applied a graph convolutional neural network (GCN). We repeated the same procedure six times as GSTCAN and then applied it to the fully connected layer. To improve the role of non-connected skeleton points in certain events during the spatial-temporal feature, we applied the attention model with GSTCAN, aiming to extract global and local features which are bound to impact model optimization. Finally, we applied a sigmoid function as a classifier and achieved a high accuracy of 99.93%, 99.74%, and 99.12% for ImViA, UR-Fall, and FDD datasets, respectively. The high-performance accuracy with the three datasets proved the superiority and efficiency of the proposed system. According to comparison Tables 5, 7, and 9, we can say that our all datasets can be considered balanced, because our method achieved high precision, sensitivity, and F-score as well as high-performance accuracy. The state-of-the-art comparison Tables 6, 8, and 10 demonstrated high performance of the proposed model for all three fall-event datasets compared to the existing state-of-the-art systems. In addition, the existing fall detection systems achieved lower performance accuracy with various models, which sometimes need high computational complexity. Our proposed system generated better performance accuracy than the hand-crafted feature and machine learning algorithms with lower computational complexity than the state-of-the-art systems. Based on the state-of-the-art comparison table, we can see that the high performance of our method with the three datasets proves the proposed system's superiority in terms of performance and efficiency. We can differentiate our model with the following: (a) It can effectively detect the motion of the fall events; (b) It achieved a more than 5% higher performance accuracy compared to the existing work; (c) It takes less time compared to the existing work because we efficiently used fewer GSTCAN units. We can conclude that our model is suitable for discriminating fall events from human activity-based video datasets with a small cost of average classification rate.

6. Conclusions

This paper proposed a graph-based spatial-temporal convolution and attention network (GSTCAN) model to extract intra- and inter-frame joint relationships to improve performance accuracy and efficiency to confirm whether a person has fallen. To emphasize the role of a non-skeleton joint, we employed a modified channel attention model to the GSTCAN feature for selecting the channel-wise effective feature. We achieved higher accuracy than the existing models on the two datasets. Our model achieved high-performance accuracy for the three benchmark fall event datasets. The high-performance accuracy with less complexity proved the superiority and efficiency of the proposed model. In the future, we plan to increase the number of hand-crafted features with spatial-temporal features to reduce the number of parameters of the model to achieve high performance with a low computational cost and apply this model to the field of movement disorder detection. In the future, we will train the model with human action recognition datasets, aiming to make it a pre-trained model for human action recognition as well as fall detection [42,82].

Author Contributions: Conceptualization, R.E., K.H. and A.S.M.M.; methodology, R.E., K.H., A.S.M.M., Y.T. and J.S.; investigation, R.E., A.S.M.M. and J.S.; data curation, R.E., A.S.M.M., Y.T. and J.S.; writing—original draft preparation, R.E., A.S.M.M. and J.S.; writing—review and editing, A.S.M.M., Y.T. and J.S.; visualization, R.E. and A.S.M.M.; supervision, J.S.; funding acquisition, J.S. All authors have read and agreed to the published version of the manuscript.

Funding: This work was supported by JSPS KAKENHI Grant Number JP23H03477.

Data Availability Statement: ImViA dataset is accessible at https://imvia.u-bourgogne.fr/en/database/fall-detection-dataset-2.html. UR fall dataset can be found at http://fenix.ur.edu.pl/~mkepski/ds/uf.html. FDD fall dataset is accessible at https://falldataset.com/.

Conflicts of Interest: The authors declare no conflict of interest.

References

1. United Nations. *World Population Ageing 2020: Highlights: Living Arrangements of Older Persons*; United Nations Department of Economic and Social Affairs: New York, NY, USA, 2021.
2. Zahedian-Nasab, N.; Jaberi, A.; Shirazi, F.; Kavousipor, S. Effect of virtual reality exercises on balance and fall in elderly people with fall risk: A randomized controlled trial. *BMC Geriatr.* **2021**, *21*, 509. [CrossRef] [PubMed]
3. Lord, S.R. Visual risk factors for falls in older people. *Age Ageing* **2006**, *35*, ii42–ii45. [CrossRef] [PubMed]

4. Romeo, L.; Marani, R.; Petitti, A.; Milella, A.; D'Orazio, T.; Cicirelli, G. Image-based Mobility Assessment in Elderly People from Low-Cost Systems of Cameras: A Skeletal Dataset for Experimental Evaluations. In Proceedings of the Ad-Hoc, Mobile, and Wireless Networks: 19th International Conference on Ad-Hoc Networks and Wireless, ADHOC-NOW 2020, Bari, Italy, 19–21 October 2020; Springer: Cham, Switzerland, 2020; pp. 125–130.
5. Gutiérrez, J.; Rodríguez, V.; Martin, S. Comprehensive review of vision-based fall detection systems. *Sensors* **2021**, *21*, 947. [CrossRef]
6. Lu, K.L.; Chu, E.T.H. An image-based fall detection system for the elderly. *Appl. Sci.* **2018**, *8*, 1995. [CrossRef]
7. Huang, Z.; Liu, Y.; Fang, Y.; Horn, B.K. Video-based fall detection for seniors with human pose estimation. In Proceedings of the 2018 4th international conference on Universal Village (UV), Boston, MA, USA, 21–24 October 2018; pp. 1–4.
8. Dong, S.; Wang, P.; Abbas, K. A survey on deep learning and its applications. *Comput. Sci. Rev.* **2021**, *40*, 100379. [CrossRef]
9. Miah, A.S.M.; Hasan, M.A.M.; Shin, J.; Okuyama, Y.; Tomioka, Y. Multistage Spatial Attention-Based Neural Network for Hand Gesture Recognition. *Computers* **2023**, *12*, 13. [CrossRef]
10. Miah, A.S.M.; Shin, J.; Hasan, M.A.M.; Rahim, M.A. BenSignNet: Bengali Sign Language Alphabet Recognition Using Concatenated Segmentation and Convolutional Neural Network. *Appl. Sci.* **2022**, *12*, 3933. [CrossRef]
11. Miah, A.S.M.; Shin, J.; Al Mehedi Hasan, M.; Rahim, M.A.; Okuyama, Y. Rotation, Translation And Scale Invariant Sign Word Recognition Using Deep Learning. *Comput. Syst. Sci. Eng.* **2023**, *44*, 2521–2536. [CrossRef]
12. Shin, J.; Musa Miah, A.S.; Hasan, M.A.M.; Hirooka, K.; Suzuki, K.; Lee, H.S.; Jang, S.W. Korean Sign Language Recognition Using Transformer-Based Deep Neural Network. *Appl. Sci.* **2023**, *13*, 3029. [CrossRef]
13. Rahim, M.A.; Miah, A.S.M.; Sayeed, A.; Shin, J. Hand gesture recognition based on optimal segmentation in human-computer interaction. In Proceedings of the 2020 3rd IEEE International Conference on Knowledge Innovation and Invention (ICKII), Kaohsiung, Taiwan, 21–23 August 2020; pp. 163–166.
14. Miah, A.S.M.; Hasan, M.J.S.L.H.S.J. Multi-Stream General and Graph-Based Deep Neural Networks for Skeleton-Based Sign Language Recognition. *Electronics* **2023**, *12*, 2841. [CrossRef]
15. Ren, L.; Peng, Y. Research of fall detection and fall prevention technologies: A systematic review. *IEEE Access* **2019**, *7*, 77702–77722. [CrossRef]
16. Xu, T.; Zhou, Y. Elders' fall detection based on biomechanical features using depth camera. *Int. J. Wavelets Multiresolution Inf. Process.* **2018**, *16*, 1840005. [CrossRef]
17. De Quadros, T.; Lazzaretti, A.E.; Schneider, F.K. A movement decomposition and machine learning-based fall detection system using wrist wearable device. *IEEE Sensors J.* **2018**, *18*, 5082–5089. [CrossRef]
18. Kibria, K.A.; Noman, A.S.; Hossain, M.A.; Islam Bulbul, M.S.; Rashid, M.M.; Musa Miah, A.S. Creation of a Cost-Efficient and Effective Personal Assistant Robot using Arduino & Machine Learning Algorithm. In Proceedings of the 2020 IEEE Region 10 Symposium (TENSYMP), Dhaka, Bangladesh, 5–7 June 2020; pp. 477–482. [CrossRef]
19. Rubenstein, L.Z. Falls in older people: Epidemiology, risk factors and strategies for prevention. *Age Ageing* **2006**, *35*, ii37–ii41. [CrossRef] [PubMed]
20. Chen, Y.; Li, W.; Wang, L.; Hu, J.; Ye, M. Vision-based fall event detection in complex background using attention guided bi-directional LSTM. *IEEE Access* **2020**, *8*, 161337–161348. [CrossRef]
21. Miah, A.S.M.; Hasan, M.A.M.; Shin, J. Dynamic Hand Gesture Recognition using Multi-Branch Attention Based Graph and General Deep Learning Model. *IEEE Access* **2023**, *11*, 4703–4716. [CrossRef]
22. Gasparrini, S.; Cippitelli, E.; Gambi, E.; Spinsante, S.; Wåhslén, J.; Orhan, I.; Lindh, T. Proposal and Experimental Evaluation of Fall Detection Solution Based on Wearable and Depth Data Fusion. In *Proceedings of the ICT Innovations 2015: Emerging Technologies for Better Living*; Springer: Berlin/Heidelberg, Germany, 2016; pp. 99–108.
23. Maddalena, L.; Petrosino, A. Background subtraction for moving object detection in RGBD data: A survey. *J. Imaging* **2018**, *4*, 71. [CrossRef]
24. Kreković, M.; Čerić, P.; Dominko, T.; Ilijaš, M.; Ivančić, K.; Skolan, V.; Šarlija, J. A method for real-time detection of human fall from video. In Proceedings of the 2012 Proceedings of the 35th International Convention MIPRO, Opatija, Croatia, 21–25 May 2012; pp. 1709–1712.
25. El Baf, F.; Bouwmans, T.; Vachon, B. Type-2 Fuzzy Mixture of Gaussians Model: Application to Background Modeling. In *Proceedings of Advances in Visual Computing*; Springer: Berlin/Heidelberg, Germany, 2008; Number Part I; pp. 772–781.
26. Guo, H.; Qiu, C.; Vaswani, N. An online algorithm for separating sparse and low-dimensional signal sequences from their sum. *IEEE Trans. Signal Process.* **2014**, *62*, 4284–4297. [CrossRef]
27. Dong, S.; Li, R. Traffic identification method based on multiple probabilistic neural network model. *Neural Comput. Appl.* **2019**, *31*, 473–487. [CrossRef]
28. Miah, A.S.M.; Mamunur Rashid, M.; Redwanur Rahman, M.; Tofayel Hossain, M.; Shahidujjaman Sujon, M.; Nawal, N.; Hasan, M.; Shin, J. Alzheimer's Disease Detection Using CNN Based on Effective Dimensionality Reduction Approach. In *Proceedings of the Intelligent Computing and Optimization*; Vasant, P., Zelinka, I., Weber, G.W., Eds.; Springer: Cham, Switzerland, 2021; pp. 801–811.
29. Kafi, H.M.; Miah, A.S.M.; Shin, J.; Siddique, M.N. A Lite-Weight Clinical Features Based Chronic Kidney Disease Diagnosis System Using 1D Convolutional Neural Network. In Proceedings of the 2022 International Conference on Advancement in Electrical and Electronic Engineering (ICAEEE), Gazipur, Bangladesh, 24–26 February 2022; pp. 1–5. [CrossRef]

30. Bouwmans, T.; Javed, S.; Sultana, M.; Jung, S.K. Deep neural network concepts for background subtraction: A systematic review and comparative evaluation. *Neural Netw.* **2019**, *117*, 8–66. [CrossRef]

31. Maldonado-Bascon, S.; Iglesias-Iglesias, C.; Martín-Martín, P.; Lafuente-Arroyo, S. Fallen people detection capabilities using assistive robot. *Electronics* **2019**, *8*, 915. [CrossRef]

32. Maas, A.L.; Hannun, A.Y.; Ng, A.Y. Rectifier nonlinearities improve neural network acoustic models. In Proceedings of the 30 th International Conference on Ma- chine Learning, Atlanta, GA, USA, 16–21 June 2013; Volume 30, p. 3.

33. Al Nahian, M.J.; Ghosh, T.; Al Banna, M.H.; Aseeri, M.A.; Uddin, M.N.; Ahmed, M.R.; Mahmud, M.; Kaiser, M.S. Towards an accelerometer-based elderly fall detection system using cross-disciplinary time series features. *IEEE Access* **2021**, *9*, 39413–39431. [CrossRef]

34. Yan, S.; Xiong, Y.; Lin, D. Spatial, temporal graph convolutional networks for skeleton-based action recognition. In Proceedings of the AAAI Conference on Artificial Intelligence, New Orleans, LA, USA, 2–7 February 2018; Volume 32.

35. Keskes, O.; Noumeir, R. Vision-based fall detection using st-gcn. *IEEE Access* **2021**, *9*, 28224–28236. [CrossRef]

36. Kwolek, B.; Kepski, M. Human fall detection on embedded platform using depth maps and wireless accelerometer. *Comput. Methods Programs Biomed.* **2014**, *117*, 489–501. [CrossRef] [PubMed]

37. Youssfi Alaoui, A.; Tabii, Y.; Oulad Haj Thami, R.; Daoudi, M.; Berretti, S.; Pala, P. Fall detection of elderly people using the manifold of positive semidefinite matrices. *J. Imaging* **2021**, *7*, 109. [CrossRef]

38. Charfi, I.; Miteran, J.; Dubois, J.; Atri, M.; Tourki, R. Optimised spatio-temporal descriptors for real-time fall detection: Comparison of SVM and Adaboost based classification. *J. Electron. Imaging JEI* **2013**, *22*, 17.

39. Mubashir, M.; Shao, L.; Seed, L. A survey on fall detection: Principles and approaches. *Neurocomputing* **2013**, *100*, 144–152. [CrossRef]

40. Miah, A.S.M.; Ahmed, S.R.A.; Ahmed, M.R.; Bayat, O.; Duru, A.D.; Molla, M.I. Motor-Imagery BCI Task Classification Using Riemannian Geometry and Averaging with Mean Absolute Deviation. In Proceedings of the 2019 Scientific Meeting on Electrical-Electronics & Biomedical Engineering and Computer Science (EBBT), Istanbul, Turkey, 24–26 April 2019; pp. 1–7. [CrossRef]

41. Liu, H.; Hartmann, Y.; Schultz, T. Motion Units: Generalized Sequence Modeling of Human Activities for Sensor-Based Activity Recognition. In Proceedings of the 2021 29th European Signal Processing Conference (EUSIPCO), Dublin, Ireland, 23–27 August 2021; pp. 1506–1510. [CrossRef]

42. Liu, H.; Gamboa, H.; Schultz, T. Sensor-Based Human Activity and Behavior Research: Where Advanced Sensing and Recognition Technologies Meet. *Sensors* **2023**, *23*, 125. [CrossRef]

43. Miah, A.S.M.; Rahim, M.A.; Shin, J. Motor-imagery classification using Riemannian geometry with median absolute deviation. *Electronics* **2020**, *9*, 1584. [CrossRef]

44. Miah, A.S.M.; Islam, M.R.; Molla, M.K.I. EEG classification for MI-BCI using CSP with averaging covariance matrices: An experimental study. In Proceedings of the 2019 International Conference on Computer, Communication, Chemical, Materials and Electronic Engineering (IC4ME2), Rajshahi, Bangladesh, 11–12 July 2019; pp. 1–5.

45. Joy, M.M.H.; Hasan, M.; Miah, A.S.M.; Ahmed, A.; Tohfa, S.A.; Bhuaiyan, M.F.I.; Zannat, A.; Rashid, M.M. Multiclass mi-task classification using logistic regression and filter bank common spatial patterns. In Proceedings of the Computing Science, Communication and Security: First International Conference, COMS2 2020, Gujarat, India, 26–27 March 2020; Revised Selected Papers; Springer: Berlin/Heidelberg, Germany, 2020; pp. 160–170.

46. Zobaed, T.; Ahmed, S.R.A.; Miah, A.S.M.; Binta, S.M.; Ahmed, M.R.A.; Rashid, M. Real time sleep onset detection from single channel EEG signal using block sample entropy. *IOP Conf. Ser. Mater. Sci. Eng.* **2020**, *928*, 032021. [CrossRef]

47. Miah, A.S.M.; Islam, M.R.; Molla, M.K.I. Motor imagery classification using subband tangent space mapping. In Proceedings of the 2017 20th International Conference of Computer and Information Technology (ICCIT), Dhaka, Bangladesh, 22–24 December 2017; pp. 1–5. [CrossRef]

48. Kabir, M.H.; Mahmood, S.; Al Shiam, A.; Musa Miah, A.S.; Shin, J.; Molla, M.K.I. Investigating Feature Selection Techniques to Enhance the Perfor-mance of EEG-Based Motor Imagery Tasks Classification. *Mathematics* **2023**, *11*, 1921. [CrossRef]

49. Miah, A.S.M.; Mouly, M.A.; Debnath, C.; Shin, J.; Sadakatul Bari, S. Event-Related Potential Classification Based on EEG Data Using xDWAN with MDM and KNN. In *Proceedings of the International Conference on Computing Science, Communication and Security*; Springer: Berlin/Heidelberg, Germany, 2021; pp. 112–126.

50. Miah, A.S.M.; Shin, J.; Hasan, M.A.M.; Molla, M.K.I.; Okuyama, Y.; Tomioka, Y. Movie Oriented Positive Negative Emotion Classification from EEG Signal using Wavelet transformation and Machine learning Approaches. In Proceedings of the 2022 IEEE 15th International Symposium on Embedded Multicore/Many-core Systems-on-Chip (MCSoC), Penang, Malaysia, 19–22 December 2022; pp. 26–31.

51. Miah, A.S.M.; Shin, J.; Islam, M.M.; Molla, M.K.I. Natural Human Emotion Recognition Based on Various Mixed Reality (MR) Games and Electroencephalography (EEG) Signals. In Proceedings of the 2022 IEEE 5th Eurasian Conference on Educational Innovation (ECEI), Taipei, Taiwan, 10–12 February 2022; pp. 408–411.

52. Daniela, M.; Marco, M.; Paolo, N. UniMiB SHAR: A Dataset for Human Activity Recognition Using Acceleration Data from Smartphones. *Appl. Sci.* **2017**, *7*, 1101.

53. Wang, J.; Zhang, Z.; Li, B.; Lee, S.; Sherratt, R.S. An enhanced fall detection system for elderly person monitoring using consumer home networks. *IEEE Trans. Consum. Electron.* **2014**, *60*, 23–29. [CrossRef]

54. Desai, K.; Mane, P.; Dsilva, M.; Zare, A.; Shingala, P.; Ambawade, D. A novel machine learning based wearable belt for fall detection. In Proceedings of the 2020 IEEE International Conference on Computing, Power and Communication Technologies (GUCON), Greater Noida, India, 2–4 October 2020; pp. 502–505.
55. Xu, T.; Zhou, Y.; Zhu, J. New advances and challenges of fall detection systems: A survey. *Appl. Sci.* **2018**, *8*, 418. [CrossRef]
56. Tian, Y.; Lee, G.H.; He, H.; Hsu, C.Y.; Katabi, D. RF-based fall monitoring using convolutional neural networks. *Proc. ACM Interactive Mobile Wearable Ubiquitous Technol.* **2018**, *2*, 1–24. [CrossRef]
57. Xue, T.; Liu, H. Hidden Markov Model and Its Application in Human Activity Recognition and Fall Detection: A Review. In *Proceedings of the Communications, Signal Processing, and Systems*; Liang, Q., Wang, W., Liu, X., Na, Z., Zhang, B., Eds.; Springer: Berlin/Heidelberg, Germany, 2022; pp. 863–869.
58. Zerrouki, N.; Houacine, A. Combined curvelets and hidden Markov models for human fall detection. *Multimed. Tools Appl.* **2017**, *77*, 6405–6424. [CrossRef]
59. Chua, J.L.; Chang, Y.C.; Lim, W.K. A simple vision-based fall detection technique for indoor video surveillance. *Signal Image Video Process.* **2015**, *9*, 623–633. [CrossRef]
60. Cai, X.; Li, S.; Liu, X.; Han, G. Vision-based fall detection with multi-task hourglass convolutional auto-encoder. *IEEE Access* **2020**, *8*, 44493–44502. [CrossRef]
61. Harrou, F.; Zerrouki, N.; Sun, Y.; Houacine, A. An integrated vision-based approach for efficient human fall detection in a home environment. *IEEE Access* **2019**, *7*, 114966–114974. [CrossRef]
62. Han, Q.; Zhao, H.; Min, W.; Cui, H.; Zhou, X.; Zuo, K.; Liu, R. A two-stream approach to fall detection with MobileVGG. *IEEE Access* **2020**, *8*, 17556–17566. [CrossRef]
63. Yao, L.; Yang, W.; Huang, W. An improved feature-based method for fall detection. *Teh. Vjesn.* **2019**, *26*, 1363–1368.
64. Tsai, T.H.; Hsu, C.W. Implementation of fall detection system based on 3D skeleton for deep learning technique. *IEEE Access* **2019**, *7*, 153049–153059. [CrossRef]
65. Zheng, H.; Liu, Y. Lightweight fall detection algorithm based on AlphaPose optimization model and ST-GCN. *Math. Probl. Eng.* **2022**, *2022*, 9962666. [CrossRef]
66. Tran, T.T.H.; Le, T.L.; Morel, J. An analysis on human fall detection using skeleton from Microsoft Kinect. In Proceedings of the 2014 IEEE Fifth International Conference on Communications and Electronics (ICCE), Danang, Vietnam, 30 July–1 August 2014; pp. 484–489.
67. Adhikari, K.; Bouchachia, H.; Nait-Charif, H. Activity recognition for indoor fall detection using convolutional neural network. In Proceedings of the 2017 Fifteenth IAPR International Conference on Machine Vision Applications (MVA), Nagoya, Japan, 8–12 May 2017; pp. 81–84. [CrossRef]
68. Pathak, D.; Bhosale, V. Fall Detection for Elderly People in Indoor Environment using Kinect Sensor. *Nternational J. Sci. Res.* **2015**, *6*, 1956–1960.
69. Hwang, S.; Ahn, D.; Park, H.; Park, T. Maximizing accuracy of fall detection and alert systems based on 3D convolutional neural network. In Proceedings of the Second International Conference on Internet-of-Things Design and Implementation, Pittsburgh, PA, USA, 18–21 April 2017; pp. 343–344.
70. Fakhrulddin, A.H.; Fei, X.; Li, H. Convolutional neural networks (CNN) based human fall detection on body sensor networks (BSN) sensor data. In Proceedings of the 2017 4th International Conference on Systems and informatics (ICSAI), Hangzhou, China, 1–13 November 2017; pp. 1461–1465.
71. Fang, H.S.; Xie, S.; Tai, Y.W.; Lu, C. Rmpe: Regional multi-person pose estimation. In Proceedings of the IEEE International Conference on Computer Vision, Venice, Italy, 22–29 October 2017; pp. 2334–2343.
72. Xiu, Y.; Li, J.; Wang, H.; Fang, Y.; Lu, C. Pose Flow: Efficient online pose tracking. *arXiv* **2018**, arXiv:1802.00977.
73. Fang, H.S.; Li, J.; Tang, H.; Xu, C.; Zhu, H.; Xiu, Y.; Li, Y.L.; Lu, C. Alphapose: Whole-body regional multi-person pose estimation and tracking in real-time. *IEEE Trans. Pattern Anal. Mach. Intell.* **2022**, *45*, 7157–7173. [CrossRef]
74. Wu, Z.; Pan, S.; Chen, F.; Long, G.; Zhang, C.; Philip, S.Y. A comprehensive survey on graph neural networks. *IEEE Trans. Neural Networks Learn. Syst.* **2020**, *32*, 4–24. [CrossRef] [PubMed]
75. Hu, J.; Shen, L.; Sun, G. Squeeze-and-excitation networks. In Proceedings of the IEEE Conference on Computer Vision and Pattern Recognition, Salt Lake City, UT, USA, 18–23 June 2018; pp. 7132–7141.
76. Tieleman, T.; Hinton, G. Lecture 6.5-rmsprop: Divide the Gradient by a Running Average of Its Recent Magnitude. *COURSERA Neural Netw. Mach. Learn.* **2012**, *17*, 26–31.
77. Paszke, A.; Gross, S.; Massa, F.; Lerer, A.; Bradbury, J.; Chanan, G.; Killeen, T.; Lin, Z.; Gimelshein, N.; Antiga, L.; et al. Pytorch: An imperative style, high-performance deep learning library. *Adv. Neural Inf. Process. Syst.* **2019**, *32*, 1–12.
78. Gollapudi, S. *Learn Computer Vision Using OPENCV*; Springer: Berlin/Heidelberg, Germany, 2019.
79. Dozat, T. Incorporating Nesterov Momentum into Adam. 2016. Available online: https://openreview.net/pdf?id=OM0jvwB8jIp57ZJjtNEZ (accessed on 21 June 2023).
80. Wang, B.H.; Yu, J.; Wang, K.; Bao, X.Y.; Mao, K.M. Fall detection based on dual-channel feature integration. *IEEE Access* **2020**, *8*, 103443–103453. [CrossRef]

81. Chamle, M.; Gunale, K.; Warhade, K. Automated unusual event detection in video surveillance. In Proceedings of the 2016 International Conference on Inventive Computation Technologies (ICICT), Coimbatore, India, 26–27 August 2016; Volume 2, pp. 1–4.
82. Hartmann, Y.; Liu, H.; Schultz, T. Interactive and Interpretable Online Human Activity Recognition. In Proceedings of the 2022 IEEE International Conference on Pervasive Computing and Communications Workshops and other Affiliated Events (PerCom Workshops), Pisa, Italy, 21–25 March 2022.

Article

Usability of Inexpensive Optical Pulse Sensors for Textile Integration and Heartbeat Detection Code Development

Niclas Richter [1], Khorolsuren Tuvshinbayar [1], Guido Ehrmann [2] and Andrea Ehrmann [1,*]

[1] Faculty of Engineering Sciences and Mathematics, Bielefeld University of Applied Sciences, 33619 Bielefeld, Germany
[2] Virtual Institute of Applied Research on Advanced Materials (VIARAM)
* Correspondence: andrea.ehrmann@fh-bielefeld.de

Abstract: Low-cost sensors and single circuit boards such as Arduino and Raspberry Pi have increased the possibility of measuring biosignals by smart textiles with embedded electronics. One of the main problems with such e-textiles is their washability. While batteries are usually removed before washing, single-board computers and microcontrollers, as well as electronic sensors, would ideally be kept inside a user-friendly smart garment. Here, we show results of washing tests with optical pulse sensors, which can be used in smart gloves not only for hospitalized patients, and ATtiny85 as an example of a single-board microcontroller, sewn onto different cotton fabrics. We report that even without any encapsulation, all tested sensors and microcontrollers endured 10 washing cycles at 30–60 °C without defects. For easier garment integration, we suggest using an ESP8266 with integrated Wi-Fi functionality and offer a new program code to measure beats per minute (BMP) with optimized accuracy.

Keywords: smart textiles; smart clothes; e-textiles; biosignal; vital signal; pulse sensor; microcontroller; washability; display

Citation: Richter, N.; Tuvshinbayar, K.; Ehrmann, G.; Ehrmann, A. Usability of Inexpensive Optical Pulse Sensors for Textile Integration and Heartbeat Detection Code Development. *Electronics* **2023**, *12*, 1521. https://doi.org/10.3390/electronics12071521

Academic Editor: Stefano Ricci

Received: 27 February 2023
Revised: 18 March 2023
Accepted: 22 March 2023
Published: 23 March 2023

1. Introduction

The increasing amount of cardiac diseases in many countries suggests the implementation of biosignal measurements in smart textiles [1,2]. Usually, the electrocardiogram (ECG) of a person is measured as one of the most important ways to detect potential irregularities, which is why several research groups developed single-lead or even 12-lead ECG garments with textile electrodes [3–6]. However, such textile electrodes necessitate a sufficient skin contact to transfer the very small ECG voltages from the skin reliably. This is normally achieved by applying a relatively high pressure [7–10], which may result in an undesired feeling and potentially even skin irritations.

In many cases, however, only the pulse needs to be monitored, offering not only the pure pulse rate but also the pulse variability as a measure of the wealth of the heart and the potential to register missing beats, if the applied software is suitable for this. While the pulse can be measured by the aforementioned conductive textile ECG electrodes, it can also be detected optically [11–13]. Especially when patients are examined by magnetic resonance imaging (MRI), optical pulse measurements can avoid any potential hazard, which may occur due to the wires of conventional ECG equipment [14].

In optical pulse measurements, the photoplethysmographic (PPG) waveform is derived by the time-dependent light absorption in the local tissue [15] and is strongly correlated with the ECG, if measured with high accuracy. An example of the correlation between both measures in the case of cardiac arrhythmia is shown in Figure 1 [15]. The pulse rate is usually measured between neighboring peaks or troughs or at the mean values [15]. In a highly resolved signal, a small minimum between the systole part, starting from the trough, and the subsequent diastole part can be visible [16].

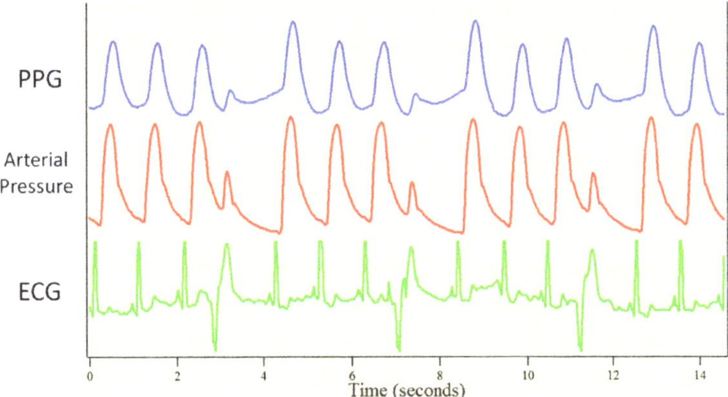

Figure 1. The effect of cardiac arrhythmia (PVCs) on the photoplethysmographic (PPG) waveform. From [15], copyright (2014), with permission from Elsevier.

Pulse oximeters are used in many clinical areas, but their measurements can be negatively influenced by patients' behavior, as well as technical errors [17]. Another problematic feature of common pulse oximeters is the rigid material of the clip, used to fix it on a finger, especially in long-term measurements [18]. On the other hand, smart watches containing optical pulse sensors may work properly for healthy people [19], but are again not suitable for 24/7 monitoring of patients in hospitals where correct placement cannot be regularly checked, and data transfer cannot be fit to the respective IT infrastructure. Most important, however, is the finding that pulse measurements are most sensitive in the carotid region and lowest in the wrist region, where smart watches are usually worn [20].

While the idea to use optical pulse measurements instead of electrical ECG measurements is, as discussed above, not new and has long been introduced into common medical technology in hospitals, the advent of steadily miniaturized microcontrollers and microcomputers nowadays offers new possibilities to integrate optical pulse sensors and the equipment for their evaluation into garments, which will not damage a patient's skin even during long-term measurements and allow defining the optimum position(s) of one or more pulse sensors [21,22]. Nevertheless, most researchers integrate only the sensors themselves into medical smart textiles and do not examine the integration of microcontrollers or microcomputers into the fabric [23–26]. One of the main reasons for this is probably the necessary washability of all electronics which are constantly embedded in textile fabrics.

Here, we thus investigate the washability of low-cost optical pulse sensors from different producers as well as ATtiny85 as one potential microcontroller small enough to be fully embedded into a medical textile. During our study, we recognized some unnecessarily complicated parts and mathematical inaccuracies of the commonly used Arduino code for such low-cost pulse sensors and thus developed a new code for Arduino, also suitable for ESP 32 which we found to be highly suitable for such measurements as well as for textile integration.

2. Materials and Methods

The following pulse sensors were purchased for the washing tests: Funduino heart rate sensor "R13-B-6-2", Berrybase pulse sensor, IDUINO SE050 (purchased from Conrad Electronic, Hirschau, Germany), Keyestudio XD-58C pulse sensor (purchased from Eckstein Komponente, Clausthal-Zellerfeld, Germany), and Joy-it KY-039 (DEBO Sens Heart) pulse sensor (purchased from Reichelt Elektronik, Sande, Germany). Moreover, the microcontroller Digispark Rev.3 ATtiny85 (AZ Delivery, Germany) was used in the washing tests.

While the Funduino sensor was washed separately in pre-tests for ten washing cycles, the other sensors and ATtiny85 were sewn on five different cotton woven fabrics, as depicted in Figure 2 (n = 5 for all pulse sensors and the microcontroller).

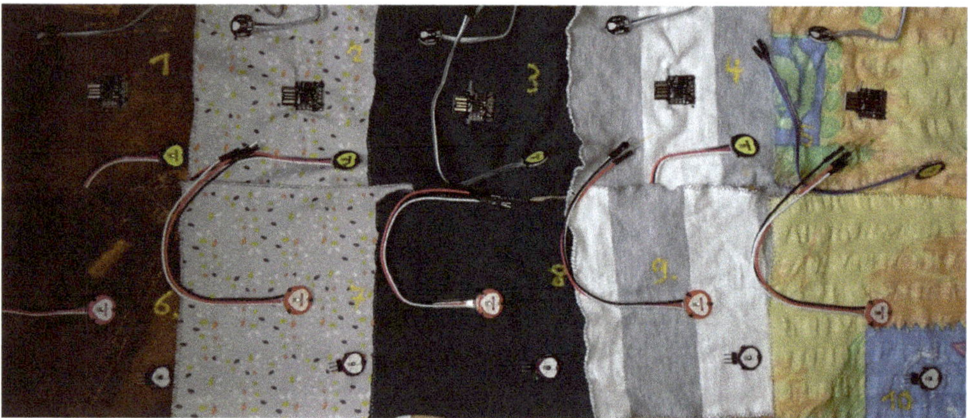

Figure 2. Samples for washing tests.

Washing was performed in a household washing machine Miele WWA028 WPS Active White with heavy-duty detergent in liquid form (30 °C, 40 °C) or in powder form (60 °C). The washing parameters for the main tests are depicted in Table 1. Washing was always performed in a laundry bag to avoid contacting the residual laundry with the multi-pin connectors of the sensors.

Table 1. Washing parameters.

Washing Cycle No.	Temperature/°C	Softener	Spin Cycle/min^{-1}
1	60	Yes	1200
2	40	Yes	1200
3	30	No	600
4	60	Yes	1200
5	30	No	900
6	40	Yes	1200
7	40	Yes	1200
8	40	Yes	1200
9	30	No	900
10	60	Yes	1200

For the setup to be included in medical garments, the following parts were tested: A small microcontroller ESP8266 D1 mini with Wi-Fi functionality via ESPNOW, enabling unobtrusive integration into smart clothes and data transfer towards a receiver; a larger microcontroller ESP 32 as receiver, and a 1.8″ display to allow for showing the pulse signal without an additional computer monitor.

3. Results

After a brief comparison of the different pulse sensors, the results of the washing tests are depicted and discussed before improved software and hardware solutions for the future integration of a pulse sensor into medical smart clothes are provided.

3.1. Sensor Comparison

Generally, all of the aforementioned pulse sensors allow measuring the pulse, with more or less resolution. The first measurements were performed based on the code from

Pulsesensor.com [27], allowing the pulse signal in the serial plotter of the Arduino IDE to be depicted. The signal of the 10-bit Arduino input is given in the value range of 0–1023, corresponding to the voltage range of 0–5 V. The base line of the signal is usually found in the middle of this value range, i.e., around 511. Figure 3 depicts measurements of different fingers from two probands, taken by the Funduino heart rate sensor, indicating the high dependence of these measurements on positioning and pressure on the skin.

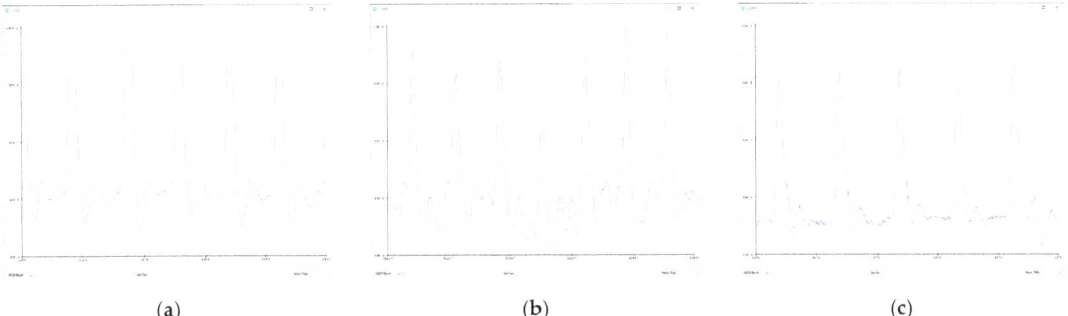

(a) (b) (c)

Figure 3. Pulse measurements of (**a**) proband A, left little finger (y-scale 200–1000); (**b**) proband A, left little finger, measured on the next day (400–720); (**c**) proband B, left little finger (450–650).

Generally, for a single sensor, signal heights between approx. 100 and 500 were found, depending on the pressure between the sensor and skin, the sensor position, the blood pressure, etc. In most cases, only digital noise was observed, no power grid noise or the like which may occur due to the USB connection to the laptop [28]. For signal heights of around 100, the digital noise of ±1 corresponds to a signal variation of $\pm1\%$, which is negligible for data evaluation. Although being low-cost, this sensor apparently enables sufficiently resolved pulse measurements.

The other pulse sensors showed strongly varying signals, partly with additional noise, partly with much smaller signal heights than in the pretest with the Funduino sensor. Figure 4 depicts exemplary measurements taken on proband B's left little finger with different sensors. While there is again a large variability of the signal height upon changes in the measurement conditions, it was also found that the maximum signal height is not identical for all sensors, i.e., with some sets of sensors from the same producer, only signal heights much smaller than 100 could be reached (e.g., Figure 4a).

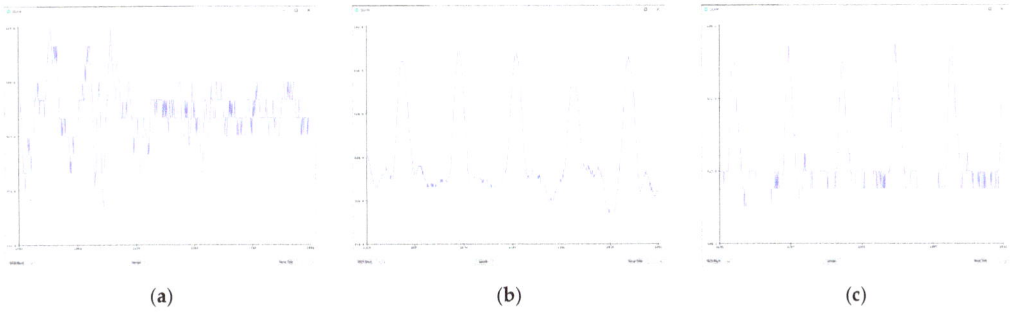

(a) (b) (c)

Figure 4. *Cont.*

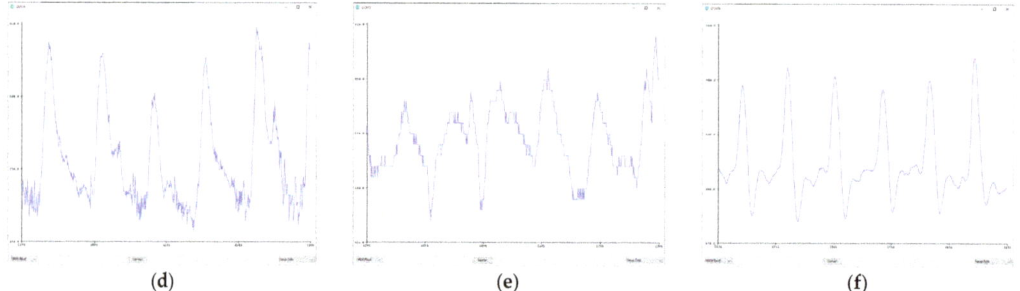

(d)   (e)   (f)

Figure 4. Pulse measurements of proband B, left little finger, with a sensor from (**a**) Berrybase (y-scale 501–513); (**b**) Iduino (490–540); (**c**) Iduino (508–520); (**d**) Keyestudio (500–560); (**e**) Joy-it (505–525); (**f**) Joy-it (420–660).

3.2. Washing Tests

For a future integration of sensors and microcontrollers into smart clothes, it can be expected that microcontrollers and soldering points will be carefully encapsulated. However, it is not fully clear whether this is also possible for the optical pulse sensor, without disturbing its function. Here, washing tests were thus performed for the worst possible case, i.e., without any encapsulation.

In the pre-tests, one ATtiny85 was programmed to blink when inserted into a USB port and washed 10 times, without any reduction of this function. In the main test series, five microcontrollers sewn on textile fabrics (cf. Figure 2) were washed 10 times (cf. Table 1), again showing no inhibition of their function. Apparently, these microcontrollers are washable even without encapsulation.

The results of the sensor washing tests are given in Table 2.

Table 2. Sensor washing tests. "-" indicates detached connector cables, "ok" means that a pulse signal could be measured. Connector cables were originally attached at Keyestudio, Joy-it and Iduino sensors, while multi-pin connectors were soldered to Berrybase sensors.

Washing Cycle No.	Sensor	Sample Number									
		1	2	3	4	5	6	7	8	9	10
1	Keyestudio	ok	ok	-	-	-					
1	Joy-it	ok	ok	ok	ok	ok					
1	Iduino						ok	-	ok	ok	ok
1	Berrybase						ok	ok	ok	ok	ok
2	Keyestudio	ok	-	-	-	-					
2	Joy-it	-	ok	ok	-	ok					
2	Iduino						ok	-	ok	ok	ok
2	Berrybase						ok	ok	ok	ok	ok
3	Keyestudio	ok	-	-	-	-					
3	Joy-it	-	ok	ok	-	ok					
3	Iduino						-	-	ok	ok	ok
3	Berrybase						ok	ok	ok	ok	ok
4	Keyestudio	ok	-	-	-	-					
4	Joy-it	-	ok	-	-	-					
4	Iduino						-	-	ok	-	-
4	Berrybase						ok	ok	ok	ok	ok
5	Keyestudio	ok	-	-	-	-					
5	Joy-it	-	ok	-	-	-					
5	Iduino						-	-	-	-	-
5	Berrybase						ok	ok	ok	ok	ok

Table 2. *Cont.*

Washing Cycle No.	Sensor	Sample Number									
		1	2	3	4	5	6	7	8	9	10
6	Keyestudio	ok	-	-	-	-					
6	Joy-it	-	ok	-	-	-					
6	Iduino						-	-	-	-	-
6	Berrybase						ok	ok	ok	ok	ok
7	Keyestudio	ok	-	-	-	-					
7	Joy-it	-	-	-	-	-					
7	Iduino						-	-	-	-	-
7	Berrybase						ok	ok	ok	ok	ok
8	Keyestudio	-	-	-	-	-					
8	Joy-it	-	-	-	-	-					
8	Iduino						-	-	-	-	-
8	Berrybase						ok	ok	ok	ok	ok
9	Keyestudio	-	-	-	-	-					
9	Joy-it	-	-	-	-	-					
9	Iduino						-	-	-	-	-
9	Berrybase						ok	ok	ok	ok	ok
10	Keyestudio	-	-	-	-	-					
10	Joy-it	-	-	-	-	-					
10	Iduino						-	-	-	-	-
10	Berrybase						ok	ok	ok	ok	ok

While the connector cables were mostly detached after a few washing cycles, measuring pulse signals in a similar quality as before washing (cf. Figure 3) was possible for all 40 sensors after re-establishing a connection at the sensors that had connector cables attached in their original state. The self-soldered multi-pin connectors of the Berrybase sensors were not impacted by washing.

These results are promising for the planned integration of such low-cost sensors and microcontrollers into smart clothes, even if it turns out that the optical sensor unit cannot be encapsulated. However, ATtiny85 may not be the ideal choice of a small microcontroller since programming is unnecessarily complicated due to pins with double functions, and no Wi-Fi functionality is embedded. This is why the next sub-section describes a more sophisticated hardware setup, including ESP8266 D1 mini as a sender and ESP 32 as a receiver, as well as a new lean code usable on Arduino, ESP, and other microcontrollers.

3.3. Hard- and Software Development

The hardware suggested for textile integration consists of the four parts depicted in Figure 5. The heart rate sensor (upper part in Figure 5a) emits green light and measures the reflection from the skin by a photodiode, which is transferred into an analog voltage signal. The pulse signal (as visible in Figure 5b) thus shows the blood circulation since reflection is reduced with a higher amount of red hemoglobin, which absorbs the green light. The ESP8266 D1 mini (lower part of Figure 5a) reads the analog signal and transmits it by ESPNOW (without the necessity of a local Wi-Fi network) to a receiver. Another reason for using the ESP8266 D1 mini is its very low energy consumption. In the c state of the project, the signal is sent in defined time intervals together with the heart frequency measured from the raw pulse signal, as described below.

ESP32 serving as the receiver (Figure 5b, right part) first saves the transmitted sensor value in an array, sets the display to fully white, and then plots the graph together with the average heart frequency (Figure 5b, left part).

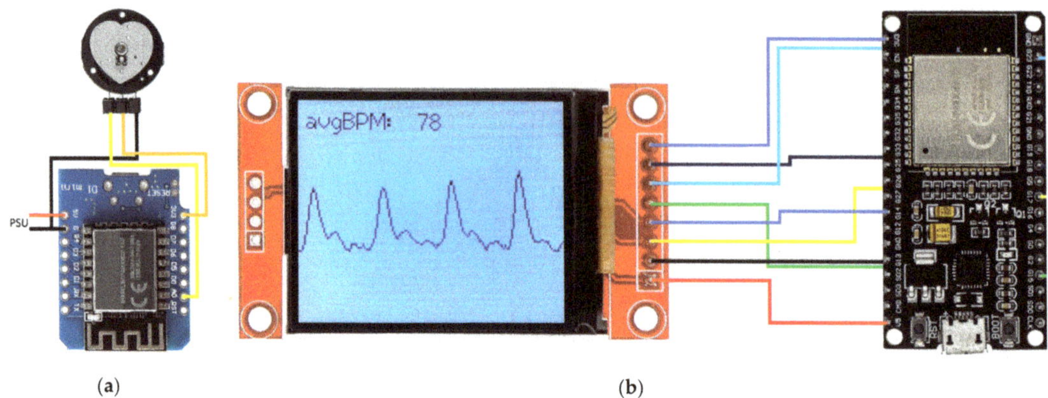

(a)   (b)

Figure 5. Hardware for textile-integrated pulse measurement: (**a**) pulse sensor and ESP8266 D1 mini to be embedded in the medical clothes, PSU = power supply unit; (**b**) ESP32 and 1.8″ display receiving and depicting the measured data outside the garment.

The code that was developed within the current project is based on mathematical criteria which must be fast, reliable, and repeatable. In order to determine the time between two heartbeats as accurately as possible, the time window in which these criteria are fulfilled must be very small. Currently, two criteria are used to detect a main chamber contraction. One is whether the analog signal $S(t)$ exceeds a certain threshold α, which is further referred to as criterion C_1. However, this condition can be fulfilled over a certain period of time, which makes it difficult to determine the heart rate accurately.

$$C_1(S) = \begin{cases} true & if\ S(t) > \alpha \\ false & if\ S(t) \le \alpha \end{cases} \tag{1}$$

Another criterion would be the rate of change over time. By numerically deriving the signal value using a backward differencing scheme (BDS), it can be determined whether a rising edge is currently being measured. Currently, the BDS is first order, so only the signal value one time step before is needed. Similar to the first criterion, it is also checked whether the derivative exceeds a certain value β. The disadvantage of this criterion is that it may be fulfilled several times in the cycle of a heartbeat.

$$\partial_t S(t) = \frac{S(t_n) - S(t_{n-1})}{\Delta t} \tag{2}$$

$$C_2(S) = \begin{cases} true & if\ \partial_t S(t) > \beta \\ false & if\ \partial_t S(t) \le \beta \end{cases} \tag{3}$$

By combining the two criteria, the described disadvantages can be completely circumvented. If suitable values for α and β are selected, a reliable criterion is formulated. Equation (4) shows the combination of the presented criteria in the form as it is implemented in the code.

$$C_{1\wedge2}(S) = \begin{cases} true & if\ S(t) > \alpha\ \wedge\ \partial_t S(t) > \beta \\ false & if\ S(t) \le \alpha\ \vee\ \partial_t S(t) \le \beta \end{cases} \tag{4}$$

Figure 6 illustrates schematically the previously described properties and problems of the individual criteria and time windows. When looking at $Cycle_1$, one can see the large time window in which the criterion C_1 would have detected a heartbeat. This is sufficient for pure pulse detection, but too inaccurate for the determination of the time between subsequent heartbeats. In $Cycle_2$, the criterion C_2 was applied. As can be seen, the rate of change within a cycle is very high in the three areas. That is, it is difficult to distinguish between the contraction of the atrium [·], the contraction of the main chamber [:], or the relaxation of the entire myocardium [∴]. $Cycle_3$ finally shows the combination of the two criteria and that both criteria complement each other well for their strengths. The time window was significantly reduced, and multiple detection was also excluded.

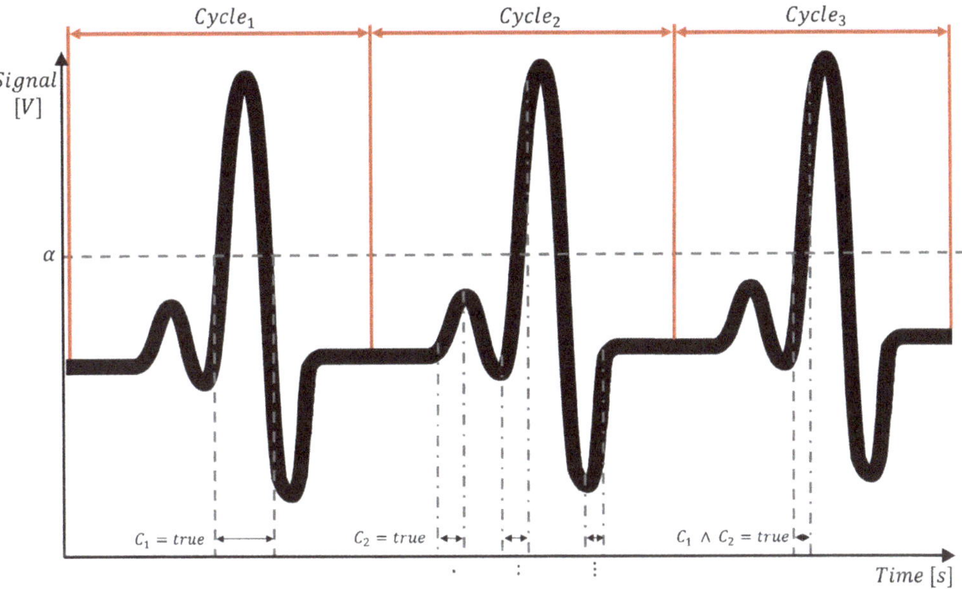

Figure 6. Schematic representation of pulse detection criteria.

Now that a reliable method for detecting a heartbeat has been implemented, the time between two beats must be determined. For this, two more variables are needed first. One is the time Δt_{loop} which determines the delay between the individual measurements. In addition, a counter n_{loop} is needed, which counts the number of loops between two heartbeats. The time between the heartbeats Δt_{HB} and the beats per minute (BPM) can now be calculated:

$$\Delta t_{HB} = n_{loop} \cdot \Delta t_{loop} \tag{5}$$

$$BPM = \frac{60}{\Delta t_{HB}} \tag{6}$$

The following flowchart (Figure 7) simplifies the basic idea of calculating beats per minute. If no heartbeat is detected, the counter is increased by one in each loop. When a heartbeat is detected, the quantities are calculated based on the counter and the counter is set back to zero. To the original condition $C_{1\wedge2}$, the condition for the counter n_{loop} was added additionally.

The resulting BPM signal is not forwarded to the receiver unit unfiltered, but averaged for the time being. The goal is an average value over a series of measurement data. Especially in this case, one would like to calculate BPM_{mean}. To avoid unnecessary operations inside loop conditions, instead of the BPM value, the average Δt_{HB} will be calculated first. If the mean value of the time between two heartbeats is determined, the corresponding BPM_{mean} can be calculated once before sending the information to the receiver unit. The span over which the average is taken is $n_{samples}$. So, a list of Δt_{HB} values containing $n_{samples}$ elements is needed. This list is created as an array $\Delta t_{HB,Arr}$ where the individual elements of the array can be addressed and used with the help of their indices. In this case, we decided to add the new elements at the end of the list. So, when adding a new element at the end, the first element is "pushed out" by the second one. For example, if $\Delta t_{HB,Arr}[n_{samples}]$ is selected, the most recent value is always returned. If $\Delta t_{HB,Arr}[1]$ is selected, this returns the oldest value.

$$\Delta t_{HB,Arr} = \left(\Delta t_{HB,1}, \ldots, \Delta t_{HB, n_{samples}} \right) \tag{7}$$

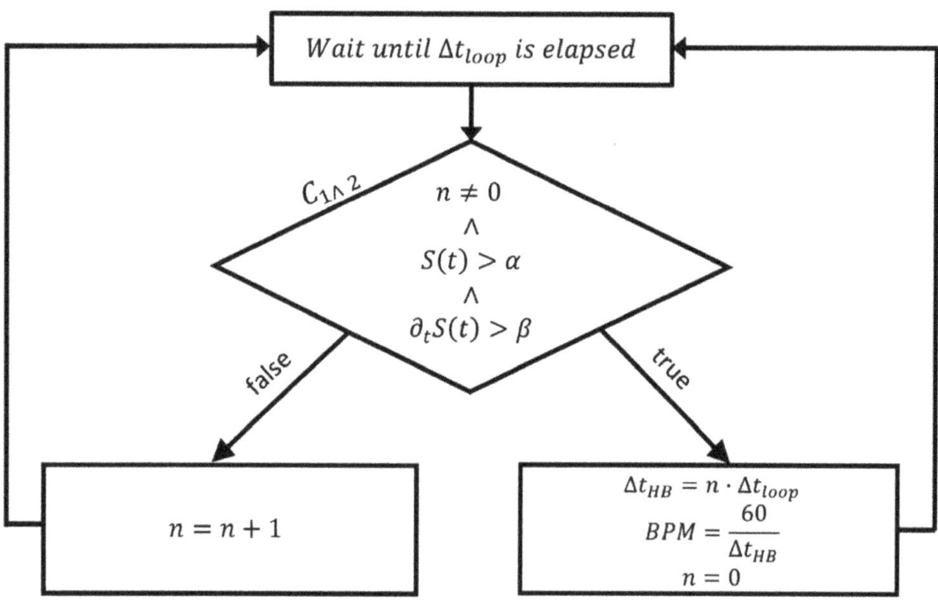

Figure 7. Flowchart of beats per minute (BPM) calculation. The other parameters are described in the main text.

As mentioned before, when adding a new Δt_{HB} value, the oldest one at the beginning of the list falls out. This gives the mean value the characteristic of a running mean over the measured values. The calculation can now be carried out using the following expression:

$$BPM_{mean} = 60 \cdot \left(\frac{1}{n_{samples}} \sum_{i=1}^{n_{samples}} \Delta t_{HB,Arr}[i] \right)^{-1} \tag{8}$$

This procedure is illustrated by an extended flowchart (Figure 8), which was supplemented by the averaging and the construction and sending of the data package. In the first loop after the Δt_{HB} calculation, the populating of the array can be observed. The first value in the array takes the value of the second one, the second one takes the value of the third one, etc., until the penultimate element. Then, the last element is assigned the currently calculated Δt_{HB} value. If the list is completely populated, the sum is calculated over this array and within the calculation of BPM_{mean} it is divided by the number of samples.

This code version was tested and found to work well for healthy probands, while the detection of cardiac arrhythmia necessitates further development, as described in Section 4.

The full code for the sensor unit is given in Appendix A, while the code for the receiver unit is provided in Appendix B.

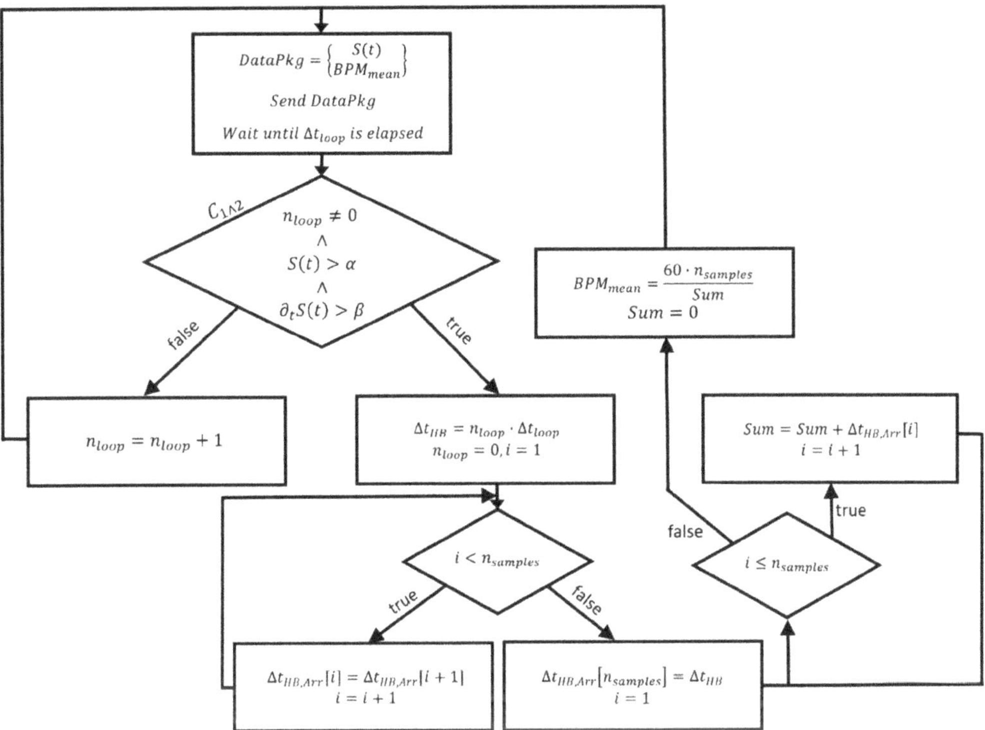

Figure 8. Flowchart of the average beats per minute (BPM) calculation.

4. Future Developments

In the next development stages, the operation of the sensor unit with battery is to be prepared. On the hardware side, this will be achieved by an even more compact ESP and a suitable battery for the sensor unit, and on the software side by restructuring the program flow chart. As described above, all data are currently sent continuously to the receiver unit. This would mean unnecessarily high energy consumption in battery mode.

Nevertheless, the balance between detecting emergency events and low energy consumption must be defined, possibly in different ways for different patients. Especially, the definition of "sleep" for the sensor unit must be defined. If sleep means the D1 mini is still analyzing and recording the signal from the sensor itself, it is easy to code a criterion for the special disease, which can be detected from the signal of an optical pulse-sensor and trigger a "SendAlarm" function to alert the medical staff. Energy saving is achieved by not sending the data to the receiver unit via ESP-Now continuously, only when a specific irregularity is detected. Additionally, the D1 mini has a large onboard storage, which is unused for now. For example, this storage can be used to save signal history and timestamps until they are retrieved.

While all sensors and microcontrollers were undamaged by different temperatures and detergents during the first ten washing cycles, and only mechanical damage of the connection lines occurred, which must be avoided in future developments, it is nevertheless necessary to develop a suitable encapsulation so that especially no heavy damages occur in case of a battery forgotten to be taken out of the equipment before washing.

When measuring and sending continuously in the moment, in the case of an arrhythmia, the algorithm would filter out the measured value, because it would be classified as implausible. Extending the algorithm with the ability to distill medical information from the signal is another point for future development. Especially cardiac arrhythmia can be detected by the time between two heartbeats, and if the time is greater than a defined variable, it can trigger an alarm, for example. The greatest advantage of the newly developed algorithm is the precise detection of heartbeats and, consequently, the time between two heartbeats, without skipping a single one. This is the reason why very precise timing is so important and why this was one of the first steps in the project.

On the programming side, the following sequence is intended:

- Put the sensor it into an energy saving mode;
- Wake up the receiver unit wirelessly;
- Acquire data over a certain period of time;
- Send data to the receiver unit;
- Return to the energy saving mode.

Another advantage is that multiple sensor units can thus be managed and queried via a central station.

One criticism of the current method would be the accuracy of calculating the time between two heartbeats. If the tasks in the loop in which the counter is counted between two heartbeats take too much time, this distorts the calculated time between two heartbeats, which in the recent tests provided a negligible error of less than 0.1%. Essentially, the clock frequency of the chip or on-board timers in general should be used.

The aspect of data protection and security should also be mentioned. Compliance with security standards when sending and receiving sensitive data still has to be implemented.

In the future, the individual functionalities are to be made available in the form of a library and further functions, such as heart rate variability and maxima/minima, are to be added.

5. Conclusions

A recent project showed that washing optical pulse sensors from different companies as well as microcontroller ATtiny85 in a common household washing machine for ten cycles did not damage any of them, suggesting their integration into smart clothes for pulse monitoring. Using an ESP8266 D1 mini as a transmitter and an ESP32 with 1.8″ display as a receiver, a stand-alone pulse measuring system was set up and equipped with an improved code for Arduino and ESP. These tests and developments serve as the base for a fully textile-integrated pulse measurement with mobile data storage and depiction.

Author Contributions: Conceptualization, N.R. and A.E.; methodology, all authors; software, N.R.; validation, N.R. and A.E.; investigation, all authors; writing—original draft preparation, N.R. and A.E.; writing—review and editing, K.T. and G.E.; visualization, N.R. and A.E. All authors have read and agreed to the published version of the manuscript.

Funding: This research was partly funded by the German Federal Ministry of Education and Research (BMBF), grant number 13GW0202C.

Institutional Review Board Statement: Not applicable.

Informed Consent Statement: Informed consent was obtained from all subjects involved in the study.

Data Availability Statement: All data are included in the paper and the Appendices A and B. The original sketches are available from https://github.com/NiclasRichter/PulseSensorTextileTechnologies (accessed on 26 February 2023).

Appendix A. Code for the Sensor Unit

```
/* Sensor-Unit */

#include <Arduino.h>
#include <ESP8266WiFi.h>
#include <espnow.h>

#define analogPin A0                              // Pin for reading sensor signal S(t)

int       PS_value      =0;                       // PulseSensor-Value from analogPin
const int diffOrder     =1;                       // Order of discretization for further developments
int       PS_signal[diffOrder+1];                 // Array for calculating the derivative
int       delta_t_loop  =20;                      // Loop-delay in ms
int       millimin      =60000;                   // Milliseconds per minute
double    alpha         =570.0;                   // Threshold for C_1, alpha
double    beta          =0.2;                      // Threshold for C_2, beta
double    dt_PS_signal;                           // Rate of change for Criteria C_2
int       n_loop        =0;                       // Loops between two heartbeats
int       delta_t_HB    =1000;                    // Time between two heartbeats
double    avgBPM        =0.0;                      // Average beats per minute
const int n_samples     =5;                       // Number of samples for calculating average BPM
double    delta_t_HB_Arr[n_samples];              // Storage for calculating average BPM
double    sum;                                    // Summed up timeBetweenHeartbeats for averaging

uint8_t peer1[] = {0x24, 0x0A, 0xC4, 0x5A, 0x05, 0x74};   // Define the Receiver-Unit via MAC-address

typedef struct message {
  int PS_value;
  double avgBPM;
};
struct message DataPkg;

void setup() {
  Serial.begin(115200);
  WiFi.mode(WIFI_STA);
  // Get Mac Add
  Serial.print("Mac Address: ");
  Serial.print(WiFi.macAddress());
  Serial.println("ESP-Now Sender");

  if (esp_now_init() != 0){
    Serial.println("Problem during ESP-NOW init");
    return;
  }
  esp_now_set_self_role(ESP_NOW_ROLE_CONTROLLER);
  esp_now_add_peer(peer1, ESP_NOW_ROLE_SLAVE, 1, NULL, 0);
}

void loop() {

PS_value = analogRead(analogPin);                 // Read analog value from Sensor and store it in S(t)
PS_signal[0]=PS_signal[1];                        // Push data through array with 2 elements
PS_signal[1]=PS_value;                            // -> 2, because order of differencing scheme is one
dt_PS_signal=1.0*(PS_signal[1]-PS_signal[0])/delta_t_loop;   // Backward differencing scheme first order
```

```cpp
if (n_loop != 0 && PS_value>=alpha && dt_PS_signal>=beta){   // Detect the heartbeat with the presented Criteria
C_1&2
    delta_t_HB=(double)n_loop*(double)delta_t_loop;          // If heartbeat detected, calculate time between heart-
beat

    for(int i=0;i<n_samples-1;i++){                          // Push data through array, until second last element
        delta_t_HB_Arr[i]=delta_t_HB_Arr[i+1];
    }

    if(delta_t_HB>350){                                      // Little bit of cosmetics to the calculation of the mean-
value
        delta_t_HB_Arr[n_samples-1]=delta_t_HB;              // If time between two heartbeats is long enough, take it
for the calculation
    }
    else{delta_t_HB_Arr[n_samples-1]=delta_t_HB_Arr[n_samples-2];} // If time is too short, take the last plausible one

    sum=0.0;                                                 // Sum for calculation mean-value
    for(int i=0;i<=(int)n_samples-1;i++){                    // Add up the whole array
        sum=sum+delta_t_HB_Arr[i];
    }

    avgBPM=1.0*n_samples*millimin/sum;                       // Calculate avgBPM
    n_loop=0;                                                // Set n_loop to 0 again
}
else {                                                       // Count up number of loops to calculate time between heartbeats
    n_loop++;
}

DataPkg.PS_value = PS_value;                                 // Construct DataPkg
DataPkg.avgBPM = avgBPM;
esp_now_send(NULL, (uint8_t *) &DataPkg, sizeof(DataPkg)); // Send DataPkg

delay(delta_t_loop);
}
```

Appendix B. Code for the Receiver Unit

```
/* Receiver-Unit */
#include <Arduino.h>
#include <WiFi.h>
#include <esp_now.h>
#include <TFT_eSPI.h>
#include <SPI.h>
TFT_eSPI tft = TFT_eSPI();

typedef struct message
{
   int PS_Value;                                    // S(t)
   double avgBPM;                                   // BPM_mean
};
struct message DataPkg;

int PS_Value;
double avgBPM;

const int displayLength = 160;                      // Length of Display in Pixel
const int displayHeight = 128;                      // Height of Display in Pixel
int graph[displayLength];                           // Storage for Plotting S(t)

void onDataReceiver(const uint8_t * mac, const uint8_t *incomingData, int len) // Do this {...}, when receiving DataPkg
{
  memcpy(&DataPkg, incomingData, sizeof(DataPkg));

  PS_Value=( DataPkg.PS_Value);                     // Assign the Elements of the package to the variables
  avgBPM=(DataPkg.avgBPM);

for(int i=0;i<displayLength-1;i++){                 // "Push the Data through the Array"
    graph[i]=graph[i+1];                            // First element gets value from second, second from third, ...
}                                                   // until second last element,
graph[displayLength-1]=map(PS_Value,500,650,128,0); // because last one gets the value mapped from DataPkg / S(t)

tft.fillScreen(TFT_WHITE);                          // Reset screen to white
tft.drawString("avgBPM:", 5, 5, 2);                 // Draw BPM_mean string on display
tft.drawNumber(avgBPM, 70, 5, 2);                   // Draw BPM_mean value on display

for(int i=0;i<displayLength-1;i++){                 // Draw black line from one coordinate to another
    tft.drawLine(i,graph[i],i+1,graph[i+1], TFT_BLACK); // Coordinates are the single S(t)-values which are mapped to
  }                                                 // the pixel range of the Display and stored in graph[i]
}

void setup() {
Serial.begin(115200);
WiFi.mode(WIFI_STA);
// Get Mac Add
Serial.print("Mac Address: ");
Serial.print(WiFi.macAddress());
Serial.println("\nESP-Now Receiver");

if (esp_now_init() != 0) {                          // Initializing ESP-NOW
    Serial.println("Problem during ESP-NOW init");
```

```
    return;
  }

tft.setTextColor(TFT_BLACK, TFT_WHITE);
tft.init();
tft.setRotation(1);
tft.fillScreen(TFT_WHITE);

  for(int i=0;i<displayLength;i++){            // Initialize graph[i] as a straight line in the display middle
     graph[i]=64;
  }

  for(int i=0;i<displayLength;i++){            // Draw line
     tft.drawPixel(i,graph[i], TFT_BLACK);
  }
  tft.drawString("avgBPM: No Signal!", 5, 5, 2);
  esp_now_register_recv_cb(onDataReceiver);
}

  void loop() {
  }
```

References

1. Ritchie, H.; Roser, M. Causes of Death. Our World in Data. 2018. Available online: https://ourworldindata.org/causes-of-death (accessed on 7 January 2022).
2. Ahmad, F.B.; Anderson, R.N. The leading causes of death in the US for 2020. *JAMA* **2021**, *325*, 1829–1830. [CrossRef] [PubMed]
3. Aumann, S.; Trummer, S.; Brücken, A.; Ehrmann, A.; Büsgen, A. Conceptual design of a sensory shirt for fire-fighters. *Text. Res. J.* **2014**, *84*, 1661–1665. [CrossRef]
4. Trummer, S.; Ehrmann, A.; Büsgen, A. Development of underwear with integrated 12 channel ECG for men and women. *AUTEX Res. J.* **2017**, *17*, 344–349. [CrossRef]
5. An, X.; Stylios, G.K. A hybrid textile electrode for electrocardiogram (ECG) measurement and motion tracking. *Materials* **2018**, *11*, 1887. [CrossRef] [PubMed]
6. Matsouka, D.; Vassiliadis, S.; Tao, X.; Koncar, V.; Bahadir, S.K.; Kalaoglu, F.; Jevsnik, S. Electrical connection issues on wearable electronics. *IOP Conf. Ser. Mater. Sci. Eng.* **2018**, *459*, 012017. [CrossRef]
7. Acar, G.; Ozturk, O.; Golparvar, A.J.; Elboshra, T.A.; Böhringer, K.; Yapici, M.K. Wearable and Flexible Textile Electrodes for Biopotential Signal Monitoring: A review. *Electronics* **2019**, *8*, 479. [CrossRef]
8. Nigusse, A.B.; Malengier, B.; Mengistie, D.A.; Tseghai, G.B.; van Langenhove, L. Development of Washable Silver Printed Textile Electrodes for Long-Term ECG Monitoring. *Sensors* **2020**, *20*, 6233. [CrossRef]
9. Euler, L.; Guo, L.; Persson, N.-K. Textile Electrodes: Influence of Knitting Construction and Pressure on the Contact Impedance. *Sensors* **2021**, *21*, 1578. [CrossRef]
10. Uz Zaman, S.; Tao, X.Y.; Cochrane, C.; Koncar, V. Smart E-Textile Systems: A Review for Healthcare Applications. *Electronics* **2022**, *11*, 99. [CrossRef]
11. Wang, J.Y.; Liu, K.W.; Sun, Q.Z.; Ni, X.L.; Ai, F.; Wang, S.M.; Yan, Z.J.; Liu, D.M. Diaphragm-based optical fiber sensor for pulse wave monitoring and cardiovascular diseases diagnosis. *J. Biophotonics* **2019**, *12*, e201900084. [CrossRef]
12. Tanima; Saini, I.; Saini, B.S. Physiological Characteristics Classification by Optical Pulse Sensor using Arterial Pulse Waves. In Proceedings of the 2020 2nd International Conference on Innovative Mechanisms for Industry Applications (ICIMIA), Bangalore, India, 5–7 March 2020; pp. 676–679.
13. Singh, M.; Li, J.K.; Sigel, G.H.; Amory, D. Fiber optic pulse sensor for noninvasive cardiovascular applications. In Proceedings of the Sixteenth Annual Northeast Conference on Bioengineering, State College, PA, USA, 26–27 March 1990; pp. 103–104.
14. Henning, M.R.; Gerdt, D.W.; Spraggins, T.A. Using a fiber-optic pulse sensor in magnetic resonance imaging. *Proc. SPIE* **1991**, *1420*, 34–40. [CrossRef]
15. Alian, A.A.; Sheeley, K.H. Photoplethysmography. *Best Pract. Res. Clin. Anaesthesiol.* **2014**, *28*, 395–406. [CrossRef] [PubMed]
16. Mark, J.B. *Atlas of Cardiovascular Monitoring*; Churchill Livingstone Inc.: New York, NY, USA, 1998.
17. Milner, Q.J.W.; Mathews, G.R. An assessment of the accuracy of pulse oximeters. *Anaesthesia* **2012**, *67*, 396–401. [CrossRef]
18. Kim, J.H.; Kim, N.Y.; Kwon, M.J.; Lee, J.H. Attachable Pulse Sensors Integrated with Inorganic Optoelectronic Devices for Monitoring Heart Rates at Various Body Locations. *ACS Appl. Mater. Interfaces* **2017**, *9*, 25700–25705. [CrossRef]

19. Bachmann, A.; Klebsattel, C.; Schankin, A.; Riedel, T.; Beigl, M.; Reichert, M.; Santangelo, P.; Ebner-Priemer, U. Leveraging smartwatches for unobtrusive mobile ambulatory mood assessment. In *UbiComp/ISWC'15 Adjunct, Proceedings of the 2015 ACM International Joint Conference on Pervasive and Ubiquitous Computing, Osaka, Japan, 7–11 September 2015*; ACM: New York, NY, USA, 2015.
20. Lee, Y.J.; Shin, H.G.; Choi, H.J.; Kim, C.S. Can pulse check by the photoplethysmography sensor on a smart watch replace carotid artery palpation during cardiopulmonary resuscitation in cardiac arrest patients? A prospective observational diagnostic accuracy study. *BMJ Open* **2019**, *9*, e023627. [CrossRef] [PubMed]
21. Ehrmann, G.; Blachowicz, T.; Homburg, S.V.; Ehrmann, A. Measuring biosignals with single circuit boards. *Bioengineering* **2022**, *9*, 84. [CrossRef] [PubMed]
22. Ehrmann, G.; Ehrmann, A. Suitability of common single circuit boards for sensing and actuating in smart textiles. *Commun. Dev. Assem. Text. Prod.* **2020**, *1*, 170–179. [CrossRef]
23. Oldfrey, B.; Jackson, R.; Smitham, P.; Miodownik, M. A deep learning approach to non-linearity in wearable stretch sensors. *Front. Robot. AI* **2019**, *6*, 27. [CrossRef]
24. Nuramdhani, I.; Jose, M.; Samyn, P.; Adriaensens, P.; Malengier, B.; Deferme, W.; de Mey, G.; van Langenhove, L. Charge-discharge characteristics of textile energy storage devices having different PEDOT:PSS ratios and conductive yarns configuration. *Polymers* **2019**, *11*, 345. [CrossRef]
25. Li, E.; Lin, X.Y.; Seet, B.-C.; Joseph, F.; Neville, J. Low profile and low cost textile smart mat for step pressure sensing and position mapping. In Proceedings of the IEEE Instrumentation and Measurement Technology Conference 2019, Auckland, New Zealand, 20–23 May 2019; pp. 1564–1568.
26. Anbalgan, A.; Sundarsingh, E.F.; Ramalingam, V.S. Design and experimental evaluation of a novel on-body textile antenna for unicast applications. *Microw. Opt. Technol. Lett.* **2020**, *62*, 789–799. [CrossRef]
27. Pulsesensor.com. Available online: https://pulsesensor.com/ (accessed on 25 February 2023).
28. Tuvshinbayar, K.; Ehrmann, G.; Ehrmann, A. 50/60 Hz Power Grid Noise as a Skin Contact Measure of Textile ECG Electrodes. *Textiles* **2022**, *2*, 265–274. [CrossRef]

Article

Lidom: A Disease Risk Prediction Model Based on LightGBM Applied to Nursing Homes

Feng Zhou [1], Shijing Hu [1], Xin Du [1,*], Xiaoli Wan [2,*], Zhihui Lu [1] and Jie Wu [1]

[1] School of Computer Science, Fudan University, Shanghai 200438, China
[2] Information Center, Zhejiang International Business Group Hangzhou, Hangzhou 310003, China
* Correspondence: jsjduxin@163.com (X.D.); wanxl@zibchina.com (X.W.)

Abstract: With the innovation of technologies such as sensors and artificial intelligence, some nursing homes use wearable devices to monitor the movement and physiological indicators of the elderly and provide prompts for any health risks. Nevertheless, this kind of risk warning is a decision based on a particular physiological indicator. Therefore, such decisions cannot effectively predict health risks. To achieve this goal, we propose a model Lidom (A LightGBM-based Disease Prediction Model) based on the combination of the LightGBM algorithm, InterpretML framework, and sequence confrontation network (SeqGAN). The Lidom model first solves the problem of uneven samples based on the sequence confrontation network (SeqGAN), then trains the model based on the LightGBM algorithm, uses the InterpretML framework for analysis, and finally obtains the best model. This paper uses the public dataset MIMIC-III, subject data, and the early diabetes risk prediction dataset in UCI as sample data. The experimental results show that the Lidom model has an accuracy rate of 93.46% for disease risk prediction and an accuracy rate of 99.8% for early diabetes risk prediction. The results show that the Lidom model can provide adequate support for the prediction of the health risks of the elderly.

Keywords: wearable sensors; machine learning; health monitoring; disease risk prediction; Internet of Things; big data

Citation: Zhou, F.; Hu, S.; Du, X.; Wan, X.; Lu, Z.; Wu, J. Lidom: A Disease Risk Prediction Model Based on LightGBM Applied to Nursing Homes. *Electronics* **2023**, *12*, 1009. https://doi.org/10.3390/electronics12041009

Academic Editor: Gabriella Olmo

Received: 21 December 2022
Revised: 31 January 2023
Accepted: 13 February 2023
Published: 17 February 2023

1. Introduction

The rapid development of sensors, the Internet of Things, and artificial intelligence have made earth-shaking changes in people's lives, and realizes the usage of vital signs and physiological indicators. The ageing of the population makes people pursue better health, and at the same time, the demand for real-time health monitoring and early warning is increasing, which has motivated the emergence of wearable health monitoring and testing equipment, for example, wristbands, blood four index monitors, blood oxygen meters [1], etc. Nursing homes use these wearable health monitoring and testing equipment to monitor and detect individuals' daily exercise, sleep quality, and physiological indicators in real time, such as steps, blood oxygen, blood pressure, heart rate [2], blood sugar, sweat, tears, etc.

At present, nursing homes can only provide early warnings for risk factors that affect health based on individual values obtained from the monitoring and testing of these wearable devices. For example, a health risk warning will be given when the detected diastolic blood pressure is more significant than 90 mmHg or the systolic blood pressure is greater than 140 mmHg, when the detected total cholesterol is greater than 6.2 mmol/L or triglyceride is greater than 2.3 mmol/L, when the detected fasting blood sugar is higher than 7.0 mmol/L, when the detected low-density lipoprotein is greater than 3.12 mmol/L, etc. However, these early warnings are single decision-making outputs based on a single indicator, not health risk predictions, after a comprehensive analysis of the obtained physical symptoms and physiological values [3]. Therefore, this single decision making lacks comprehensiveness in the early warning of health risks.

In order to realize comprehensive and practical health warnings for the elderly in nursing homes, we first build electronic health records for the elderly; secondly, we increase the equipment for physiological index monitoring and detection to collect more comprehensive physiological index data; and then use the public dataset based on American intensive care, the medical information database MIMIC-III (Medical Information Mart for Intensive Care-III), and subject data and the early diabetes risk prediction dataset in the UCI machine learning database of the University of California, Irvine, are used as sample data to train the disease risk prediction model Lidom (A LightGBM-based Disease Prediction Model); finally, we use the Lidom model to make a comprehensive health risk prediction based on the physiological values obtained from the monitoring and detection of wearable health devices. Its application scenario is shown in Figure 1.

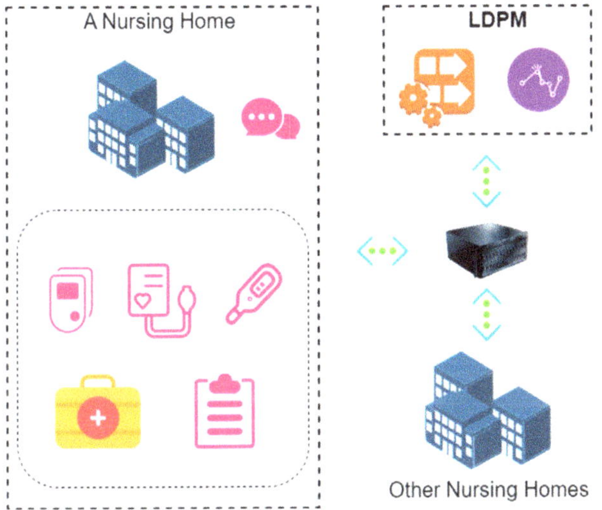

Figure 1. Application Scenarios of Disease Risk Prediction.

In Figure 1, a nursing home with informatization and intelligence capabilities is used to illustrate the application scenario of disease risk prediction. Nursing homes use IoT devices to conduct health checks on the physiological indicators of the elderly. After the health inspection, the electronic health records and physiological index values are sent to the background server. After the server completes the data reception, the Lidom model predicts the disease risk of the received data and pushes the predicted results to the relevant terminals of the nursing home in real time.

The goal of this research is that the disease risk prediction model, Lidom, can provide a comprehensive and effective early warning of the health of the elderly in nursing homes. In this paper, the public dataset MIMIC-III1.4 (V1.4), the subject data, and UCI's early diabetes risk prediction dataset were used as sample data for model training and evaluation. The MIMIC-III1.4 (V1.4) dataset includes vital signs, scores, laboratory tests, imaging reports, length of hospital stay, and survival data. The investigators of this study completed the "Protecting Human Research Participants" course on the website of the National Institutes of Health NIH (National Institutes of Health) and obtained the right to use the data (certificate number: 48686713). The data attributes in the early diabetes risk prediction dataset include age, sex, polyuria, polydipsia, sudden weight loss, weakness, polyphagia, genital thrush, visual blurring, itching, irritability, delayed healing, partial paresis, muscle stiffness, alopecia, obesity and class.

In order to achieve a comprehensive and effective early warning of the health of the elderly in nursing homes, we first extracted patient data, disease data, and outpatient examination data from the dataset MIMIC-III and combined them with the physiological data of the elderly in nursing homes and the physiological data of young, healthy subjects. We trained a sample generation model based on the sequence confrontation network (SeqGAN) with mixed sample data. Then, based on LightGBM (Light Gradient Boosting Machine), we trained the classification model for sample feature selection of the disease risk prediction model. Secondly, we used the machine learning interpretable framework InterpretML to analyze each feature pair. The contribution value of the prediction results was analyzed, and the sample characteristics were adjusted according to the analysis results; then the optimized samples were used to complete the training of the disease risk prediction model based on the LightGBM algorithm, and the parameters of the model were set based on the grid search method (GridSearchCV). Finally, we realized the comprehensive and effective early warning of the health of the elderly in nursing homes. In this paper, to further improve the accuracy of disease risk prediction, we used the early diabetes risk prediction dataset in the UCI machine learning database as sample data to predict the risk of early diabetes and achieved an accuracy rate of 99.8%.

The main contribution of this paper is as follows:

- We propose a sample generation model for disease risk prediction based on sequence adversarial networks (SeqGAN) to solve the problem of uneven samples.
- We propose a sample feature impact analysis model based on the InterpretML framework and the LightGBM algorithm to analyse the contribution value of the prediction result.
- We propose a disease risk prediction model based on the LightGBM algorithm to obtain the best prediction for disease risk.

There are six sections in this article. Section 1 introduces the current business status of disease risk prediction. The Section 2 introduces the relevant status of disease risk prediction. Section 3 describes the design of the disease risk prediction model. The Section 4 elaborates on the realization process of the disease risk prediction model. The Section 5 compares and analyzes the experimental results. The Section 6 summarizes the research results of this paper and looks forward from the results.

2. Related Works

Many studies have been devoted to disease prediction based on machine learning in recent years. Machine learning has also produced good results in predicting diseases and more clinical applications. Green et al. proposed a prediction model for acute coronary syndrome based on regression and neural networks [4]. Experiments showed that the model based on a neural network achieved better prediction results. Das et al. proposed a complex neural network model that uses the neural network model as the basic model and combines multiple neural network models to predict valvular heart disease [5]. Atkov et al. proposed a prediction model for coronary heart disease based on a neural network [6]. In constructing the model, features such as genetic factors in the disease were added to the training. Experiments showed that after adding more information, the prediction effect improved greatly.

Hatim Guermah et al. proposed an ontology-based architecture using linear SVM to predict chronic diseases [7], such as chronic renal failure. Experimental results show that the accuracy of prediction reaches 93.3%. Ching-seh (Mike) Wu et al. found through experiments that when constructing a model for predicting the survival rate of heart disease [8], using the random forest as the base classifier achieved a higher accuracy than using decision trees, logistic regression, and the Bayesian method. In order to promote the clinical application of the disease prediction model [9], Shao Rongqiang et al. proposed to improve the interpretability of the model, simplify the model training and call process, use business parameters to allow doctors to participate in the training, and improve the experience of the system to realize the clinical application of the disease prediction model. Quick application of the Artificial Immune System (AIS) achieved 98.08% accuracy on the breast cancer dataset and 70% accuracy on the heart disease dataset [10].

Graph convolutional neural networks are also beginning to be used in disease prediction [11], but using graphs to represent real-life applications will complicate the design of graph convolutional neural networks. Building graph convolutional neural networks on large-scale graphs is also an enormous challenge. Electronic health data containing semantic information such as demographics, clinical diagnoses and measurements, health behaviors, laboratory results, prescriptions, and care provide convenience for disease prediction models. Yichen He et al. used the Gibbs sampling method and the ARIMA model to construct a parametric–nonparametric forecasting model [12]. By analyzing the time series, infectious diseases were predicted. The model's predicted value showed that it was consistent with the actual results.

Use of the Adaptive Neural Fuzzy Inference System (ANFIS) in the field of heart disease prediction, can draw more apparent conclusions based on relevant information [13]. Aiming at the problems of low data trust and uncontrollable data sharing in telemedicine, Meiquan Wang et al. proposed a blockchain-based secure medical data management and a neural-network-based multi-label disease prediction model [14]. However, due to the low recall rate of some disease predictions, it cannot be used in clinical applications. In order to predict the number of potential patients with a specific disease at a particular time and place in the future, Junyi Gao et al. proposed a real-time population-level disease prediction model PopNet (real-time population-level disease prediction with data latency) [15], which realized a 47% root mean square error and 24% mean absolute error.

Dahiwade et al. proposed a general disease prediction based on patient symptoms and used K-nearest neighbor (KNN) and convolutional neural network (CNN) machine learning algorithms to predict the disease [16]. The experimental results showed that CNN's accuracy in predicting prevalent diseases reached 84.5%. Ambesange et al. used the Indian Liver Patient Dataset (ILPD) to build a disease prediction model based on the random forest (RF) algorithm [17], and the model prediction accuracy reached 100%. Zhao et al. used the liver function records of 573 patients and proposed the W-LR-XGB algorithm for the prediction model of liver disease patients [18]. The experimental results show that the prediction model established by the W-LR-XGB algorithm has an accuracy rate of 83% and a recall rate of 100% and 82%, and the F1 score reached 81%.

Islam et al. used a hybrid genetic algorithm (HGA) for data clustering when predicting early heart disease [19]. The experimental results showed that the prediction accuracy reached 94.06%. Nguyen et al. proposed a method based on known lncRNA-disease associations and known disease-miRNA associations and verified lncRNA-miRNA interactions to construct a lncRNA-disease-miRNA tripartite in order to find the potential function of lncRNAs in the disease [20]. The figure was used to predict the lncRNA-disease association, and the prediction accuracy reached 98.4%. Thummala et al. compared random forest and KNN in terms of heart disease prediction [21]. The experimental results show that the performance of the random forest algorithm is significantly better than KNN.

Chen et al. developed a heart disease prediction system (HDPS) based on neural networks [22], which can input clinical data, display ROC curves, and display prediction performance (execution time, accuracy, sensitivity, specificity, and prediction results). Kumari et al. tried seven machine learning algorithms to predict heart disease [23]. They tried ensemble methods such as AdaBoost and voting ensemble methods to improve the accuracy of poorly performing algorithms. Rubaiyat et al. built a disease prediction model based on three machine learning algorithms: logistic regression [24], random forest classifier, and KNN, and used a dataset of 310 orthopedic disease patients for training. The experimental results show that the prediction accuracy rate of the prediction model based on the random forest algorithm reaches 89%. Ambekar et al. proposed using structured data for heart disease prediction and built a model based on convolutional neural networks [25], and achieved an accuracy rate of more than 65%.

Reddy et al. built a heart disease prediction model based on dynamic KNN and SVM, respectively [26]. The accuracy of the prediction model based on dynamic KNN was 84.44%, and the accuracy of the prediction model based on SVM was 67.21%. Xue et al. established a classification prediction model by mining the physical index data of heart disease patients and optimizing the optimal parameters with particle swarm optimization [27]. The accuracy rate was 1.33% higher than SVM, and the prediction efficiency was also improved. Li et al. proposed a novel prediction model LRWHLDA based on local random walks to predict potential associations between human lncRNAs and diseases [28].

In order to predict the severity of heart disease, Yuan et al. proposed a prediction model based on machine learning [29], which can realize the prediction of two-class and multi-class heart disease at the same time. Experimental results show the model's accuracy and stability in binary and multi-class predictions. Bhoyar et al. proposed a prediction model based on a multi-layer perceptron (MLP) neural network [30], which achieved an accuracy of 87.30% in the cardiovascular disease dataset. Latha et al. proposed a heart disease prediction model based on a partially observable Markov decision process (POMDP) [31]. In an emergency, doctors can send alerts to patients through fog computing.

In order to improve the accuracy of predicting heart disease, Yadav et al. used genetic algorithms to optimize the characteristics of the dataset and built prediction models based on SVM, KNN, naive Bayes, and random forest, among which the naïve-Bayes-based prediction model achieved a prediction accuracy of 96% [32].

It can be seen from Table 1 that the existing disease prediction research has both single disease risk and comprehensive disease risk prediction. It can be seen from the above paragraphs that in the field of disease prediction, the accuracy rate of chronic renal failure prediction has reached 93.3%, the accuracy rate of liver disease prediction has reached 100%, the accuracy rate of heart disease prediction has reached 94.06%, and the accuracy rate of cardiovascular disease prediction has reached 93.3%. The accuracy rate reached 87.30%. However, the basic algorithms mainly used are random forest, CNN, KNN, SVM, etc. When the random forest algorithm has many decision trees, the time and space required for training is more prominent, and the interpretability of the algorithm itself is not high; the interpretability of the features extracted by the convolutional neural network CNN (convolutional neural network) model is poor; the K-nearest neighbor algorithm KNN (K-nearest neighbors) and the support vector machine SVM (support vector machine) algorithms have a low operating efficiency when the sample size is large. In summary, the existing research cannot meet the current demand for disease risk prediction in nursing homes, and the Lidom model we proposed can be well applied to the current application scenario of disease risk prediction in nursing homes.

Table 1. Related Research Statistics.

Researcher	Name	Single Disease Risk Prediction	Integrated Disease Risk Prediction	Algorithm
Green [4]	Coronary Syndrome	√		Regression Models, Neural Networks
Das [5] Ching-seh [8] Islam [19] Thummala [21] Kumari [23] Ambekar [25] Reddy [26] Xue [27] Yuan [29] Latha [31] Yadav [32]	Heart Disease	√		Neural Networks, Random Forests, HGA, KNN, AdaBoost, CNN, SVM, Markov Decision, Naive Bayes
Atkov [6]	Coronary Heart Disease	√		Neural Networks
Hatim [7]	Chronic Renal Failure	√		SVM
Junyi [15]	Population Disease Prediction		√	Attention network, Neural network
Dahiwade [16]	Early Disease Prediction		√	KNN, CNN
Ambesange [17] Zhao [18]	Liver Disease		√	Random Forest, XGBoost
Rubaiyat [24]	Orthopedic Disease		√	Logistic Regression, Random Forest Classifier, KNN
Bhoyar [30]	Cardiovascular Diseases		√	Neural Networks

3. Model Design

This chapter introduces the sequence confrontation network (SeqGAN), LightGBM (Light Gradient Boosting Machine) algorithm, and InterpretML, an interpretable framework for machine learning. Finally, we describe training a disease risk prediction model based on the LightGBM algorithm.

3.1. Sequential Adversarial Network

The generative adversarial network (GAN) proposed by Ian Goodfellow et al. has achieved good results in data generation [33]. However, certain limitations exist when generating discrete sequences, such as text data. Mainly because of the discrete sequence generated by the GAN generation model, the gradient update cannot be transferred from the discriminative model to the generative model. To solve this problem, Lantao Yu et al. proposed the sequence adversarial network (SeqGAN) framework [34], which models the generator as a stochastic reinforcement learning (RL) policy using gradient policy updates to bypass the generator differentiation problem. The reward of reinforcement learning is based on the judgment of the GAN discriminant model on the complete sequence.

As shown in Figure 2, the existing dot is State [34], and the dot will be generated in action. Since the discriminant model D needs to judge a complete sequence, use MCTS (Monte Carlo Search) to complete the multiple possibilities of each action. After the discriminant model D generates rewards for these complete sequences, update this way to the generative model G using reinforcement learning.

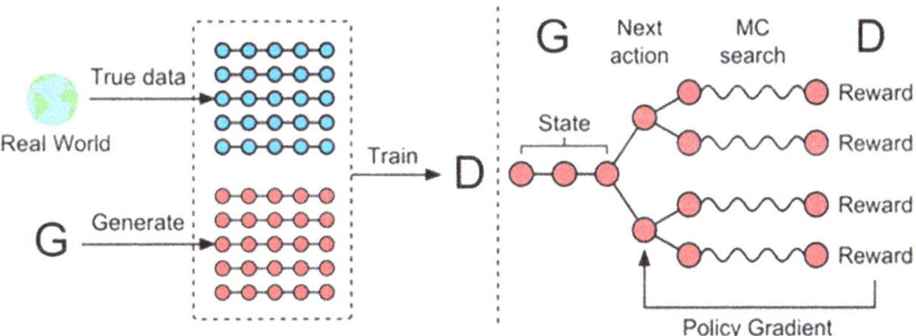

Figure 2. The illustration of SeqGAN [34].

3.2. Feature Classification Base Model

The residual value output from the previous iteration is used as the input value of the next iteration and is the principle of GBDT (gradient boosting decision tree). Since the input value of each round depends on the output value of the previous round, the training samples will be traversed all the time during the training process, increasing the training time and storage space. The XGBoost algorithm proposed by Chen et al. is an integrated algorithm with high controllability and high efficiency [35]. Its basic principle is to be used pre-sorted. The method mines the optimal split point of the feature. However, due to the need to store eigenvalues and sorting results, it is also necessary to calculate the split gain value when traversing split points. That will also increase the training time and storage space.

The basic model used in this paper when constructing the feature classification model is the LightGBM algorithm proposed by Ken et al. [36]. Methods such as growth strategy, GOSS (Gradient-based One-Side Sampling), and EFB (Exclusive Feature Bundling) have effectively solved the problem of massive data processing.

As shown in Figure 3, the flow of the histogram algorithm is first to discretize the floating-point eigenvalues into integers and construct a histogram; secondly, the index, when traversing the samples, is a discrete value. To carry out cumulant statistics, the third step is to traverse the samples according to the discrete values of the histogram and search for the best split point [36].

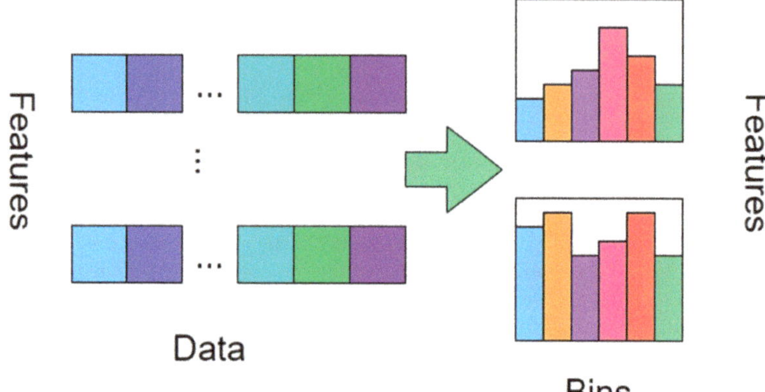

Figure 3. Histogram algorithm.

As shown in Figure 4, histogram difference acceleration means that calculating the histogram of a leaf node is obtained by making the difference between the histogram of the parent node and the histogram of the leaf node at the same level. Conventionally, the leaf nodes of the large histogram are obtained by calculating the leaf nodes of the small histogram. This operation method improves training efficiency.

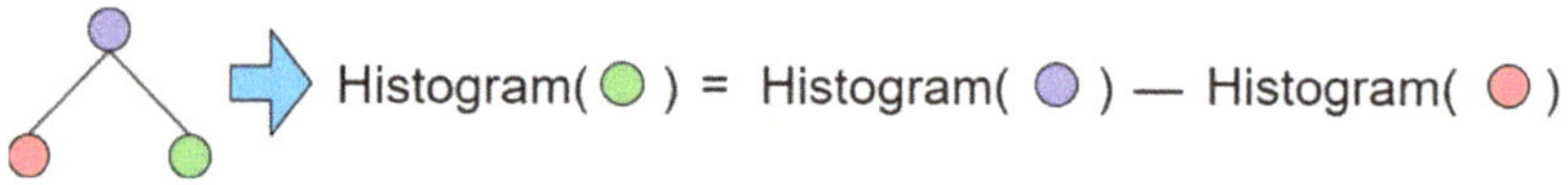

Figure 4. Histogram difference acceleration.

As shown in Figure 5, since the leaf-wise tree growth strategy is to split from the leaf node with the most significant split gain when traversing the leaf nodes, the error of the leaf-wise tree growth strategy is small. However, too many splits are prone to overfitting. Therefore, when the LightGBM algorithm uses the leaf-wise tree growth strategy, it avoids overfitting by increasing the upper limit of the number of splits.

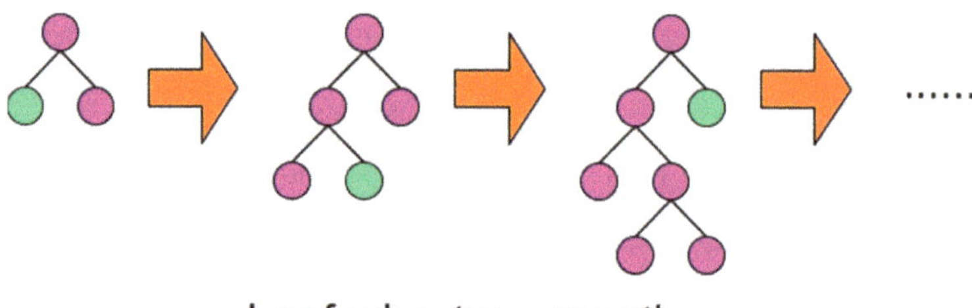

Figure 5. Tree structure of leaf node increase method.

The GOSS (Gradient-based One-Side Sampling) algorithm reduces the sample size by deleting small gradient samples and using the remaining samples to calculate the gain. Suppose $\{x_1, \ldots, x_n\}$ represents the sample, x_i represents the s-dimensional vector of X^s space, and $\{g_1, \ldots, g_n\}$ represents the gradient of the loss function, in the gradient boosting decision tree algorithm GBDT, the split gain calculation of the training set O feature j at point d is shown in Formula (1) [36].

$$V_{j|O}(d) = \frac{1}{n_O}\left(\frac{\left(\sum_{\{x_i \in O: x_{ij} \le d\}} g_i \right)^2}{n^j_{l|O}(d)} + \frac{\left(\sum_{\{x_i \in O: x_{ij} > d\}} g_i \right)^2}{n^j_{r|O}(d)} \right) \tag{1}$$

In Formula (1), $n_O = \sum I[x_i \in O]$, $n^j_{l|O}(d) = \sum I[x_i \in O : x_{ij} \le d]$, $n^j_{r|O}(d) = \sum I[x_i \in O : x_{ij} > d]$. In the GOSS (Gradient-based One-Side Sampling) algorithm, the absolute value of the gradient of the samples is first arranged in descending order. Then the large gradient sample set A and the small gradient sample set A^C is obtained, and a subset of $b \times |A^C|$ size is randomly extracted B. Finally, split on the variance gain $\widetilde{V}_j(d)$ is calculated by $A \cup B$. Specifically, it is shown in Formula (2) below [36].

$$\widetilde{V}_j(d) = \frac{1}{n}\left(\frac{\left(\sum_{x_i \in A_l} g_i + \frac{1-a}{n} \sum_{x_i \in B_l} g_i \right)^2}{n^j_l(d)} + \frac{\left(\sum_{x_i \in A_r} g_i + \frac{1-a}{n} \sum_{x_i \in B_r} g_i \right)^2}{n^j_r(d)} \right) \tag{2}$$

In Formula (2), $A_l = \{x_i \in A : x_{ij} \le d\}$, $A_r = \{x_i \in A : x_{ij} > d\}$, $B_l = \{x_i \in B : x_{ij} \le d\}$, $B_r = \{x_i \in B : x_{ij} > d\}$.

The essential condition of the mutually exclusive feature merging EFB (Exclusive Feature Bundling) algorithm is the sparsity of the feature space. Ken et al. used the feature scanning algorithm to find the features with the same feature histogram and then added a bias constant to the feature value to combine these features. They were bundling by reducing the feature dimension in such a way without compromising accuracy.

However, in practical applications, we found that when building an application model based on the LightGBM algorithm, we could not obtain the influence of each feature in the sample on the prediction result. If the model misjudges some sample characteristics, it will have a particular impact on the effect of the model. Therefore, if we can obtain the contribution value of each feature to the prediction result, it is of great significance for analyzing the model's effect and the sample's characteristics. The Lidom model proposed in this paper uses the LightGBM algorithm as the base model to build a sample feature classification model. When analyzing the impact of features on the prediction results, use the machine learning interpretable framework InterpretML to interpret the contribution value of each feature.

3.3. Explainable Framework for Machine Learning

In order to gain insight into the impact of each feature value on the prediction results, we used the InterpretML framework proposed by Harsha Nori et al. to analyze the behavior of the model we built and the logic of individual predictions, such as model debugging, feature engineering, fairness detection, human–machine interaction, compliance, etc. [37].

The Interpretable Enhancement Machine (EBM) algorithm proposed by the InterpretML framework is a generalized additive model (GAM), and the form of the algorithm is shown in Formula (3) [37].

$$g(E[\mathcal{Y}]) = \beta_0 + \sum f_j(x_j) \tag{3}$$

In Equation (3), g is the function that adapts the generalized additive model (GAM) to different settings, and $f_j(x_j)$ is the feature function of sample j. Since the Interpretable Boosting Machine (EBM) algorithm is an additive model, the form of the algorithm to automatically detect the interaction term is shown in Equation (4) [37].

$$g(E[\mathcal{Y}]) = \beta_0 + \sum f_j(x_j) + \sum f_{i_j}(x_i, x_j) \tag{4}$$

The Interpretable Enhancement Machine (EBM) algorithm uses the form in Formula (4), which improves the accuracy of model interpretation while ensuring the clarity of model interpretation.

3.4. Lidom Model

The design process of the disease risk prediction Lidom model proposed in this paper is shown in Figure 6.

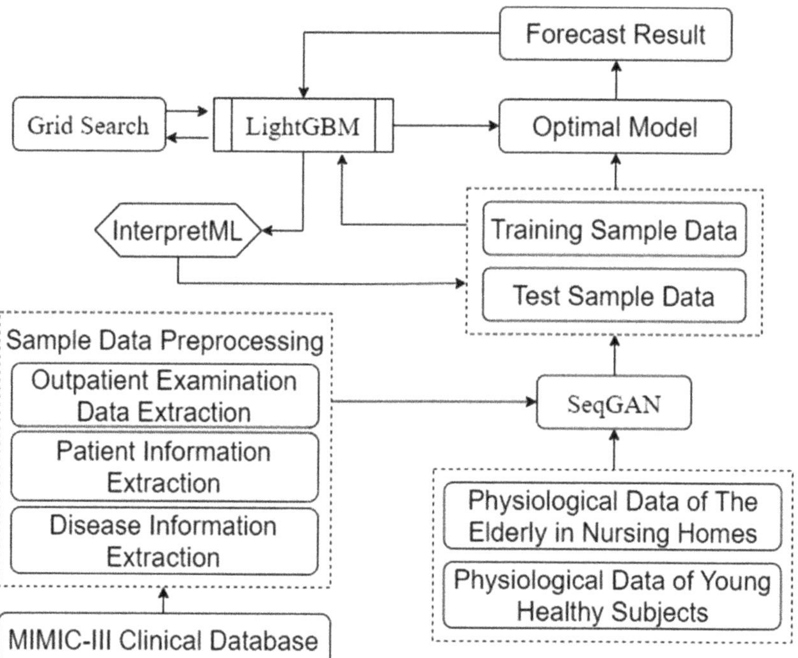

Figure 6. Disease risk prediction model process.

The detailed process of the design process of the disease risk prediction model in Figure 6 is as follows:

1. Preprocess the disease information, patient information, and outpatient examination data extracted from the MIMIC-III Clinical Database.
2. The mixed sample data comprise the preprocessed MIMIC-III data, the physiological data of the elderly in nursing homes, and the physiological data of young, healthy subjects.
3. Based on the sequence confrontation network (SeqGAN), the mixed sample data and the sample imbalance in the early diabetes risk prediction dataset are correlated. The training sample data and the test sample data are formed.
4. Using LightGBM as the base model, train the Lidom model, use the InterpretML framework to analyze the prediction results, and then adjust the features in the sample data according to the analysis results.
5. The parameters of the model settings are optimized using the grid search method (GridSearchCV).
6. Use the adjusted sample data to test the Lidom model, and adjust the relevant parameters of the Lidom model according to the test results.
7. Iterate repeatedly until the model accuracy rate reaches the preset threshold of 93%, the precision rate (Precision) reaches the preset threshold of 50%, and the recall rate (Recall) reaches the preset threshold of 45%. The false positive rate (False Positive Rate,

FPR) reaches the preset threshold of 100%, and the F1 value (F1-Measure) reaches the preset threshold of 45%.

The code flow of the Lidom model algorithm is shown in Algorithm 1.

Algorithm 1: Lidom model algorithm

Input: MIMIC-III data, physiological data of elderly in nursing homes, physiological data of young, healthy subjects.
Output: Lidom best model
S: training sample data
T: test sample data

1	Analysis of data in the MIMIC-III Clinical Database
2	According to the requirements of the Lidom model, the disease data, patient data, and outpatient examination data in the MIMIC-III Clinical Database are extracted and preprocessed.
3	The preprocessed MIMIC-III data, the physiological data of the elderly in nursing homes, and the physiological data of young, healthy subjects are used to form mixed sample data.
4	The sample data imbalance processing logic is realized based on SeqGAN.
5	Start sample data processing
6	Divide the sample data into five parts, four parts for training and one part for testing, to obtain training sample data S and test sample data T
7	Load Lidom model
8	Load the InterpretML framework
9	Interpreting the InterpretML Contribution Report
10	Update sample data S
11	for epochs do
12	Preset Lidom parameter values
13	Enable Lidom model
14	Enable parameter optimization model
15	Execute Lidom
16	According to the scoring results of relevant parameters in the Lidom model, update the relevant parameter values of the Lidom model
17	Use the test sample data T to test the accuracy rate, precision rate, recall rate, false positive rate, and F1 value
18	if accuracy $\geq$ *Acc* and precision $\geq$ *Pre* and recall $\geq$ *Rec* and false positive rate $\geq$ *FPR* and F1 value $\geq$ *F1*:
19	Output Lidom best model
20	End

In Algorithm 1, epochs represent the number of iterations to build the Lidom model. Acc represents the accuracy threshold, which is preset to 93.3% in this paper. Pre represents the accuracy threshold, which is preset to 50% in this article. Rec represents the recall rate threshold, which is preset to 45% in this article. FPR represents the false positive rate threshold, which is preset to 100% in this paper. F1 represents the F1 value threshold, which is preset as 45% in this paper.

3.5. Evaluation Metrics

In this paper, the evaluation indicators of the intelligent disease prediction model Lidom are accuracy, precision, recall, false positive rate, and $F1$ value.

$$\text{Accuracy} = \frac{TP + TN}{TP + TN + FP + FN} \times 100\% \tag{5}$$

$$\text{Precision} = \frac{TP}{TP + FP} \times 100\% \tag{6}$$

$$\text{Recall} = \frac{TP}{TP + FN} \times 100\% \tag{7}$$

$$FPR = \frac{FP}{FP + TN} \times 100\% \tag{8}$$

$$F1 = \frac{2TP}{2TP + FP + FN} \times 100\% \tag{9}$$

In the above formula, TP means true positive, TN means true negative, FP means false positive, and FN means false negative.

In this chapter, we train the Lidom model based on the sequence confrontation network (SeqGAN), LightGBM, and InterpretML frameworks. We describe the related algorithms, frameworks, model implementation processes, and evaluation indicators.

4. Model Realization

In this chapter, we mainly introduce the implementation process of the disease risk prediction model Lidom applied to nursing homes. We first analyzed the data in the MIMIC-III Clinical Database according to the requirements of the Lidom model [38], constructed training samples and test samples, and then trained the sample generation model based on the sequence confrontation network (SeqGAN); secondly, we trained the sample feature classification model based on the LightGBM algorithm, then used the InterpretML framework to analyze the contribution of the prediction results, and then updated the sample features according to the analysis results. Then we trained the Lidom model based on the LightGBM algorithm, and, finally, according to the accuracy (Accuracy), precision (Precision), and recall (Recall). The five evaluation index values of the false positive rate (FPR) and F1 measure (F1-Measure) output the best model parameters. The software platforms used in this experiment were: Ubuntu Server 18.04, Python 3.7.6, Keras 2.6.0, LightGBM 3.3.2, interpret 0.3.0, TensorFlow-GPU 2.6.0, XGBoost 1.6.1, Numpy 1.18.5, Pandas 1.3.5, and sklearn 1.1.

4.1. Sample Data

The sample data used in this article came from the public dataset MIMIC-III1.4(V1.4) (Medical Information Mart for Intensive Care-III1.4 (V1.4)), and consisted of terminology dictionary class, patient information, outpatient treatment, and ICU treatment; these four tables were composed, and the detailed information is shown in Table 2 [39].

Table 2. Dataset Data Distribution.

Term Dictionary	Patient Information	Outpatient Treatment	ICU Treatment
D_CPT	PATIENTS	CPTEVENTS	CAREGIVERS
D_ICD_DIAGNOSES	ADMISSIONS	DIAGNOSES_ICD	CHARTEVRNTS
D_ICD_PROCEDURES	CALLOUT	DRGCODES	DATETIMEEVENTS
D_ITEMS	ICUSTAYS	LABEVENTS	INPUTEVENTS
D_LABITEMS	TRANSFERS	MICROBIOLOGYEVENTS	INPUTEVENTS_MV
	SERVICES	PRESCRIPTIONS	NOTEEVENTS
			OUTPUTEVENTS
			PROCEDUREEVENTS_MV
			PROCEDURES_ICD

In Table 2, D_CPT is the glossary of medical services, D_ICD_DIAGNOSES is the dictionary table of ICD disease diagnosis, D_ICD_PROCEDURES is the dictionary table of ICD medical process, D_ITEMS is the dictionary table of ICU laboratory tests, D_LABITEMS is the dictionary table of outpatient laboratory tests, PATIENTS is the dictionary table of patients, ADMISSIONS is a table for recording patient hospitalization information, CALLOUT is a table for recording patient discharge information, ICUSTAYS is an ICU record table, TRANSFERS is a patient ward transfer table, SERVICES is a patient service table, CPTEVENTS is a medical service record table used by patients, DIAGNOSES_ICD is the diagnosis information table, DRGCODES is the diagnosis code table, LABEVENTS is the patient outpatient examination record table, MICROBIOLOGYEVENTS is the patient microbial detection record table, PRESCRIPTIONS is the prescription information table, CAREGIVERS is the medical staff information table, CHARTEVRNTS is the laboratory test record table, DATETIMEEVENTS is the date event table, INPUTEVENTS is the injection event table (CV), INPUTEVENTS_MV is the injection event table (MV), NOTEEVENTS is the text record event table, OUTPUTEVENTS is the patient excretion record table, PROCE-DUREEVENTS_MV is the patient medical process event table, and PROCEDURES_ICD is the ICD operation record [39].

After our analysis, the extraction of sample data required by the Lidom model in this paper was used to extract the patient number SUBJECT_ID INT, medical record number HADM_ID INT, preliminary DIAGNOSIS, and other data from the ADMISSIONS table, and extract the patient number SUBJECT_ID INT and medical record number HADM_ID from the LABEVENTS table INT, item identifier ITEMID, measurement VALUE, and other data, and extract item identifier ITEMID, item LABEL, and other data from the D_LABITEMS table.

In the LABEVENTS table, up to 726 items were examined for a single patient. In this paper, we extracted 21 kinds of inspection items: oxygen saturation, temperature, potassium, whole blood, sodium, alanine aminotransferase (ALT), alpha-fetoprotein, aspartate aminotransferase (AST), cholesterol, total cholesterol, HDL, LDL, measured creatinine, homocysteine, triglycerides, uric acid, platelet count, red blood cells, white blood cells, CD4, lymphs, bilirubin, and prostatic acid phosphatase.

There were 15,654 preliminary diagnoses recorded in the ADMISSIONS table. This paper extracted 56 such as coronary artery disease, bowel obstruction, etc.

4.2. Feature Preprocessing

The sample data in this paper use mixed sample data consisting of the MIMIC-III Clinical Database, physiological data of the elderly in nursing homes, and physiological data of young, healthy subjects. The preprocessing steps are as follows:

1. Uniformly complete the rows with missing attribute values and convert the text into values [40].
2. OrdinalEncoder encodes character features and classification labels [41].

4.3. Lidom Model Training and Prediction Results' Contribution to Value Analysis

In this chapter, we design and train the Lidom model based on LightGBM (Light Gradient Boosting Machine) and use the machine learning interpretable framework InterpretML to analyze the influence of features on the prediction results. The specific process is divided into several steps.

1. Construct Lidom and enable it.
2. Load the machine learning interpretable framework InterpretML.
3. Train and parse Glassbox models.
4. Train and parse the Blackbox model.
5. Interpret the contribution of each feature, and adjust the features.
6. Analyze and predict relevant evaluation indicators and optimize them.

4.4. Parameter Optimization

The method of 50-fold cross-validation is used to divide the sample data into five parts, where four parts are used for training, and one part is used for testing. After completing five experiments, the average value is calculated. This article uses the five-fold cross-validation method for training sample data. Before the model is loaded [42], we first set a range value for the basic parameters, parameters that affect accuracy, overfitting, and training speed, and then use the grid search method (GridSearchCV) to obtain the values of related parameters. For scoring, after iterative training, we obtain the optimal values of the relevant parameters; finally, the optimal model of the Lidom model is obtained.

In this chapter, we describe and explain the process of sampling data based on the MIMIC-III clinical database, the preprocessing of sample data features, Lidom model training and the interpretation of feature contribution values, and the process of Lidom parameter optimization.

5. Experimental Results and Analysis

In this chapter, we mainly introduce the prediction results of the disease risk prediction model Lidom applied to nursing homes, the contribution analysis of prediction results, model optimization, and early diabetes risk prediction.

5.1. Forecast Result

The artificial neural network (ANN), K-nearest neighbor classification (KNN), support vector machine (SVM), logistic regression (LR), distributed gradient boosting framework based on decision tree algorithm LightGBM (Light Gradient Boosting Machine) are standard classification prediction algorithms. We compared the prediction accuracies of the models built by different algorithms, as shown in Figure 7a. We made statistics based on the five indicators of the Lidom model, namely accuracy, precision, recall, false positive sate (FPR), and F1-measure, as shown in Figure 7b.

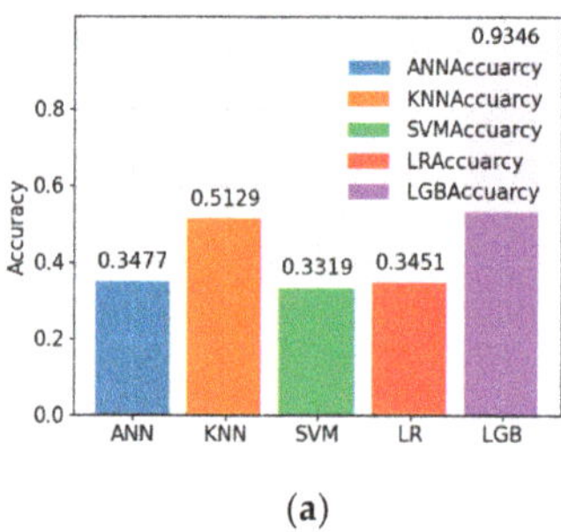
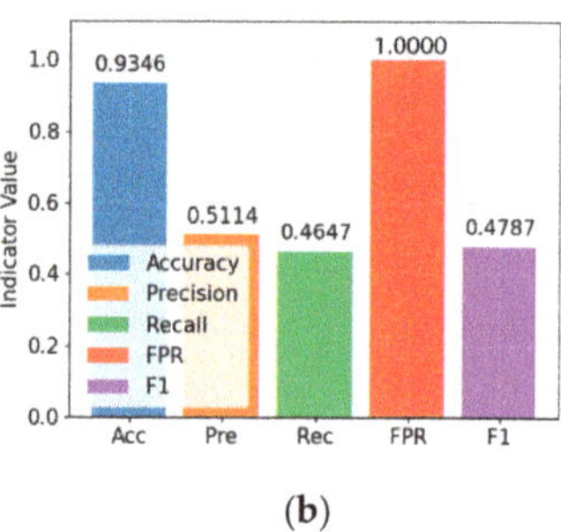

Figure 7. Training prediction results; (**a**): prediction accuracy; (**b**): Lidom model evaluation results.

From Figure 7a, it can be concluded that the accuracies of the models based on artificial neural network ANN, K-nearest neighbor classification KNN (K-nearest neighbors), support vector machine SVM (support vector machines), and logistic regression LR (logistic regression) algorithm were lower than 60%. The accuracy rate of the model built based on the LightGBM algorithm reached 93.46%.

From Figure 7b, it can be concluded that the precision rate of the model based on the LightGBM algorithm reached 51.14%, the recall rate reached 46.47%, the false positive rate reached 100%, and the F1 value reached 47.87%.

We used the precision rate (Precision), recall rate (Recall), F1 value (F1-Measure), and false positive rate FPR (False Positive Rate) for the models artificial neural network ANN, K-nearest neighbor classification KNN (K-nearest neighbors), and support vector. The results of the models constructed by the five algorithms of machine SVM (support vector machines), logistic regression LR (logistic regression), and LightGBM (Light Gradient Boosting Machine) were further compared and are shown explicitly in Figures 8 and 9.

From Figure 8, it can be concluded that the precision rate (Precision) and recall rate (Recall) are low. The highest are the precision rate (Precision) and recall rate (Recall) of the model built using the LightGBM algorithm.

From Figure 9, it can be concluded that the F1 value (F1-Measure) is lower. The F1 value (F1-Measure) of the model constructed using the LightGBM algorithm is the highest.

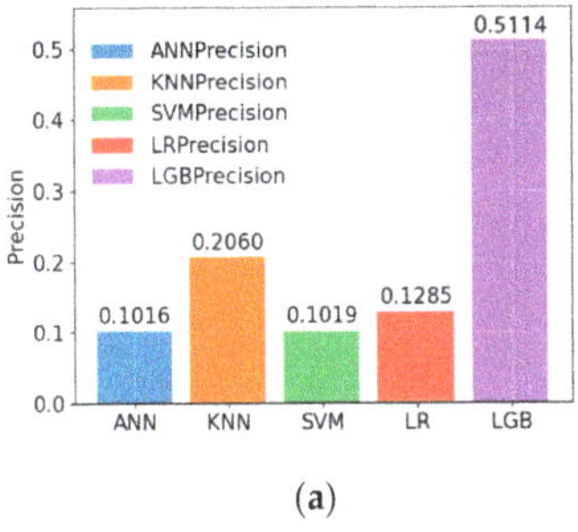

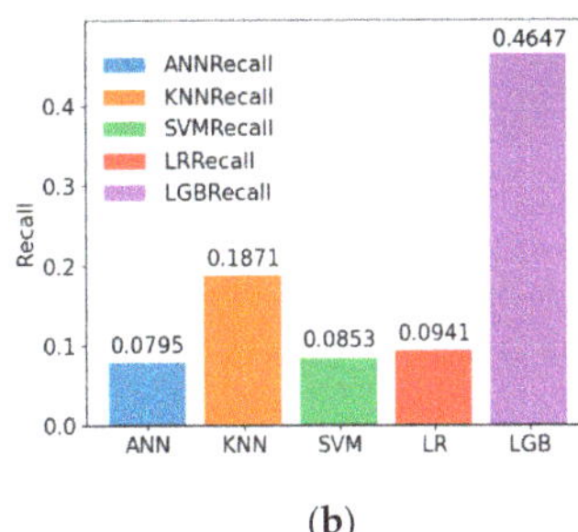

Figure 8. Comparison of precision and recall; (**a**): comparison of precision of five algorithms; (**b**): comparison of recall of five algorithms.

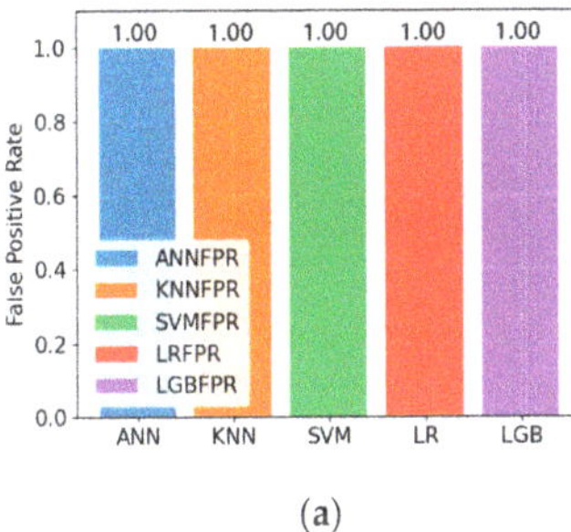

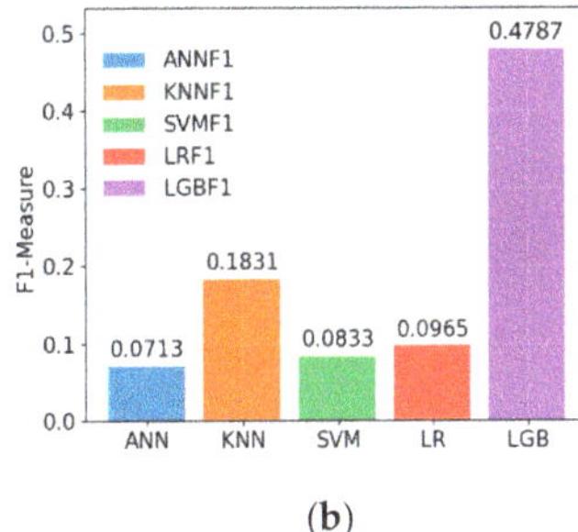

Figure 9. Comparison of FPR (False Positive Rate) and F1-value (F1-Measure); (**a**): comparison of FPR (False Positive Rate) of five algorithms; (**b**): F1-value (F1-Measure) of five algorithms—measure comparison.

The contribution of this paper is to innovatively apply the algorithm to the application scenario of disease risk prediction in nursing homes. The above experimental results show that the accuracy rate (Accuracy), precision rate (Precision), recall rate (Recall), and F1 value (F1-Measure) of the Lidom model are the highest. Therefore, the Lidom model can effectively support the prediction of disease risk.

5.2. Contribution Analysis of Prediction Results

In this paper, we use the machine learning interpretable framework InterpretML to calculate the contribution value of each feature in the sample data on each sample, as shown in Figure 10.

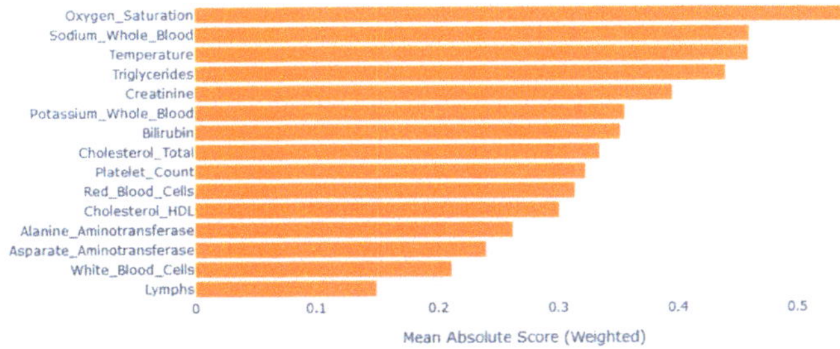

Figure 10. Global Feature Importance.

As seen in Figure 10, 15 features significantly impact the results of disease risk prediction in the sample data. To continue to understand the impact of each feature on the disease risk prediction results, we looked at the contribution of each feature to the disease risk prediction results. Figure 11 is a breakdown of the contribution of the Oxygen_Saturation feature to the disease risk prediction results.

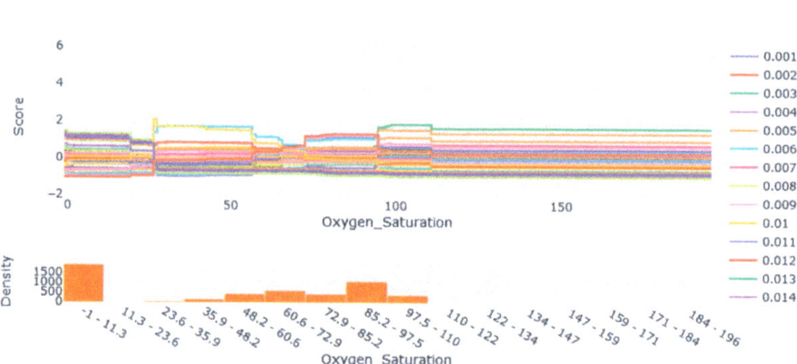

Figure 11. The influence of a single feature on the prediction result.

In Figure 11, density represents the distribution of the Oxygen_Saturation feature on the training sample data. The score represents log odds, and log odds can obtain the probability through the logit function, as shown in Equation (10).

$$\log odds = Logit(P) = ln\frac{p}{1-p} \tag{10}$$

From Figure 12, we can see the local interpretation of some features; that is, the distribution of these features in local importance. Based on the above analysis, we conclude that the features Oxygen_Saturation, Sodium_Whole_Blood, temperature, triglycerides, creatinine, Potassium_Whole_Blood, bilirubin, Cholesterol_Total, Platelet_Count, Red_Blood_Cells, Cholesterol_HDL, Alanine_Aminotransferase, and Aspartate_aminotranspherase have a more significant contribution to the prediction results of the Lidom model. Among them, the Oxygen_Saturation index has the most significant impact on disease risk prediction. Therefore, it is indispensable to monitor the blood oxygen index of the elderly in nursing homes in time.

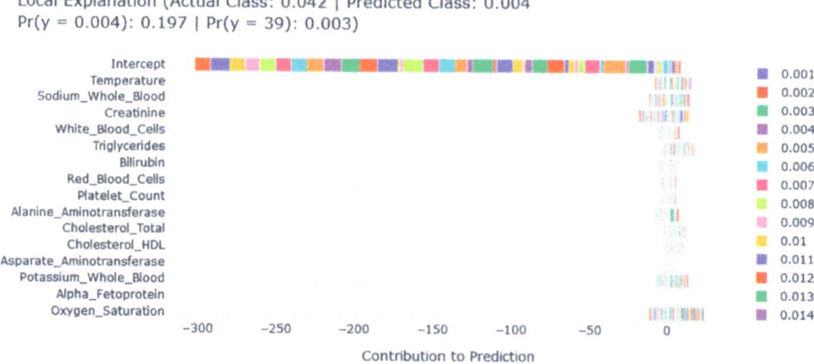

Figure 12. The local importance of features.

5.3. Model Optimization

The work in this paper on model optimization mainly consists of sample optimization and parameter optimization. When optimizing the sample, we used the method described in the previous chapter to interpret each feature's influence on the prediction result's contribution value and obtain the features that significantly impact the IDPM model we constructed. We adjusted the training sample data and constructed the sample data using features that significantly impact the IDPM results. This paper also uses the five evaluation indicators of accuracy, precision, recall, FPR, and F1-Measure to compare the samples before and after adjustment. The details are shown in Figure 13a.

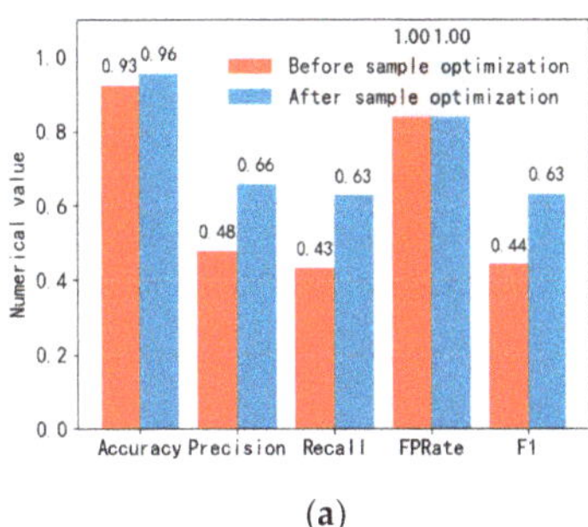

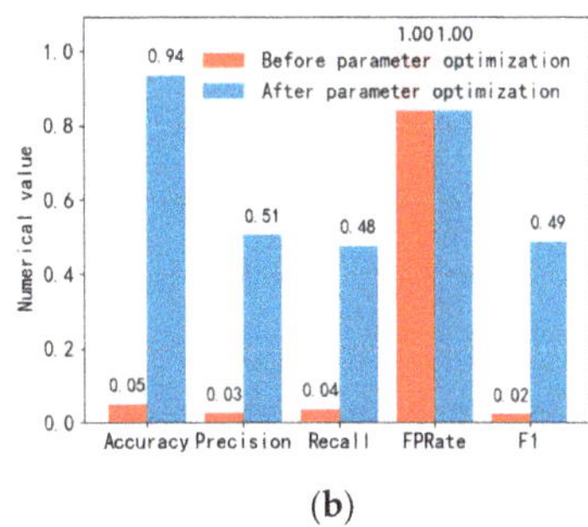

Figure 13. Comparison before and after model optimization; (**a**): comparison of forecast results before and after sample adjustment; (**b**): comparison of forecast results before and after parameter adjustment.

The parameters of the LightGBM algorithm are composed of model structure parameters, training accuracy parameters, and overfitting parameters. The model structure parameters are the parameters of the base classifier structure. The more complex the structure, the better the fitting ability of the model will be, but the phenomenon of overfitting will occur. This parameter is the maximum number of leaf nodes in a single base classifier. Within a specific range, the larger the value, the more accurate the classification. For the parameter of the maximum depth of the tree in a single base classifier, the larger the value, the stronger the fitting ability. The smaller the value, the easier it is to underfit.

For training accuracy parameters such as the number of trees within a specific range, the higher the value, the higher the accuracy. However, after exceeding the range, the higher the value, the lower the accuracy; the learning rate parameter directly affects the training efficiency. Overfitting parameters are those such as regularization parameters, sampling ratio parameters of training samples, etc. When optimizing the model parameters, we use GridSearchCV to optimize the model parameters according to the above parameter optimization ideas. This article uses the five evaluation indicators of accuracy, precision, recall, FPR, and F1-Measure to compare parameters before and after adjustment. The details are shown in Figure 13b.

Figure 13a shows that the Lidom model trained using the training sample data after feature adjustment has higher prediction accuracy, precision, and recall. The F1 value has improved to a certain extent.

From Figure 13b, we can observe that all relevant evaluation indicators improve after the parameters are adjusted. This is because, in the LightGBM algorithm, the default parameter values have less control over model-related indicators. Therefore, the relevant evaluation indicators all improve after the parameters are adjusted.

5.4. Early Diabetes Risk Prediction

Due to the low precision rate, recall rate, and F1 value of the disease risk prediction results of the Lidom model, in order to further improve the accuracy rate, precision rate, recall rate, and F1 value of the Lidom model for disease risk prediction, and to understand the differences in the early diabetes risk predictions of the five algorithms, artificial neural network (ANN), K-nearest neighbor classification (KNN), support vector machine (SVM), logistic regression (LR), and LightGBM, we used the early diabetes risk prediction dataset in the UCI machine learning database as sample data. The prediction results are shown in Figures 14–16.

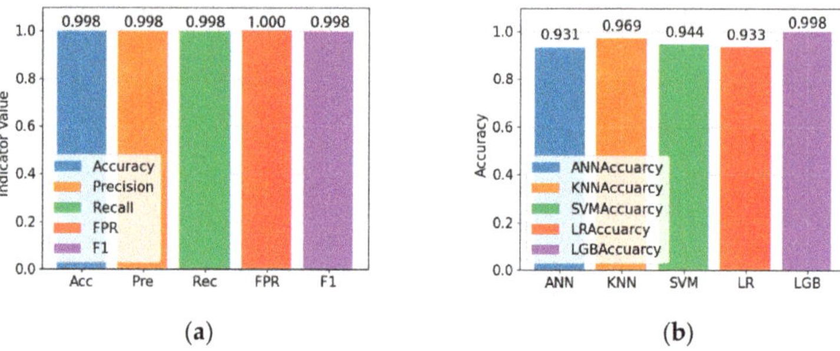

(a) (b)

Figure 14. Training prediction results; (**a**): Lidom model evaluation results; (**b**): prediction accuracy.

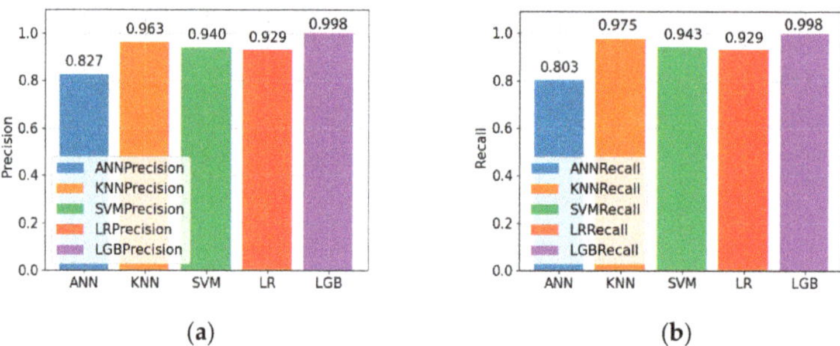

(a) (b)

Figure 15. Comparison of precision and recall of early diabetes risk prediction; (**a**): Comparison of precision of five algorithms for early diabetes risk prediction; (**b**): recall of five algorithms for early diabetes risk prediction Rate (Recall) comparison.

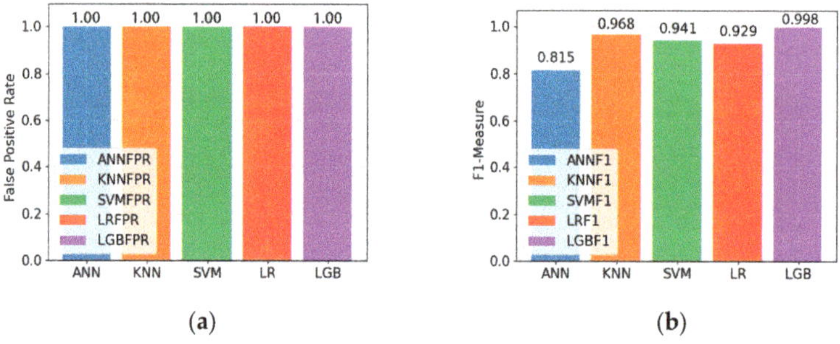

(a) (b)

Figure 16. Comparison of false positive rate FPR (False Positive Rate) and F1 value (F1-Measure) in early diabetes risk prediction; (**a**): Comparison of false positive rate FPR (False Positive Rate) in early diabetes risk prediction of five algorithms; (**b**): Comparison of five algorithms for early diabetes risk prediction F1 value (F1-Measure).

It can be seen from Figure 14a that the accuracy, precision, recall, and F1 value of Lidom in the prediction of early diabetes risk all reached 99.8%.

It can be seen from Figure 14b that the accuracy rates of the models constructed based on artificial neural network ANN, K-nearest neighbor classification KN, support vector machine SVM, and logistic regression LR algorithm all exceeded 93%. The accuracy rate of the model built based on the LightGBM algorithm reached 99.8%.

It can be concluded from Figure 15 that among the models constructed using the five algorithms of ANN, KNN, SVM, LR, and LightGBM, the precision and recall rates of the model constructed using the LightGBM algorithm were the highest.

It can be concluded from Figure 16 that among the models constructed using the five algorithms of ANN, KNN, SVM, LR, and LightGBM, the F1-Measure rates of the model constructed using the LightGBM algorithm were the highest.

We made statistics on the accuracy of some recent models for diabetes risk prediction, and the statistical results are shown in Table 3.

Table 3. Comparison of model accuracy on the same disease prediction topic.

Deep NN [43]	TSGB [44]	DT [45]	RF [45]	GBC [45]	Lidom
98%	75.08%	95.962%	95.19%	98.65%	99.8%

From Table 3, we can conclude that the accuracy rates of the proposed models based on deep neural network, multi-task learning, gradient boosting decision tree, decision tree, random forest, and gradient boosting classifier exceeded 95%. Based on the model Lidom proposed by LightGBM, the prediction accuracy rate reached 99.8%.

During the experiment, to understand the changes in the loss and accuracy of the trained model more clearly, we intercepted the loss and accuracy of some models and outputted them, as shown in Figure 17.

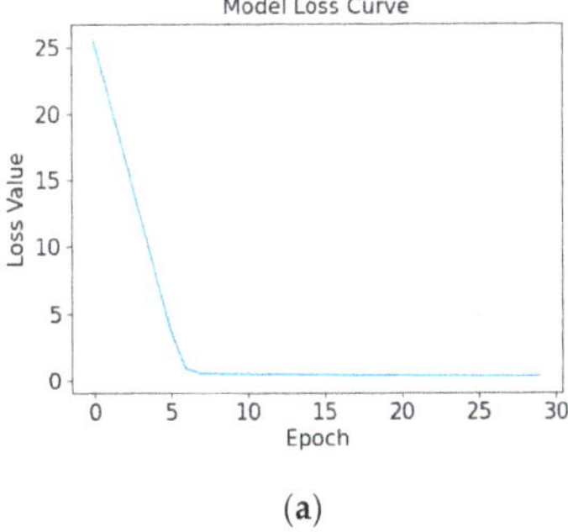

(a)

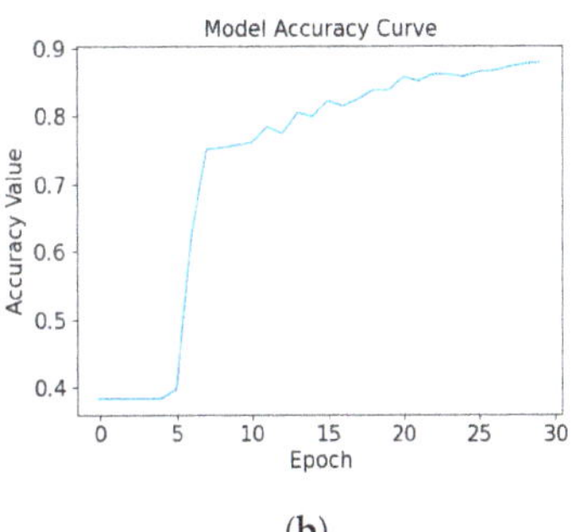

(b)

Figure 17. The model's loss curve and accuracy rate curve; (**a**): the loss curve of the model; (**b**): the accuracy rate curve of the model.

Figure 17 clearly shows that with the reduction in model training loss, the model's accuracy also improved accordingly. To understand the model's superiority based on LightGBM more clearly, we compared the AUC (Area under the Curve of ROC) value and KS (Kolmogorov–Smirnov) value of the models proposed based on KNN, SVM, LR, and LightGBM, respectively, as shown in Figure 18.

Figure 18a show the ROC (Receiver Operating Characteristic), and the abscissa indicates the false positive rate, and the ordinate indicates the true positive rate. The total area enclosed by the ROC curve, the abscissa, and the ordinate give the AUC value. The model is more realistic when the AUC value is closer to 1.0. In Figure 18b, the abscissa represents the threshold value, and the ordinate represents the difference between the TPR and FPR. The maximum difference between the TPR and the FPR is the KS value. The larger the KS value, the stronger the discrimination ability of the model. From Figure 18, we can see that the classification accuracy based on the model Lidom proposed by LightGBM is the highest.

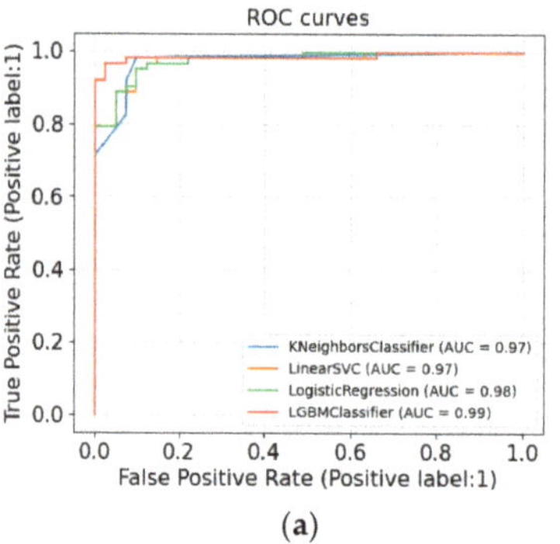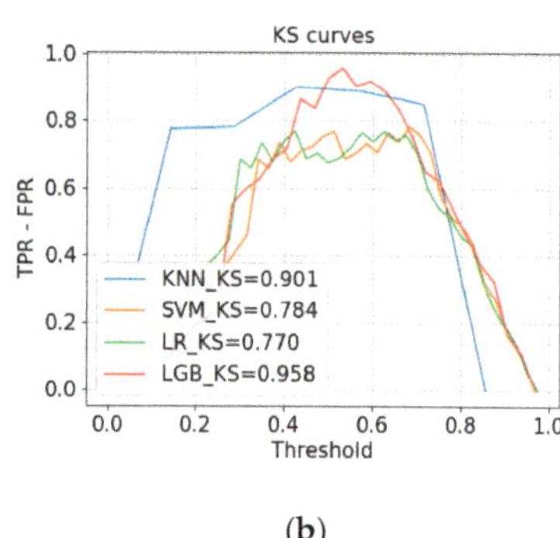

(a)   (b)

Figure 18. ROC curve and KS curve of each model; (**a**): ROC curve of each model; (**b**): KS curve of each model.

In this chapter, we compared the IDPM models proposed in this paper based on ANN, KNN, SVM, LR, and LightGBM, using relevant evaluation indicators. We adjusted the characteristics of the training sample data and the parameters of the model based on the analysis of the contribution of the characteristics to the prediction results. In order to improve the accuracy, precision, recall, and F1 value of Lidom model prediction, we predicted the risk of early diabetes. The experimental results again show that the Lidom model can effectively support the prediction of disease risk in nursing homes.

6. Conclusions

In this paper, we use MIMIC-III, subject data, and UCI's early diabetes risk prediction dataset as sample data, train the Lidom model, apply a disease risk prediction model to nursing homes based on the LightGBM algorithm, and use the InterpretML framework to contribute to the prediction results from the accuracy of the Lidom model for disease risk prediction, which reached 93.46%. The accuracy rate for early diabetes risk prediction reached 99.8%. The effective prediction of the health risks of the elderly in nursing homes was realized. However, the samples used in this article only selected 21 features from MIMIC-III, and only 56 diseases were selected. In the future, we aim to increase the number of features of the samples and the number of predicted diseases. In addition, this article's proportion of subject data is relatively small, and it is hoped to increase subject data in the future in order to further enhance the support of the Lidom model for the health risk prediction of the elderly in nursing homes.

Author Contributions: F.Z., S.H. and X.D. wrote the main manuscript text, and Z.L. provided the idea. X.W. and J.W. prepared the data and figures. All authors reviewed the manuscript. The authors read and approved the final manuscript. All authors have read and agreed to the published version of the manuscript.

Funding: The work of this paper is supported by the National Key Research and Development Program of China (2019YFB1405000), and the National Natural Science Foundation of China under Grant (No. 61873309, 92046024, 92146002).

Data Availability Statement: The MIMIC-III Clinical Database dataset is from https://physionet.org/content/mimiciii/1.4/ (accessed on 5 September 2022).

Acknowledgments: Thanks to Xiong Yu, chief physician of the Cardiology Department of Wuhu Hospital of Traditional Chinese Medicine; Sun Jun, deputy chief physician of the Cardiology Department; Cheng Hui, deputy chief physician of the Orthopedics Department; Du Qingqing, director of the Infection Prevention Office, and Cao Jin, deputy chief physician of the Department of Anesthesiology for their guidance and coordination in clinical medicine.

Conflicts of Interest: The authors declare no conflict of interest.

References

1. Luo, J.; Li, J.-W.; Wei, M.; Li, Y.; Yan, Z.; Wang, Y.; Guo, Y.; Gan, Z. Dynamic Blood Oxygen Saturation Monitoring based on a New IPPG Detecting Device. In Proceedings of the 2021 11th International Conference on Biomedical Engineering and Technology (ICBET '21), Tokyo, Japan, 17–20 March 2021; Association for Computing Machinery: New York, NY, USA, 2021; pp. 92–99. [CrossRef]
2. Dissanayake, V.; Elvitigala, D.S.; Zhang, H.; Weerasinghe, C.; Nanayakkara, S. CompRate: Power Efficient Heart Rate and Heart Rate Variability Monitoring on Smart Wearables. In Proceedings of the 25th ACM Symposium on Virtual Reality Software and Technology (VRST '19), Parramatta, Australia, 12–15 November 2019; Association for Computing Machinery: New York, NY, USA, 2019; pp. 1–8. [CrossRef]
3. Du, X.; Tang, S.; Lu, Z.; Gai, K.; Wu, J.; Hung, P.C.K. Scientific Workflows in IoT Environments: A Data Placement Strategy Based on Heterogeneous Edge-Cloud Computing. *ACM Trans. Manage. Inf. Syst.* **2022**, *13*, 42. [CrossRef]
4. Green, M.; Bjork, J.; Forberg, J.; Ekelund, U.; Edenbrandt, L.; Ohlsson, M. Comparison between neural networks and multiple logistic regression to predict acute coronary syndrome in the emergency room. *Artif. Intell. Med.* **2006**, *38*, 305–318. [CrossRef] [PubMed]
5. Das, R.; Turkoglu, I.; Sengur, A. Diagnosis of valvular heart disease through neural networks ensembles. *Comput. Methods Programs Biomed.* **2009**, *93*, 185–191. [CrossRef] [PubMed]
6. Atkov, O.Y.; Gorokhova, S.G.; Sboev, A.G.; Generozov, E.V.; Muraseyeva, E.V.; Moroshkina, S.Y.; Cherniy, N.N. Coronary heart disease diagnosis by artificial neural networks including genetic polymorphisms and clinical parameters. *J. Cardiol.* **2012**, *59*, 190–194. [CrossRef] [PubMed]
7. Guermah, H.; Fissaa, T.; Guermah, B.; Hafiddi, H.; Nassar, M. Using Context Ontology and Linear SVM for Chronic Kidney Disease Prediction. In Proceedings of the International Conference on Learning and Optimization Algorithms: Theory and Applications (LOPAL '18), Rabat, Morocco, 2–5 May 2018; Association for Computing Machinery: New York, NY, USA, 2018; pp. 1–6. [CrossRef]
8. Wu, C.-S.M.; Badshah, M.; Bhagwat, V. Heart Disease Prediction Using Data Mining Techniques. In Proceedings of the 2019 2nd International Conference on Data Science and Information Technology (DSIT 2019), Seoul, Republic of Korea, 19–21 July 2019; Association for Computing Machinery: New York, NY, USA, 2019; pp. 7–11. [CrossRef]
9. Rongqiang, S.; Yan, C.; Chang, D.; Qingyue, G. Research on promoting the application of disease prediction system based on machine learnin. In Proceedings of the 2020 International Symposium on Artificial Intelligence in Medical Sciences (ISAIMS 2020), Beijing, China, 11–13 September 2020; Association for Computing Machinery: New York, NY, USA, 2020; pp. 45–50. [CrossRef]
10. Günay, M.; Orman, Z. Disease Prediction Using Weighted Artificial Immune System. In Proceedings of the 2019 3rd International Symposium on Computer Science and Intelligent Control (ISCSIC 2019), Amsterdam, The Netherlands, 25–27 September 2019; Association for Computing Machinery: New York, NY, USA, 2020; pp. 1–5. [CrossRef]
11. Xiaoai, G.; Yujing, X.; Lin, L.; Lin, T. An Overview of Disease Prediction based on Graph Convolutional Neural Network. In Proceedings of the 2021 6th International Conference on Intelligent Information Processing (ICIIP 2021), Bucharest, Romania, 29–31 July 2021; Association for Computing Machinery: New York, NY, USA, 2021; pp. 27–32. [CrossRef]
12. He, Y.; Liu, H.; Xie, X.; Gu, W.; Mao, Y.; Luo, W. Infectious Disease Prediction and Analysis Based on Parametric-nonparametric Hybrid Model. In Proceedings of the 2021 2nd International Conference on Artificial Intelligence and Information Systems (ICAIIS 2021), Chongqing, China, 28–30 May 2021; Association for Computing Machinery: New York, NY, USA, 2021; pp. 1–6. [CrossRef]
13. Song, S.; Chen, T.; Antoniou, G. ANFIS Models for Heart Disease Prediction. In Proceedings of the 2021 the 5th International Conference on Innovation in Artificial Intelligence (ICIAI 2021), Xiamen, China, 5–8 March 2021; Association for Computing Machinery: New York, NY, USA, 2021; pp. 32–35. [CrossRef]
14. Wang, M.; Zhang, H.; Wu, H.; Li, G.; Gai, K. Blockchain-based Secure Medical Data Management and Disease Prediction. In Proceedings of the Fourth ACM International Symposium on Blockchain and Secure Critical Infrastructure (BSCI '22), Nagasaki, Japan, 30 May 2022; Association for Computing Machinery: New York, NY, USA, 2022; pp. 71–82. [CrossRef]
15. Gao, J.; Xiao, C.; Glass, L.M.; Sun, J. PopNet: Real-Time Population-Level Disease Prediction with Data Latency. In Proceedings of the ACM Web Conference 2022 (WWW '22), Lyon, France, 25–29 April 2022; Association for Computing Machinery: New York, NY, USA, 2022; pp. 2552–2562. [CrossRef]
16. Dahiwade, D.; Patle, G.; Meshram, E. Designing Disease Prediction Model Using Machine Learning Approach. In Proceedings of the 2019 3rd International Conference on Computing Methodologies and Communication (ICCMC), Erode, India, 27–29 March 2019; pp. 1211–1215. [CrossRef]
17. Ambesange, S.; Vijayalaxmi, A.; Uppin, R.; Patil, S.; Patil, V. Optimizing Liver disease prediction with Random Forest by various Data balancing Techniques. In Proceedings of the 2020 IEEE International Conference on Cloud Computing in Emerging Markets (CCEM), Bengaluru, India, 6–7 November 2020; pp. 98–102. [CrossRef]
18. Zhao, R.; Wen, X.; Pang, H.; Ma, Z. Liver disease prediction using W-LR-XGB Algorithm. In Proceedings of the 2021 International Conference on Computer, Blockchain and Financial Development (CBFD), Nanjing, China, 23–25 April 2021; pp. 245–248. [CrossRef]

19. Islam, T.; Rafa, S.R.; Kibria, G. Early Prediction of Heart Disease Using PCA and Hybrid Genetic Algorithm with k-Means. In Proceedings of the 2020 23rd International Conference on Computer and Information Technology (ICCIT), Dhaka, Bangladesh, 19–21 December 2020; pp. 1–6. [CrossRef]

20. Nguyen, V.T.; Le, T.T.K.; Tran, D.H. A new method on lncRNA-disease-miRNA tripartite graph to predict lncRNA-disease associations. In Proceedings of the 2020 12th International Conference on Knowledge and Systems Engineering (KSE), Can Tho City, Vietnam, 12–14 November 2020; pp. 287–293. [CrossRef]

21. Thummala, G.S.R.; Baskar, R. Prediction of Heart Disease using Decision Tree in Comparison with KNN to Improve Accuracy. In Proceedings of the 2022 International Conference on Innovative Computing, Intelligent Communication and Smart Electrical Systems (ICSES), Antalya, Turkey, 10–13 November 2022; pp. 1–5. [CrossRef]

22. Chen, A.H.; Huang, S.Y.; Hong, P.S.; Cheng, C.H.; Lin, E.J. HDPS: Heart disease prediction system. In Proceedings of the 2011 Computing in Cardiology, Hangzhou, China, 18–21 September 2011; pp. 557–560.

23. Kumari, A.; Mehta, A.K. A Novel Approach for Prediction of Heart Disease using Machine Learning Algorithms. In Proceedings of the 2021 Asian Conference on Innovation in Technology (ASIANCON), Pune, India, 27–29 August 2021; pp. 1–5. [CrossRef]

24. Rubaiyat, N.; Apsara, A.I.; Al Farabe, A.; Ishtiak, I. Classification and prediction of Orthopedic disease based on lumber and pelvic state of patients. In Proceedings of the 2019 IEEE International Conference on Electrical, Computer and Communication Technologies (ICECCT), Coimbatore, India, 20–22 February 2019; pp. 1–4. [CrossRef]

25. Ambekar, S.; Phalnikar, R. Disease Risk Prediction by Using Convolutional Neural Network. In Proceedings of the 2018 Fourth International Conference on Computing Communication Control and Automation (ICCUBEA), Pune, India, 16–18 August 2018; pp. 1–5. [CrossRef]

26. Reddy, K.S.K.; Kanimozhi, K. Novel Intelligent Model for Heart Disease Prediction using Dynamic KNN (DKNN) with improved accuracy over SVM. In Proceedings of the 2022 International Conference on Business Analytics for Technology and Security (ICBATS), Dubai, United Arab Emirates, 16–17 February 2022; pp. 1–5. [CrossRef]

27. Xue, T.; Jieru, Z. Application of Support Vector Machine Based on Particle Swarm Optimization in Classification and Prediction of Heart Disease. In Proceedings of the 2022 7th International Conference on Intelligent Computing and Signal Processing (ICSP), Xi'an, China, 15–17 April 2022; pp. 857–860. [CrossRef]

28. Li, J.; Zhao, H.; Xuan, Z.; Yu, J.; Feng, X.; Liao, B.; Wang, L. A Novel Approach for Potential Human LncRNA-Disease Association Prediction Based on Local Random Walk. *IEEE/ACM Trans. Comput. Biol. Bioinform.* **2019**, *18*, 1049–1059. [CrossRef] [PubMed]

29. Yuan, X.; Chen, J.; Zhang, K.; Wu, Y.; Yang, T. A Stable AI-Based Binary and Multiple Class Heart Disease Prediction Model for IoMT. *IEEE Trans. Ind. Inform.* **2021**, *18*, 2032–2040. [CrossRef]

30. Bhoyar, S.; Wagholikar, N.; Bakshi, K.; Chaudhari, S. Real-time Heart Disease Prediction System using Multilayer Perceptron. In Proceedings of the 2021 2nd International Conference for Emerging Technology (INCET), Belagavi, India, 21–23 May 2021; pp. 1–4. [CrossRef]

31. Latha, R.; Vetrivelan, P. Blood Viscosity based Heart Disease Risk Prediction Model in Edge/Fog Computing. In Proceedings of the 2019 11th International Conference on Communication Systems & Networks (COMSNETS), Bengaluru, India, 7–11 January 2019; pp. 833–837. [CrossRef]

32. Yadav, D.; Saini, P.; Mittal, P. Feature Optimization Based Heart Disease Prediction using Machine Learning. In Proceedings of the 2021 5th International Conference on Information Systems and Computer Networks (ISCON), Mathura, India, 22–23 October 2021; pp. 1–5. [CrossRef]

33. Goodfellow, I.J.; Pouget-Abadie, J.; Mirza, M. Generative adversarial nets. In Proceedings of the International Conference on Neural Information Processing Systems, Cambridge 2014, 2672–2680, Montreal, QU, Canada, 8–13 December 2014.

34. Yu, L.; Zhang, W.; Wang, J.; Yu, Y. SeqGAN: Sequence Generative Adversarial Nets with Policy Gradient. In Proceedings of the AAAI Conference on Artificial Intelligence 2017, San Francisco, CA, USA, 4–9 February 2017; p. 31. [CrossRef]

35. Chen, T.; Guestrin, C. XGBoost: A Scalable Tree Boosting System. In Proceedings of the 22nd ACM SIGKDD International Conference on Knowledge Discovery and Data Mining (KDD '16), San Francisco, CA, USA, 13–17 August 2016; Association for Computing Machinery: New York, NY, USA, 2016; pp. 785–794. [CrossRef]

36. Ke, G.; Meng, Q.; Finley, T.; Wang, T.; Chen, W.; Ma, W.; Ye, Q.; Liu, T. Y Lightgbm: A highly efficient gradient boosting decision tree. *Adv. Neural Inf. Process. Syst.* **2017**, *30*, 52.

37. Nori, H.; Jenkins, S.; Koch, P.; Caruana, R. InterpretML: A Unified Framework for Machine Learning Interpretability. *arXiv* **2019**, arXiv:1909.09223.

38. Du, X.; Liu, Y.; Lu, Z.; Duan, Q.; Feng, J.; Wu, J.; Chen, B.; Zheng, Q. A Low-Latency Communication Design for Brain Simulations. *IEEE Netw.* **2022**, *36*, 8–15. [CrossRef]

39. Johnson, A.; Pollard, T.; Mark, R. MIMIC-III Clinical Database (version 1.4). *PhysioNet* **2016**, *101*, e215–e220. [CrossRef]

40. Zhou, F.; Du, X.; Li, W.; Lu, Z.; Wu, J. NIDD: An intelligent network intrusion detection model for nursing homes. *J. Cloud. Comp.* **2022**, *11*, 91. [CrossRef]

41. Du, X.; Chen, X.; Lu, Z.; Duan, Q.; Wang, Y.; Wu, J. BIECS: A Blockchain-based Intelligent Edge Cooperation System for Latency-Sensitive Services. In Proceedings of the 2022 IEEE International Conference on Web Services (ICWS), Barcelona, Spain, 11–15 July 2022; pp. 367–372. [CrossRef]

42. Meng, H.; Lei, T.; Li, M.; Li, K.; Xiong, N.; Wang, L. *Advances in Natural Computation, Fuzzy Systems and Knowledge Discovery*; Springer Nature: Berlin/Heidelberg, Germany, 2021; Volume 88.

43. Daanouni, O.; Cherradi, B.; Tmiri, A. Predicting diabetes diseases using mixed data and supervised machine learning algorithms. In Proceedings of the 4th International Conference on Smart City Applications (SCA '19), Casablanca, Morocco, 2–4 October 2019; Association for Computing Machinery: New York, NY, USA, 2019; pp. 1–6. [CrossRef]

44. Chen, M.; Wang, Z.; Zhao, Z.; Zhang, W.; Guo, X.; Shen, J.; Qu, Y.; Lu, J.; Xu, M.; Xu, Y.; et al. Task-wise Split Gradient Boosting Trees for Multi-center Diabetes Prediction. In Proceedings of the 27th ACM SIGKDD Conference on Knowledge Discovery & Data Mining (KDD '21), Virtual Event, Singapore, 14–18 August 2021; Association for Computing Machinery: New York, NY, USA, 2021; pp. 2663–2673. [CrossRef]

45. Moniruzzaman; Zaman, A.G.M.; Tasnia, R.; Biswas, S.; Khanam, M. Early-Stage Diabetes Prediction using Data Mining Algorithms. In Proceedings of the 2nd International Conference on Computing Advancements (ICCA '22), Dhaka, Bangladesh, 10–12 March 2022; Association for Computing Machinery: New York, NY, USA, 2022; pp. 240–248. [CrossRef]

Article

MDP-Based MAC Protocol for WBANs in Edge-Enabled eHealth Systems

Haoru Su [1], Meng-Shiuan Pan [2], Huamin Chen [1,*] and Xiliang Liu [1]

[1]   Faculty of Information Technology, Beijing University of Technology, Beijing 100124, China
[2]   Department of Electronic Engineering, National Taipei University of Technology, Taipei 10608, Taiwan
*   Correspondence: chenhuamin@bjut.edu.cn

Abstract: In recent years, eHealth systems based on the Internet of Things (IoT) have attracted considerable attention. The wireless body area network (WBAN) is an essential technology of eHealth systems. A major challenge in WBAN is the design of the medium access control (MAC) protocol, which plays a significant role in avoiding collisions, enhancing the energy efficiency, maximizing the network life, and improving the quality of service (QoS) as well as the quality of experience (QoE). In this study, we apply the mobile edge computing (MEC) network architecture to an eHealth system and design a multi-channel MAC protocol for WBAN based on the Markov decision process (MDP). In this protocol, the channel condition and the reward value are considered. By continuously interacting with the environment, the optimal channel resource allocation strategy is generated. Simulation results indicate that the proposed WBAN MAC protocol can adaptively assign different channels to the sensor nodes for data transmission, thereby reducing the collision rate, decreasing the energy consumption, improving the channel utilization, and enhancing the system throughput and QoE.

Keywords: medium access control protocol; channel allocation; wireless body area networks; energy efficiency; quality of service

Citation: Su, H.; Pan, M.-S.; Chen, H.; Liu, X. MDP-Based MAC Protocol for WBANs in Edge-Enabled eHealth Systems. *Electronics* **2023**, *12*, 947. https://doi.org/10.3390/electronics12040947

Academic Editor: José Santa

Received: 5 January 2023
Revised: 7 February 2023
Accepted: 9 February 2023
Published: 14 February 2023

1. Introduction

The development of wearable sensing devices and communication technologies has facilitated diverse applications in many fields such as healthcare. The traditional healthcare system is under severe pressure due to the growth of the aging population and the increase in chronic diseases [1]. Therefore, eHealth systems based on wearable body sensing devices and the Internet of Things (IoT) are considered as promising technologies for realizing remote healthcare services [2,3]. Wireless sensor networks (WSN) have evolved into a technology suited to human mobility through the development of wearable or implanted sensors. The main goal of these networks is to collect data and transfer the data to a remote server [4]. This technology is referred to as the wireless body area network (WBAN), which consists of several sensors and a central base station or coordinator [5]. These heterogeneous, miniature, energy-constrained sensors capture the physiological information of the human body, such as the body temperature, heartbeat, ECG, and EEG, and then send the physiological data to the central coordinator or base station [6]. Besides healthcare, WBANs are employed in various fields, such as sports, entertainment, military, and emergency rescue.

Traditional eHealth systems based on WBANs generally employ the cloud-based network architecture [7]. Data are gathered by the sensor node, transmitted to the coordinator, and then sent to the remote cloud data center [8]. The time delay is longer than that in the local process. However, emergencies such as heart attacks require a rapid response. The data transmission delay between the user and the remote cloud server can be quite long, which may result in a late response. Recently, there has been a trend of moving computing closer to the users [9]. Mobile edge computing (MEC) has emerged as a popular research topic in the field of communication and networking. In the edge-enabled network architecture, MEC servers are deployed near the nodes of WBANs [10].

The communication between sensing devices and the coordinator is supported by the IEEE 802.15.4 and IEEE 802.15.6 standard protocol stacks. The IEEE 802.15.4 standard was designed for low-rate wireless personal area networks (LR-WPANs) [11], and it can also support WBANs. The IEEE 802.15.6 standard is geared toward applications of WBANs. It defines the physical (PHY) and medium access control (MAC) layers [12]. There are three types of PHY layers: human body communications (HBC), ultra-wideband (UWB), and narrowband (NB). In the MAC layer, the communication channels are separated into superframes. Beacon frames are used for synchronization in the beacon-enabled mode.

The MAC layer protocol is in charge of the time and channel access in a network. It plays an important role in the network resource allocation and influences the network performance [13]. Energy efficiency is a critical metric to be considered for designing proper WBAN protocols. The sensing devices of WBANs are implanted in the human body and/or attached to the clothes (i.e., they are wearable). The batteries have limited capacity and it is difficult to change them. A well-designed MAC protocol can significantly improve the energy efficiency of WBANs. In energy-restricted networks, the MAC protocol is a critical aspect in terms of decreasing the power consumption and prolonging the network lifetime [13]. The responsibility of the MAC layer is to coordinate the sensing devices when they access the WBAN channels. Therefore, energy efficiency is the main property of a proper WBAN MAC protocol. If the channel and/or time slot utilization can be enhanced, data collisions and/or beacon frame collisions can be mitigated, the idle time can be decreased, and the energy consumption of WBANs can be reduced considerably.

In addition, heterogeneous wearable body sensing devices produce different types of data packets that have various requirements in terms of the network quality of service (QoS). Some applications require a high probability of success in packet delivery (i.e., high packet delivery ratio) as well as timely packet delivery (i.e., low latency). For instance, the time-delay requirement of heart-rate information is shorter than that of the body temperature for heart disease patients. Moreover, the network QoS (time delay, delivery ratio, reliability, fairness, throughput, and so on) must be enhanced. Generally, medical and healthcare applications are mission critical, and they need reliable data delivery [14]. Technologies and devices from all the protocol layers of a WBAN are responsible for reliability. In [15], transmission control protocol (TCP) is utilized. The MAC protocol for WBANs should be flexible in order to support diverse data and various applications.

Furthermore, there are contradictions among the different QoS requirements. For instance, delay-bounded and reliable data delivery may require higher energy consumption, thereby shortening the network lifetime. This problem complicates the design of the MAC protocol in terms of satisfying the QoS requirements. To satisfy the system QoS, the MAC protocol must be designed properly. In particular, the MAC protocol must achieve a trade-off between the energy efficiency and the data transmission reliability.

In this study, we apply an MEC-based network architecture to an eHealth system. We design an energy-efficient multi-channel MAC protocol based on a Markov decision process (MDP). This protocol attempts to minimize the beacon and data collisions of coordinators and sensor nodes by learning the collision patterns as well as by effectively assigning frequencies over time to ensure that neighboring nodes do not collide with one another.

The remainder of this paper is organized as follows. Section 2 reviews the related studies. Section 3 presents the MEC-based network architecture. Section 4 introduces the priority-aware MAC layer resource allocation mechanism. Section 5 discusses the performance evaluation. Finally, Section 6 concludes the paper.

2. Related Work

Recently, numerous MAC protocols have been proposed for WBANs [16]. In this section, we review some existing MAC protocols that attempt to improve the power efficiency and system QoS of wearable sensor eHealth systems.

WBAN MAC protocols can be classified into two types: contention-based protocols and contention-free protocols. In contention-based MAC protocols, the sensor nodes access the communication channels through competition. It is not necessary to establish a fixed structure; hence, such protocols are more scalable. However, data packet collisions may occur during data transmission, which may lead to retransmission and energy wastage. By contrast, in contention-free MAC protocols, the communication resources such as time and channels are assigned to each senor node. For example, the well-known time-division multiple access (TDMA) MAC protocol divides the timeline into equal-length time slots and assigns them to the sensing devices. During the assigned time slots, each sensor node transmits the physiological data to the coordinator. Subsequently, the sensor nodes go into the sleep mode to save energy. Contention-free MAC protocols can reduce not only the probability of beacon and/or data collisions but also the energy consumption due to retransmissions.

In [17], the authors proposed the in-body sensor medium access control (i-MAC) protocol for implanted sensor communication in WBANs. They designed a superframe structure that separates emergency and regular event access. Further, they proposed a scheduled access mechanism according to the node priority. EE-DCAA [18] is a channel allocation mechanism that is compatible with the IEEE 802.15.6 WBAN standard. It exploits polling and contention access mechanisms. MG-HYMAC [19] is a hybrid MAC for WBANs. It integrates a transmission scheduling scheme into duty cycle operations in order to reduce the system energy consumption and extend the network lifetime.

DeepBAN [20] is a communication framework with channel prediction using a temporal convolution network (TCN) based deep learning approach. To enhance the energy efficiency of the system, the authors proposed a joint algorithm including time slot allocation, power control, and relay node selection. In the previous study [21], a WBAN power control scheme based on reinforcement learning was proposed. The end-edge-cloud orchestrated network architecture is adopted. The transmission power level is modified according to the current channel condition to reduce energy consumption. In [22], a WBAN MAC protocol based on the priority ladders resource scheduling (PLRS) scheme was used to degrade the difference of the time slot numbers among disparate sensor nodes with similar priorities. EEEA-MAC [23] is a hybrid CSMA/CA-TDMA pattern MAC for WBANs. The authors of [24] extended the superframe structure in the IEEE 802.15.6 standard. In the dynamic superframe protocol, the time slots dedicated based on priorities are assigned by the criteria importance through inter-criteria correlation (CRITIC).

In [25], the authors proposed a bargaining-based optimal slot sharing (BOSS) mechanism to solve the optimization problem of time slot assignment for sensor nodes. The communication among the sensing devices follows the IEEE 802.15.6 standard. A cooperative game theoretic method was designed to distribute the time slots to the sensors. It employs the Nash bargaining solution (NBS). In [26], the authors used the graph coloring problem to formulate the channel assignment problem. To solve the problem, the game-theoretic approach was adopted. A distributed two-hop incomplete coloring (DTIC) mechanism was proposed. In this scheme, the two-hop information is used to reuse channels. If the color number is inadequate to color all the sensing nodes, the algorithm addresses the incomplete graph coloring problem. In addition, the authors proposed a distributed message-passing scheme to avoid collisions during data transmission as well as to share coloring information among sensor nodes.

In [27], the authors used cellular-assisted device-to-device (D2D) communications for WBANs. A radio resource allocation mechanism using the Hungarian algorithm and convex optimization was proposed. The algorithm aims to optimize the system throughput, while mitigating collisions among WBANs and meeting the QoS requirements. In [28], the authors proposed a programmable MAC scheme that adds a command to the beacon of the IEEE 802.15.6 standard beacon frame. At the start of every superframe, the coordinator broadcasts the beacon to the sensor nodes. In [29], the authors proposed a greedy algorithm to solve the time slot allocation problem in WBANs. The priority of the sensor node is based on the degree of importance, data rate, remaining energy, and overtime. In [30], the authors adopted the IEEE 802.15.4 standard for body sensor networks, in which user behaviors are detected autonomously. They proposed a cross-layer time-slotted channel hopping (TSCH) scheme.

In [31], the authors presented an optimal MAC (OMAC) protocol, which employs the multi-dimensional (MD) graph optimal mechanism. This protocol attempts to achieve a compromise between the QoS and the energy consumption of data transmission. Before data transmission, the nodes wait for a period of time for more data. The length of the waiting time is determined by the application. Aggregation is used to reduce the size of the data. In [32], the authors proposed a radio resource allocation method based on the marginal utility theory. The allocation problem was modeled as a sum-utility maximization problem to attain the throughput of the required network QoS, considering the efficiency and fairness among the sensor nodes. In [33], the authors proposed a numerical MAC optimization algorithm for WBANs based on the IEEE 802.15.4 standard beacon-enabled mode. The algorithm aims to satisfy the existing traffic and QoS requirements, as well as to optimize the network delay, reliability, and energy.

The aforementioned studies proposed various WBAN access control protocols. However, these methods focus on the time division multiple access (TDMA) mechanism. There are few access protocols based on frequency division multiple access (FDMA). The channel utilization can be improved. In addition, the superframe and time slot length in the existing protocols are mostly fixed mechanisms. As the application scenarios of WBANs are heterogeneous, multi-source data have diverse requirements in terms of the QoS and quality of experience (QoE). The network performance of the MAC protocol must be improved.

3. MEC-Based Architecture for eHealth Systems

The traditional healthcare system network is based on the cloud architecture. Figure 1 shows an example of the network architecture of cloud-based eHealth systems. In one WBAN, the coordinator (or personal devices such as smartphones or PDAs) collects the sensing data. It also provides functions such as data storage, processing, fusion, analysis, and display to the access gateway (AG). After collecting the data, the coordinator transfers the data to the AG or access point (AP). The AG or AP may link to the Internet. The collected data can be uploaded to the remote cloud server, where data are processed, visualized, exhibited, and/or stored. In some systems, there is no AG or AP. The physiological data are transmitted to the Internet by smartphones through mobile networks. The reference nodes (RNs) are equipped with GPS or preprogrammed with their locations; they are used for sensor node localization. The medical provider, such as a physician or nurse, can access the patients' data through the Internet and thus examine, monitor, and/or provide medical services to the patients. Collaboration or consultation by experts from different locations can be carried out through the Internet. The nearest ambulance can be sent out if an emergency occurs. Furthermore, the patients' health information can be stored in the cloud. Data and statistical analysis can be performed in the long term.

However, with the rapid development of eHealth applications, the aforementioned cloud-based network architecture faces some challenges. Owing to the massive increase in the number of users and the large amount of heterogeneous data from diverse application scenarios, data transmission and processing become a heavy burden on the network and remote cloud, which have limited communication and computing resources. System QoS degradation may often occur because of data congestion. Moreover, the network QoS requirements of eHealth systems based on wearable sensors, such as the time delay, delivery ratio, reliability, and fairness, differ significantly among various applications. Some vital data, such as the heartbeats of cardiac patients, must be transmitted with low latency and high reliability. Massive data generated by wearable sensors and uploaded to the cloud will lead to unpredictable delays. In the case of an emergency, the existing system may not give timely feedback owing to the return delay of the remote server. Therefore, how to improve both the efficiency of data transmission and the QoS has emerged as a major challenge for enhancing the application scope of eHealth systems based on wearable sensing devices.

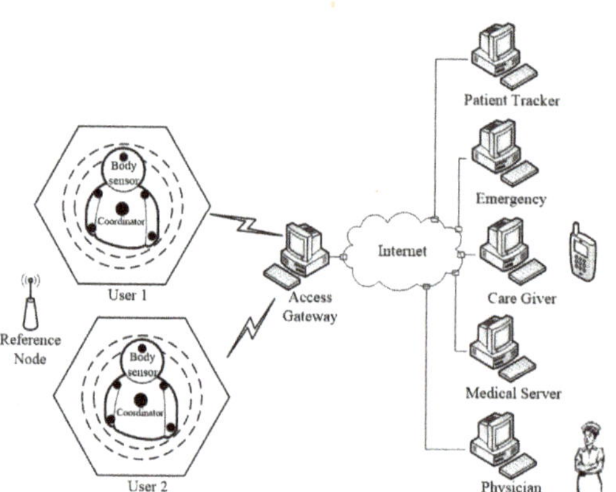

Figure 1. Network architecture of cloud-based eHealth systems.

In this study, we employ the MEC network architecture to solve the aforementioned problems. MEC technology is considered as a promising paradigm for the next-generation Internet, in which the function of clouds will progressively move toward the edge of networks [34]. The edge layer close to the user can effectively reduce the transmission delay, improve the rapid response ability, and significantly decrease the volume of data sent to the cloud [35]. The MEC servers can respond to the crucial request promptly and then contact a nearby ambulance if required. Compared with the traditional cloud-based architecture, it has many advantages, such as shorter latency, lower energy consumption, and higher QoS. Figure 2 shows an example of the network architecture of edge-cloud eHealth systems. The system has four tiers: intra-WBAN communication, inter-WBAN communication, mobile edge communication, and remote cloud server.

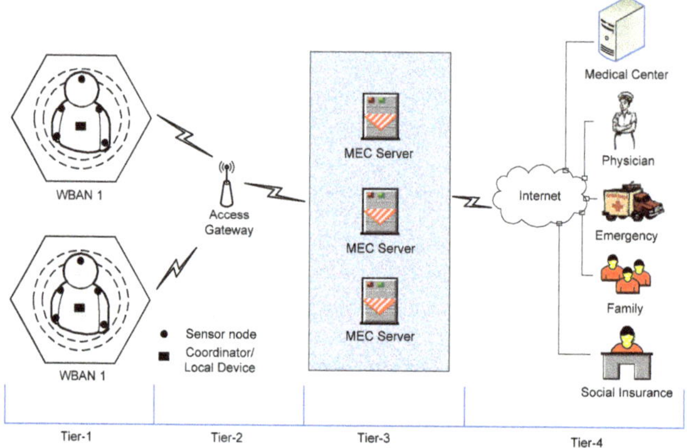

Figure 2. Network architecture of edge-cloud eHealth systems.

The first tier is intra-WBAN communication. In this tier, body sensing devices are placed inside, on, or around the human body to collect physiological data, such as the body temperature, heartbeat, ECG, EEG, and EMG. Activity sensors are positioned on the human body to detect the posture and movement, such as lying, walking, sitting, and running. The sensors are organized by short-distance wireless communication networks, such as Wi-Fi, Bluetooth, and ZigBee. Data are sent to the coordinator or local device, such as a smartphone. The gateway transfers the information to the next tier.

The second tier is inter-WBAN communication. This tier is the access gateway, which bridges the gap between the coordinator and the MEC servers. The coordinators of multiple users send data to the access gateway (AG) through a wireless or mobile network. Essentially, communication at this tier aims to connect the WBAN with other systems or networks.

The third tier is mobile edge communication, which is composed of MEC servers and the access gateway (AG). The coordinator transmits the data and requirements to the MEC server and AG. Edge computing aims to offload mobile devices to a nearby data process center, which can provide short-latency services. The MEC servers perform tasks in order of priority and data fusion. Then, the MEC servers relay the data to the cloud system and database. On account of data fusion, the data volume from the sensors to the cloud is reduced significantly. The utility efficiency of network communication and the computing capability can be increased by employing the edge computing architecture. Therefore, using the MEC architecture can improve the QoS and QoE of eHealth systems.

The fourth tier is remote cloud server. The data of patients are stored in the database. The doctors or expert systems examine the patients and then provide medical services. Through the Internet, experts from different locations can conduct a collaboration or consultation. If an emergency occurs, such as a heart attack, the eHealth system will send an alarm as well as the required information to the nearest ambulance [36]. Furthermore, the patients' health information can be stored in the cloud. Data and statistical analysis can be conducted in the long term.

4. Channel Allocation Based on MDP

In WBANs, the MAC protocol allocates the channel and time slot resources to multiple sensor nodes in the same individual area network. The MAC protocol attempts to not only avoid beacon and/or data frame collisions of the sensor nodes but also improve the transmission rate and throughput of the system. Therefore, the MAC protocol plays a critical role in improving the QoS and QoE of eHealth systems. In this section, we propose a MAC protocol based on an MDP for WBANs.

4.1. Problem Formulation Based on the Markov Decision Model

Considering the network communication of WBANs, the environment is regarded as the MDP. The process can be expressed as a five-tuple (S, A, P, γ, C), where S denotes a finite state set, A denotes a finite action set, and P denotes the transition probability. We assume that at time t, the system is in state s. Further, $P_a(s, s')$ denotes the probability of the system being in state s' at time $t + 1$ after taking action a, γ denotes the discount factor, and C denotes the reward value, which represents the reward value obtained after the system performs action a when the system is in state s. According to the property of the MDP, the conditional probability distribution of the future state of a stochastic process is only determined by the action under the current state.

The core idea of the Markov process is to find the optimal strategy π, which is the mapping from the state set to the behavior set. To evaluate the quality of the system, it is also necessary to define a value function v, which is the expected value of the reward obtained by implementing a scheme. It is defined as follows:

$$V^{\pi}(s, a) = E\left(\sum_{t=0}^{\infty} \gamma^t c_t \middle| s_0 = s\right) \tag{1}$$

where E is the mathematical expectation, $\gamma \in [0, 1)$ is the discount factor, and c_t is the immediate reward value generated at time t. The aforementioned formula is expressed recursively as follows:

$$V^\pi(s, a) = C(s, a) + \sum_{s' \in S} P_{ss'}(a) V^\pi(s') \tag{2}$$

where $C(s, a) = E[c(s, a)]$, which gives the average value of c, and $P_{ss'}(a)$ is the probability of transition from state s to state s', which indicates the next state. The optimal strategy π^* satisfies Bellman's criterion as follows:

$$V^*(s) = V^{\pi*}(s) = \min_{a \in A} \left[C(s, a) + \gamma \sum_{s' \in S} P_{ss'}(a) V^*(s') \right] \tag{3}$$

The Q function is defined on the basis of strategy π as follows:

$$Q^\pi(s, a) = C(s, a) + \sum_{s' \in S} P_{ss'}(a) V^\pi(s') \tag{4}$$

where $Q^\pi(s, a)$ is the expected reward value of state s after performing action a according to the strategy.

$$Q^*(s, a) = Q^{\pi*}(s) = C(s, a) + \gamma \left\{ \sum_{s' \in S} P_{ss'}(a) V^*(s') \right\} \tag{5}$$

where

$$V^*(s, a) = \min_{a \in A} [Q^*(s, a)] \tag{6}$$

Then, Q^* can be expressed as follows:

$$Q^*(s, a) = Q^{\pi*}(s) = C(s, a) + \gamma \left\{ \sum_{s' \in S} P_{ss'}(a) \min_{a \in A} V^*(s', a') \right\} \tag{7}$$

Then, the optimal value $Q^*(s, a)$ is found recursively. The initial value of the Q table is 0. Then, the criteria for Q-learning are as follows:

$$Q_{t+1}(s, a) = \begin{cases} Q_t(s, a) + \alpha \left\{ \left[c(s, a) + \gamma \min_{a' \in A} Q_t(s', a') \right] - Q_t(s, a) \right\}, \text{if } s = s_t, a = a_t \\ Q_t(s, a), \text{if } s \neq s_t, \text{or } a \neq a_t \end{cases} \tag{8}$$

where γ is the discount factor, which satisfies the condition $0 \leq \gamma < 1$, and α denotes the learning efficiency, which can be calculated as follows:

$$\alpha = \frac{1}{1 + m} \tag{9}$$

where m is the access number of the current state s.

If the Q value of each accessible (s, a) pair is infinitely accessible and the learning rate α is reduced to 0 appropriately, then the Q value will converge with a probability of 1. According to the learning criterion of the Q value, the optimal channel resource allocation scheme can be obtained by constantly updating the iterative Q table.

4.2. Action Selection Based on ε Greedy Strategy

The ε-greedy strategy is adopted to avoid falling into local optima in the learning process of the algorithm. The action selection is required after the execution of state s. To maximize the reward value after each action, two aspects must be considered. The first is the reward value brought by each action, and the second is the action with the largest reward value. If the reward corresponding to each action s is a certain value C, then the action with the highest reward value can be determined by performing all the possible actions. However, the reward value C of an action is derived from a probability distribution. The average reward value cannot be accurately obtained by traversing one action.

In each attempt, the probability of ε is used for exploration, i.e., an action is randomly selected with uniform probability. The action corresponding to the maximum Q value is selected by using the probability of $1-\varepsilon$. In general, ε takes a small constant value, such as 0.1. If the number of attempts is sufficient, the reward value of the action can be effectively approximated after a period of time. The unknown state is explored as much as possible. The optimal action is selected as far as possible, i.e., the optimal channel is selected.

4.3. Reward Function

The reward function is denoted by $c\,(s, a)$ in Equation (8). It is defined as the reward value of performing action a in state s at a certain time. More specifically, it is the reward when the current sensor node i selects the channel k, which is related to the service priority p and data frame length d of the WBAN node. According to the IEEE 802.15.6 standard, the service scenarios corresponding to the eight priorities are basic assurance, best effort service, excellent service, visual service, audio service, medical data or network control, high priority medical data, and emergency or medical implant events [12]. The higher the priority, the shorter is the time delay that can be tolerated, and the higher is the reliability requirement.

$$c(s, a) = (1 + p)^2 \cdot log_2(1 + d) \tag{10}$$

The following known conditions are listed: there are n sensor nodes and R available frequency resources. The channel state of sensor node i is defined as an r-dimensional vector, i.e., $U(i) = \{u_1(i), u_2(i), \ldots, u_R(i)\}, i \in \{1, 2, \ldots, N\}$, where

$$u_k(i) = \begin{cases} 1, & \text{channel } k \text{ is occupied by node } i \\ 0, & \text{others} \end{cases}, k \in \{1, 2, \ldots, R\} \tag{11}$$

4.4. Algorithm Flowchart

The objective of the algorithm is to obtain the optimal channel resource allocation mechanism to minimize the possibility of collisions when the sensor nodes transmit data. The flowchart of the MAC protocol based on the MDP is shown in Figure 3, which can also be described as follows:

Step 1: The system parameters are initialized. N denotes the number of sensor nodes, R denotes the number of available frequency resources of the system, T denotes the learning time, ε denotes the probability of the greedy strategy, γ denotes the discount factor, and α denotes the learning efficiency. The initial time t is 1.

Step 2: The system model parameters are initialized. According to the number of sensor nodes and frequency resources in the WBAN, the state space S and action space A are initialized. All the Q-function values in the table are set to 0. $Q\,(s, a) = 0$, where $s \in S$, $a \in A$. Furthermore, the channel state of sensor node i is defined as an r-dimensional vector.

Step 3: The system performs actions. A random number is generated in the current state s. If the random number is greater than ε, or if all the corresponding Q values are 0, i.e., if the state is accessed for the first time, then the action is selected randomly. If the random number is less than ε, the action corresponding to the maximum value in the Q table is selected. Thus, through this action a, a channel resource is allocated to the sensor node: $a \in \{1, 2, \ldots, R\}$. The state at a certain time is determined by i and $A(i)$, represented as $(i, A(i))$, where i denotes the current sensor node, $I \in \{1, 2, \ldots, N\}$, and $A(i)$ denotes the number of available channel resources of the current sensor node, $A(i) \in \{1, 2, \ldots, R\}$. The discount factor γ satisfies $0 \leq \gamma < 1$.

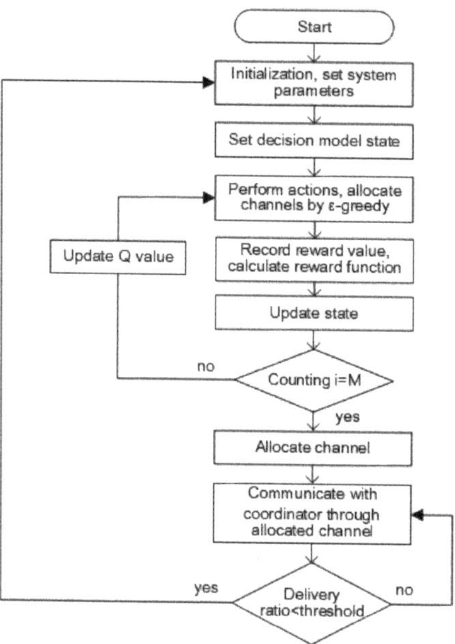

Figure 3. Flowchart of the MDP-based MAC protocol.

Step 4: Record the reward value. The reward value $c(s, a)$ appraises and calculates the effect of the actions according to Equation (10). After taking action a, the reward function value c and the next station s' are recorded.

Step 5: Update the Q table according to the reward value $c(x, k)$ and value Q by Equation (8). Update the value of $Q_{t+1}(s, a)$ accordingly. When $t < T$, go to step 3. When $t > T$, abort the learning process and accomplish the channel allocation.

Step 6: The sensor node communicates with the coordinator through the allocated channel. It maintains a transmission waiting queue with data priority.

As the link station varies with the change of the network topology and the node movement, the probability of collision during data transmission may increase, which can lead to a decrease in the delivery ratio. If the delivery ratio is lower than the threshold, the channel resources will be reallocated.

5. Performance Evaluation

This section describes the computer simulation and performance analysis of the proposed MDP-MAC protocol for WBANs introduced in the previous section. The proposed algorithm is compared with the in-body sensor medium access control (i-MAC) protocol [17], the energy efficient dynamic channel allocation algorithm (EE-DCAA) [18], and the distributed two-hop incomplete coloring (DTIC) algorithm [26]. The network performance metrics are the system energy consumption, delivery ratio, and system throughput. The quality of experience (QoE) is also defined and evaluated.

The energy consumption E is defined as the average energy consumption by each sensor.

$$E = \frac{E_t}{\sum_{i=1}^{n} T_s(i)} \tag{12}$$

where E_t denotes the system total energy consumption and $T_s(i)$ denotes the ith successful transmission. The delivery ratio D_r is defined as the percentage of the transmitted data frames that are successfully received.

$$D_r = \frac{\sum\limits_{i=1}^{n} T_s(i)}{\sum\limits_{i=1}^{n} T(i)} \tag{13}$$

where $T(i)$ denotes the ith transmission. The system throughput is calculated on the basis of the total successful transmitted packets and the total time T.

$$Thp = \frac{\sum\limits_{i=1}^{n} T_s(i)}{T} \tag{14}$$

Besides the QoS, the quality of experience (QoE) is also an important metric to evaluate network performance. QoE is the quality perceived subjectively by the users, which can be predicted and assessed by the QoS parameters [37]. In WBANs, the relationship between QoE and QoS is heterogeneous for different users. The QoE of the user is closely related to the data transmission demands. The QoE is influenced by the throughput. To evaluate the QoE precisely, it is defined as follows [38]:

$$Q_i = 5 - 5 \cdot e^{-\frac{c_i Thp_i}{Thp^i_{max}}} \tag{15}$$

where Q_i denotes user i's QoE, Thp_i is the user i's throughput, Thp^i_{max} is the maximum throughput that user i needs, which reflects the heterogeneous transmission requirements of different users, and c_i is the sensitive parameter to the throughput of user i. The system QoE, denoted by Q, is defined as the average Q_i of all the users.

In WBANs, the applications are extremely heterogeneous. The traffic data rates vary considerably. Applications transmitting simple data require a rate of a few kilobits per second. Video streams require a rate of several megabits per second. The transmission data rate may be significantly higher in a particular time period, which is called a burst. Table 1 summarizes the data rate requirements for some applications [1]. They are computed by the expected accuracy, range, and sampling rate. Overall, the user data levels are not regarded as high. Nevertheless, if the user wears many body sensors, such as ECG, temperature, EEG, and motion sensors, the system-assembled data rate can be several megabits per second, which is higher than that of commonly used radios.

Table 1. Data rate requirement of healthcare applications.

Application	Data Rate	Bandwidth (Hz)
Blood saturation	16 bps	0–1
ECG (12 leads)	288 kbps	100–1000
ECG (6 leads)	71 kbps	100–500
EEG (12 leads)	43.2 kbps	0–150
EMG	320 kbps	0–10,000
Temperature	120 bps	0–1
Motion sensor	35 kbps	0–500
Glucose monitoring	1600 bps	0–50
Cochlear implant	100 kbps	-
Voice	50–100 kbps	-
Audio	1 Mbps	-
Artificial retina	50–700 kbps	-

The sensor nodes of WBANs can be classified according to the priority of the data. There are four categories: non-constrained traffic class (NTC), delay traffic class (DTC), reliability traffic class (RTC), and critical traffic class (CTC). The NTC sensor nodes gather non-constrained data packets (NDP), which can tolerate a certain degree of losses and have a relaxed time delay requirement, such as blood pressure (BP) and temperature. The DTC sensor nodes collect delay data packets (DDP). These packets can tolerate some losses but have time delay constraints, such as telemedicine video imaging. The RTC sensor nodes collect reliability data packets (RDP). Such data have strict requirements in terms of the packet loss ratio but no time delay constraints, such as heart rate (HR) and respiratory rate (RR). The CTC sensor nodes collect critical data packets (CDP). These data packets have strict requirements in terms of the maximum loss and time delay, such as electroencephalogram (EEG) and electrocardiograph (ECG). Table 2 summarizes the four classes and priorities [2].

Table 2. Sensor classes and priorities.

Priority	Classification of Sensors	Traffic Class
0	Critical Data Packets (CDP)	Critical
1	Reliability Data Packets (RDP)	Reliability
2	Delay Data Packets (DDP)	Delay
3	Non-constrained Data Packets (NDP)	Non-constrained

The path loss (*PL*) model of WBANs can be described as follows:

$$PL(d, \theta) = a \cdot d + b + P(\theta) + N [\text{dB}] \tag{16}$$

where a denotes the gradient coefficient, which is experimentally measured as 1.92 (dB/cm); d denotes the distance; b is a constant, which has a value of 39.85 (dB) in real systems; and N denotes the stochastic fluctuation following a normal distribution $(0, \sigma_N)$, where the experimentally estimated value of σ_N is 6.59 (dB). Further, $P(\theta)$ denotes the fluctuation induced by the direction of the sensor, which can be calculated as follows:

$$P(\theta) = -20 \log_{10}(|\cos(\theta)| \cdot (1.0 - x_c) + x_c) \tag{17}$$

where θ denotes the angle between the transmitting and receiving antenna and x_c denotes the electric field direction difference between main and cross, which has a value of 0.145 in real experiments.

We simulate the performance of the proposed protocol in different scenarios with various node densities and environments. In the experiments, the sensor nodes and coordinators of the WBAN are deployed randomly in a square area of 10×10 m^2. The number of sensor nodes varies from 10 to 100. The star network topology is adopted, in which the coordinator is connected to multiple wearable sensor nodes. The WBAN coordinators remain static for 10 s after the simulation starts. Then, the coordinators begin to randomly move at a speed of [0, 1.5] (m/s). The sensing device mobility velocity is a random variable that is uniformly distributed in [0, 0.5] (m/s), which emulates the walking speed of humans. The simulation program is run 20 times to generate the average results. Every cycle of the simulation is executed independently. The average values are calculated according to the results.

In one cycle, the first data frame is generated randomly. The application traffic is assumed as a constant bit rate (CBR) distribution. The packet generation rate (PGR) is set to 1–40 packets/s for different sensors. The transport layer protocol is TCP. The data transmission rate (DTR) is set to 2–300 kbps. The maximal transmission requirement of a user is set to 10–2000 kbps. The sensitive parameter to the throughput of a user varies between 1 and 5. The buffer size (BS) is 1–4 Kbytes. The superframe length is assumed as 64 slots. The duration of a time slot is set to be 5 ms (milliseconds). The sizes of MAC header, beacon, ACK, and FCS are 27, 17, 7, and 2 (bytes), respectively. The payload length of the data frame is 30–200 bytes. The channel rate is 250 kbps. The frequency band is in the range of 2400–2483.5 MHz. The interference range of the sensors is assumed to be 2 m. The initial energy of the nodes is 100 J. The voltage supply is 3 V. Table 3 summarizes the energy consumption parameters employed in the simulation.

Table 3. Parameters of energy consumption.

State	Energy Consumption
Transmission data	15.9 mW
Reception data	22.2 mW
Idle status	18.4 mW
Sleep status	42 μW

Figure 4 shows the simulation results of the system energy consumption comparison of the proposed and existing protocols. The figure shows that the overall energy consumption increases continuously with the number of nodes. When the number of sensor nodes increases, the probability of allocating the idle channel decreases. The plot also clearly shows that the average energy consumption of each node of the proposed MDP-based MAC protocol is less than that of the existing MAC protocols. The energy consumption of the DTIC mechanism is lower than that of i-MAC and EE-DCAA.

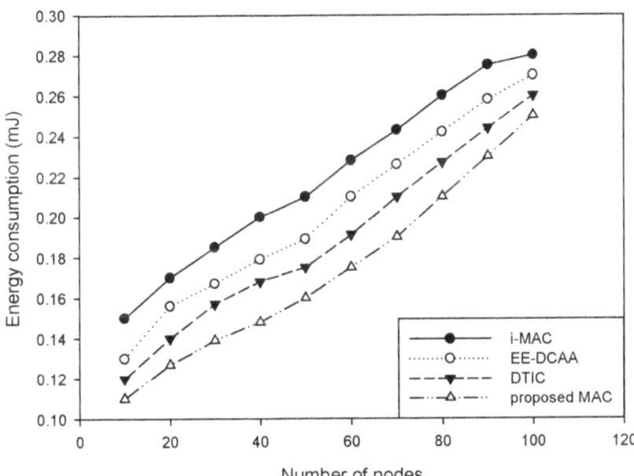

Figure 4. Simulation results of energy consumption.

For WBAN protocols, owing to the power limitation, energy efficiency is one of the most important metrics. The lifetime of WBANs can be extended by reducing the power consumption of the sensor nodes. The figure shows that by adopting the proposed MDP-based MAC protocol, the system energy consumption increases substantially. The edge-cloud network architecture reduces the data delay, thereby reducing the energy consumption. Moreover, the MDP-based MAC protocol minimizes the energy consumption while maintaining the link quality.

In the MDP-based MAC protocol, the channel utilization ratio can be decreased, which reduces not only the energy consumption of the system but also the probability of beacon and data collision. If a collision occurs, retransmission is required, which causes considerable power wastage. Therefore, the overlap area of the transmission range can be reduced. The proposed scheme can regulate the power consumption and extend the WBAN lifetime.

Figure 5 shows the WBAN delivery ratio comparison of the proposed MAC and other three protocols. Delivery ratio is another important metric for the network performance. As the number of users increases, the delivery ratio of the WBAN gradually decreases. The figure shows that the delivery ratio of the MDP-based MAC protocol is higher than that of i-MAC, EE-DCAA, and DTIC. The proposed protocol enhances the network performance in terms of the delivery ratio as it can reduce the beacon or data frame collisions. Meanwhile, this scheme also enhances the link reliability. Because of the lower collision rate, the successful transmission rate increases.

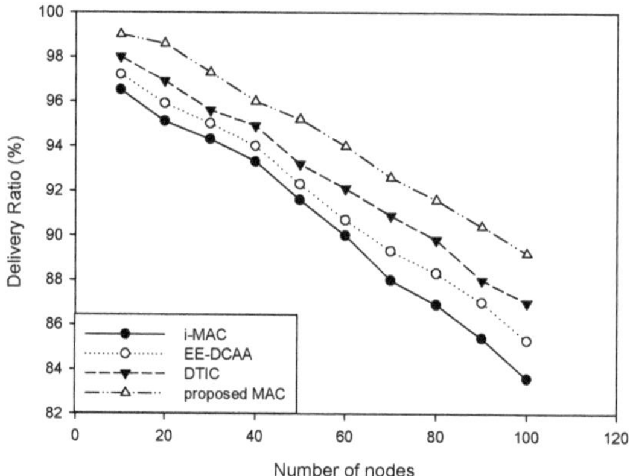

Figure 5. Simulation results of delivery ratio.

Figure 6 shows the network throughput of WBAN. The system throughput reflects the WBAN data transmission capability, which is another critical metric. As the number of nodes increases, the eHealth Internet of Things (IoT) system throughput also increases. The network throughput may be affected by a further increase in the number of sensor nodes as the probability of channel allocation decreases and the collision probability increases. Thus, the overall network throughput is affected. It is evident that the system throughput of the MDP-based MAC protocol is higher than that of i-MAC, EE-DCAA, and DTIC. The proposed protocol can reduce the beacon or data frame collisions, thereby decreasing the retransmission time. The system throughput increases as there are fewer collisions and the latency is reduced. As the retransmission rate is lower, the data delay is shorter. The channel utilization is improved, and the average time delay is reduced.

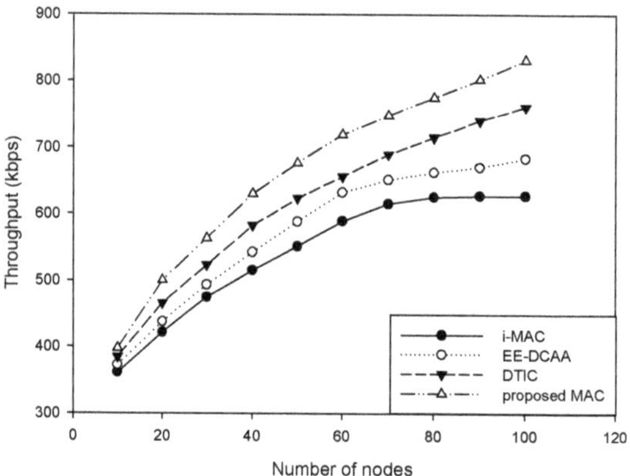

Figure 6. Simulation results of throughput.

Figure 7 demonstrates the simulation results of QoE of WBAN. The QoE reflects the average level of meeting the user's data transmission demands. As the number of nodes increases, the throughput of a user decreases, which causes a gradual decline in the QoE. It is evident that the QoE achieved using the MDP-based MAC protocol is higher than that achieved using i-MAC, EE-DCAA, and DTIC. In the proposed protocol, with the optimal strategy, the channel resources are allocated to minimize the probability of collision when the sensor nodes transmit data. This reduces the time wasted on retransmission.

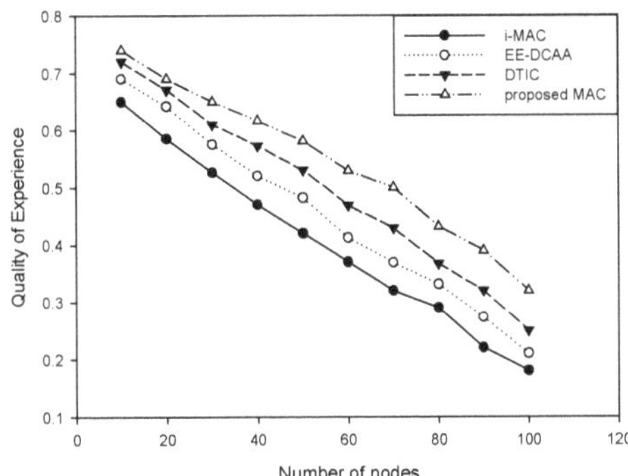

Figure 7. Simulation results of QoE.

6. Conclusions

WBANs support the development of IoT eHealth systems. Because of the limited battery capacity, the energy efficiency of sensing devices is a critical issue. Moreover, owing to the mobility of sensor nodes, the rapidly changing link state, and human body shadowing, communication reliability shares a trade-off relationship with energy consumption. To achieve a shorter task process delay, this study applied an MEC-based network architecture to an eHealth system. Further, an MDP-based MAC protocol was proposed. The energy utility model and optimization problem were solved using a learning algorithm. The performance evaluation of the proposed protocol showed that it achieves a higher system energy efficiency, delivery ratio, throughput, and QoE. In the future, we will perform some experiments to validate the performance of the proposed protocol.

Author Contributions: Conceptualization, H.S.; methodology, M.-S.P.; software, X.L.; validation, H.C.; writing—original draft preparation, H.S.; writing—review and editing, H.C. All authors have read and agreed to the published version of the manuscript.

Funding: This research was funded by the Beijing Natural Science Foundation (grant no. L222048) and NTUT-BJUT Joint Research Program (grant no. NTUT-BJUT-111-05).

Data Availability Statement: The data presented in this study are available on request from the corresponding author.

Conflicts of Interest: The authors declare no conflict of interest.

References

1. Khan, R.A.; Pathan, A.K. The state-of-the-art wireless body area sensor networks: A survey. *Int. J. Distrib. Sens. Netw.* **2018**, *14*, 23. [CrossRef]
2. Javaid, S.; Zeadally, S.; Fahim, H.; He, B. Medical sensors and their integration in Wireless Body Area Networks for Pervasive Healthcare Delivery: A Review. *IEEE Sens. J.* **2022**, *22*, 3860–3877. [CrossRef]
3. Habibzadeh, H.; Dinesh, K.; Shishvan, O.R.; Boggio-Dandry, A.; Sharma, G.; Soyata, T. A Survey of Healthcare Internet of Things (HIoT): A Clinical Perspective. *Ieee Internet Things J.* **2020**, *7*, 53–71. [CrossRef]
4. Yaghoubi, M.; Ahmed, K.; Miao, Y. Wireless Body Area Network (WBAN): A Survey on Architecture, Technologies, Energy Consumption, and Security Challenges. *J. Sens. Actuator Netw.* **2022**, *11*, 67. [CrossRef]
5. Ahmadzadeh, S.; Luo, J.; Wiffen, R. Review on Biomedical Sensors, Technologies and Algorithms for Diagnosis of Sleep Disordered Breathing: Comprehensive Survey. *IEEE Rev. Biomed. Eng.* **2022**, *15*, 4–22. [CrossRef] [PubMed]
6. Misra, S.; Roy, A.; Roy, C.; Mukherjee, A. DROPS: Dynamic Radio Protocol Selection for Energy-Constrained Wearable IoT Healthcare. *Ieee J. Sel. Areas Commun.* **2021**, *39*, 338–345. [CrossRef]

7.   Samanta, A.; Li, Y. Distributed Pricing Policy for Cloud-Assisted Body-to-Body Networks with Optimal QoS and Energy Considerations. *IEEE Trans. Serv. Comput.* **2021**, *14*, 668–682. [CrossRef]

8.   Hammood, D.; Alkhayyat, A. An overview of the Survey/Review Studies in Wireless Body Area Network. In Proceedings of the 2020 3rd International Conference on Engineering Technology and its Applications (IICETA), Najaf, Iraq, 6–7 September 2020; pp. 18–23.

9.   Mao, Y.; You, C.; Zhang, J.; Huang, K.; Letaief, K.B. A Survey on Mobile Edge Computing: The Communication Perspective. *IEEE Commun. Surv. Tutor.* **2017**, *19*, 2322–2358. [CrossRef]

10.   Bishoyi, P.K.; Misra, S. Enabling Green Mobile-Edge Computing for 5G-Based Healthcare Applications. *IEEE Trans. Green Commun. Netw.* **2021**, *5*, 1623–1631. [CrossRef]

11.   *IEEE Std 802.15.4-2011 (Revision of IEEE Std 802.15.4-2006)*; IEEE Standard for Local and metropolitan area networks—Part 15.4: Low-Rate Wireless Personal Area Networks (LR-WPANs). IEEE: New York, NY, USA, 2011; pp. 1–314. [CrossRef]

12.   *IEEE Std 802.15.6-2012*; IEEE Standard for Local and metropolitan area networks—Part 15.6: Wireless Body Area Networks. IEEE: New York, NY, USA, 2012; pp. 1–271. [CrossRef]

13.   Saboor, A.; Ahmad, R.; Ahmed, W.; Kiani, A.K.; Le Moullec, Y.; Alam, M.M. On Research Challenges in Hybrid Medium-Access Control Protocols for IEEE 802.15.6 WBANs. *Ieee Sens. J.* **2019**, *19*, 8543–8555. [CrossRef]

14.   Razzaque, M.A.; Hira, M.T.; Dira, M. QoS in Body Area Networks: A Survey. *ACM Trans. Sens. Netw.* **2017**, *13*, 46. [CrossRef]

15.   Lloret, J.; Parra, L.; Taha, M.; Tomás, J. An architecture and protocol for smart continuous eHealth monitoring using 5G. *Comput. Netw.* **2017**, *129*, 340–351. [CrossRef]

16.   Khalid, K.; Li, G.H. A Survey of MAC Protocols on Wireless Body Area Network. In Proceedings of the 4th Annual International Conference on Information Technology and Applications, Guangdong, China, 26–28 May 2017; Long, L., Li, Y., Li, X., Dai, Y., Yang, H., Eds.; ITM Web of Conferences. EDP Sciences: Paris, France, 2017; Volume 12.

17.   Misra, S.; Bishoyi, P.K.; Sarkar, S. i-MAC: In-Body Sensor MAC in Wireless Body Area Networks for Healthcare IoT. *IEEE Syst. J.* **2021**, *15*, 4413–4420. [CrossRef]

18.   Ashraf, M.; Hassan, S.; Rubab, S.; Khan, M.A.; Tariq, U.; Kadry, S. Energy-efficient dynamic channel allocation algorithm in wireless body area network. *Environ. Dev. Sustain.* **2022**, 1–17. [CrossRef]

19.   Olatinwo, D.D.; Abu-Mahfouz, A.M.; Hancke, G.P. Energy-Aware Hybrid MAC Protocol for IoT Enabled WBAN Systems. *IEEE Sens. J.* **2021**, *22*, 2685–2699. [CrossRef]

20.   Liu, K.; Ke, F.; Huang, X.; Yu, R.; Lin, F.; Wu, Y.; Ng, D.W.K. DeepBAN: A Temporal Convolution-Based Communication Framework for Dynamic WBANs. *IEEE Trans. Commun.* **2021**, *69*, 6675–6690. [CrossRef]

21.   Su, H.R.; Yuan, X.M.; Tanga, Y.J.; Tian, R.; Sun, E.C.; Yan, H.R. A Learning-based Power Control Scheme for Edge-based eHealth IoT Systems. *Ksii Trans. Internet Inf. Syst.* **2021**, *15*, 4385–4399. [CrossRef]

22.   Liang, B.; Obaidat, M.S.; Liu, X.; Zhou, H.; Dong, M. Resource Scheduling Based on Priority Ladders for Multiple Performance Requirements in Wireless Body Area Networks. *IEEE Trans. Veh. Technol.* **2021**, *70*, 7027–7036. [CrossRef]

23.   Liang, B.; Liu, X.; Zhou, H.; Leung, V.C.M.; Liu, A.; Chi, K. Channel Resource Scheduling for Stringent Demand of Emergency Data Transmission in WBANs. *IEEE Trans. Wirel. Commun.* **2021**, *20*, 2341–2352. [CrossRef]

24.   Das, K.; Moulik, S.; Chang, C.Y. Priority-Based Dedicated Slot Allocation with Dynamic Superframe Structure in IEEE 802.15.6-Based Wireless Body Area Networks. *IEEE Internet Things J.* **2022**, *9*, 4497–4506. [CrossRef]

25.   Das, K.; Moulik, S. BOSS: Bargaining-based Optimal Slot Sharing in IEEE 802.15.6-based Wireless Body Area Networks. *IEEE Internet Things J.* **2021**, *10*, 2945–2953. [CrossRef]

26.   Wu, K.J.; Hong, Y.W.P.; Sheu, J.P. Coloring-Based Channel Allocation for Multiple Coexisting Wireless Body Area Networks: A Game-Theoretic Approach. *IEEE Trans. Mob. Comput.* **2022**, *21*, 63–75. [CrossRef]

27.   Kong, P.Y. Cellular-Assisted Device-to-Device Communications for Healthcare Monitoring Wireless Body Area Networks. *IEEE Sens. J.* **2020**, *20*, 13139–13149. [CrossRef]

28.   Michaelides, C.; Pavlidou, F.N. Programmable MAC in Body Area Networks, One Command at a Time. *IEEE Sens. Lett.* **2019**, *3*, 7500604. [CrossRef]

29.   Sun, G.; Wang, K.; Yu, H.; Du, X.; Guizani, M. Priority-Based Medium Access Control for Wireless Body Area Networks with High-Performance Design. *IEEE Internet Things J.* **2019**, *6*, 5363–5375. [CrossRef]

30.   Ngo, M.V.; La, Q.D.; Leong, D.; Quek, T.Q.S.; Shin, H. User Behavior Driven MAC Scheduling for Body Sensor Networks: A Cross-Layer Approach. *IEEE Sens. J.* **2019**, *19*, 7755–7765. [CrossRef]

31.   Rasool, S.M.; Gafoor, S.A.A.A. OMAC: Optimal Medium Access Control Based Energy and QoS Compromise Techniques for Wireless Body Area Networks. *Wirel. Pers. Commun.* **2022**, *123*, 3223–3240. [CrossRef]

32.   Shen, S.; Qian, J.; Cheng, D.; Yang, K.; Zhang, G. A Sum-Utility Maximization Approach for Fairness Resource Allocation in Wireless Powered Body Area Networks. *IEEE Access* **2019**, *7*, 20014–20022. [CrossRef]

33.   Alimorad, N.; Maadani, M.; Mahdavi, M. REO: A Reliable and Energy Efficient Optimization Algorithm for Beacon-Enabled 802.15.4–Based Wireless Body Area Networks. *IEEE Sens. J.* **2021**, *21*, 19623–19630. [CrossRef]

34.   Amin, S.U.; Hossain, M.S. Edge Intelligence and Internet of Things in Healthcare: A Survey. *Ieee Access* **2021**, *9*, 45–59. [CrossRef]

35.   Ning, Z.L.; Dong, P.R.; Wang, X.J.; Hu, X.P.; Guo, L.; Hu, B.; Guo, Y.; Qiu, T.; Kwok, R.Y.K. Mobile Edge Computing Enabled 5G Health Monitoring for Internet of Medical Things: A Decentralized Game Theoretic Approach. *Ieee J. Sel. Areas Commun.* **2021**, *39*, 463–478. [CrossRef]

36. Aski, V.J.; Dhaka, V.S.; Kumar, S.; Verma, S.; Rawat, D.B. Advances on networked ehealth information access and sharing: Status, challenges and prospects. *Comput. Netw.* **2022**, *204*, 108687. [CrossRef]
37. Taha, M.; Ali, A.; Lloret, J.; Gondim, P.R.L.; Canovas, A. An automated model for the assessment of QoE of adaptive video streaming over wireless networks. *Multimed. Tools Appl.* **2021**, *80*, 26833–26854. [CrossRef]
38. Yao, C.; Jia, Y.; Wang, L. The QoE Driven Transmission Optimization Based on Cognitive Air Interface Match for Self-Organized Wireless Body Area Network. *IEEE Access* **2019**, *7*, 138203–138210. [CrossRef]

Article

Data Glove with Bending Sensor and Inertial Sensor Based on Weighted DTW Fusion for Sign Language Recognition

Chenghong Lu, Shingo Amino and Lei Jing *

Graduate School of Computer Science and Engineering, University of Aizu, Tsuruga, Ikki-machi, Aizuwakamatsu City 965-8580, Japan
* Correspondence: leijing@u-aizu.ac.jp

Abstract: There are numerous communication barriers between people with and without hearing impairments. Writing and sign language are the most common modes of communication. However, written communication takes a long time. Furthermore, because sign language is difficult to learn, few people understand it. It is difficult to communicate between hearing-impaired people and hearing people because of these issues. In this research, we built the Sign-Glove system to recognize sign language, a device that combines a bend sensor and WonderSense (an inertial sensor node). The bending sensor was used to recognize the hand shape, and WonderSense was used to recognize the hand motion. The system collects a more comprehensive sign language feature. Following that, we built a weighted DTW fusion multi-sensor. This algorithm helps us to combine the shape and movement of the hand to recognize sign language. The weight assignment takes into account the feature contributions of the sensors to further improve the recognition rate. In addition, a set of interfaces was created to display the meaning of sign language words. The experiment chose twenty sign language words that are essential for hearing-impaired people in critical situations. The accuracy and recognition rate of the system were also assessed.

Keywords: data glove; wearable device; sign language recognition; ubiquitous computing

Citation: Lu, C.; Amino, S.; Jing, L. Data Glove with Bending Sensor and Inertial Sensor Based on Weighted DTW Fusion for Sign Language Recognition. *Electronics* **2023**, *12*, 613. https://doi.org/10.3390/electronics12030613

Academic Editor: Paolo Visconti

Received: 11 December 2022
Revised: 23 January 2023
Accepted: 24 January 2023
Published: 26 January 2023

1. Introduction

In Japan, there are about 341,000 hearing impaired people [1]. The general way to communicate between a hearing person and a hearing-impaired person is communication by writing or sign language. However, communication by writing takes a lot of time. Furthermore, sign language, which hearing impaired people use, is not always familiar to hearing people or those who acquired a hearing impairment. Each of the two approaches has problems that hinder smooth communication in society.

Sign language recognition has always been a research problem that has received a lot of attention. There has been a large number of studies on sign language recognition in recent years [2–6].

Sign language recognition systems can be divided into non-wearable and wearable approaches. Non-wearable generally include vision-based [7,8], while non-wearable use WiFi signal-based [9,10] methods. Another approach is to recognize sign language with wearable sensor-based data gloves [11,12].

Due to the development of deep learning methods in visual sign language recognition, the recognition rate has been improved. However, deep learning is driven by data, and the quality of data collection greatly affects the results. Insufficient video frames and occlusions will also reduce the recognition accuracy. Gerges et al. [13] established dynamic hand recognition based on MediaPipe's Landmarks and compared the recognition accuracy of three deep learning methods: gated recurrent unit (GRU), long short term memory (LSTM), and bi-directional LSTM (BILSTM). Dataset collection requires complete characters, no occlusions, and a fixed duration. It is difficult to achieve these requirements in actual use.

Chang et al. [14] conducted research on recognizing sign language by detecting the places of nails and wrists using pictures of the hand. It recognizes sign language by a skeleton of the hand and the distribution of skin color from pictures taken of the hand. However, the systems that hearing-impaired people need to use in their daily lives can detect not only the hand shape part of sign language, but also the dynamic part of hand movement in sign language. In other vision-based methods, one approach uses color gloves and Kinect data stored by Microsoft. Shibata et al. [15] used color gloves for recognizing sign language. The color glove has every color at every finger and wrist. Furthermore, it operates by moving the distance and area of glove colors. However, in the detection step, if the background or the user's clothing are the same color as the part of the glove, it cannot be recognized in this way. Kinect can detect hand motions and hand places. Muaaz et al. [16] developed a system that can recognize American sign language with Kinect. This system has a high average recognition rate of 80%. Furthermore, this system can make easy sentences by recognizing sign language words. However, this system is also limited by the camera, and we can only use this system in limited positions without occlusion. In daily life, it can be large barrier for the hearing impaired to use this system.

Vision-based sign language recognition is limited by the nature of camera view observation and is not good at capturing complex two-handed interaction movements because of occlusion. It is also susceptible to the influence of the environment in-between the camera and the object. The method of wearable sensors and a data gloves forces users to accept some burdens. However, data gloves can collect data steadily in complex environments, without the problem of line of sight obstruction, noisy backgrounds, and inadequate light. It can even be used outdoors, and in low visibility. The camera method is subject to a variety of environmental constraints. Therefore, we plan to use wearable devices to capture the complex motion of the fingers.

In recent years, wearable sensor-based data gloves have been developed with continuous improvements in processing information technology and the miniaturization and high functionality of equipment. Wearable sensor-based data gloves have been able to operate a large amount of information and handle more complex processing.

Common wearable sensor data gloves for sign language recognition include flexible sensors [17], inertial measurement units (IMUs) [18], surface electromyography (sEMG) devices [19,20], and touch sensors [21]. EMG data have large individual variation. When using the bilinear model for classification, a new subject needs to perform at least one motion. Furthermore, the recognition rate will drop significantly without using a bilinear model.

The information directly related to the hand in sign language includes 21 degrees of freedom of the joints on the hand, and the spatial displacement and orientation of the hand. It is difficult to obtain the appropriate characteristics of complicated information through a single type of sensor. Korzeniewska et al. [17] chose Velostat to make bending sensors to collect data to identify Polish sign language and obtained a letter recognition rate of 86.5%. However, sign language generally uses words as the unit of recognition. Youngmin Na et al. [18] installed an accelerometer on the index finger to recognize static letter gestures in the Korean sign language alphabet, but sign language contains a lot of dynamic gestures, and only static gesture recognition was not enough. Jakub et al. [22] collected IMU sensor data installed on the palm and fingertips and used parallel hidden Markov model (HMM) approaches for sign language recognition. The finger shape data could be obtained by combining the IMU data on the fingertips and the IMU data on the palm. For collecting hand shape features, multiple inertial sensors are more expensive than multiple bending sensors.

Data gloves from a single type of sensor either collect much missing hand information or have a high price for implementation. Thus, multi-sensor fusion is a better solution. The use of wearable sensors and data gloves is moving toward practical applications as advanced MEMS technology sensors are being miniaturized. It also breaks down the spatial limitations of the hand, making multi-sensor data collection possible. Furthermore, among the multiple combinations, inertial sensors to collect hand motion and bending sensors to collect hand shape are the common approaches [23–26]. Faisal et al. [23], using the K-nearest neighbor (KNN), classified 14 static and 3 dynamic gestures for sign language recognition. Faisal et al. [24] collected data from 25 subjects for 24 static and 16 dynamic American sign language gestures for validating the system. Boon Giin Lee et al. [25] used the support vector machine (SVM) to classify American sign language.

The combination of inertial sensors and bending sensors helps us to obtain hand shape and motion information at low cost. However, how to rationalize multiple sensor data for sign language recognition is still a difficult problem. The execution length of the actions of sign language varies greatly due to people's habits or usage scenarios. The dynamic time warping (DTW) algorithm is a solution to compare the similarity between time series data of different lengths. However, the current research on the application of the DTW algorithm to sign language recognition is insufficient. Chu et al. [26] studied DTW for sign language recognition on seven Japanese sign language datasets, and validation was performed using the leave one out (LOO) approach, with recognition rates of 82.5%. First of all, seven recognition actions were insufficient. On the other hand, the variation between different sensors was significant in providing useful information for sign language recognition. Thus, it is necessary to propose weighted DTW. As shown in Table 1, we compared the studies of various sensors. Portability in the table refers to whether good results can be obtained without any data from new users.

Table 1. Comparison of related research (KNN: K-nearest neighbor).

Research	Sensor	Accuracy	Subject	Kinds	Portability	Algorithm	Dynamic Motion
Muaaz et al. [16]	Kinect	95.6%	5	10	○	DTW	○
Tateno et al. [19]	EMG	97.7%	20	20	×	LSTM	○
Lee et al. [21]	Touch	92%	-	36	○	Tree	×
Faisal et al. [23]	Inertial and Flex	64%	35	3	○	KNN	○
Chu et al. [26]	Inertial and FlexForce	82.5%	3	7	○	DTW	○
Ours	Inertial and Flex	85.21%	8	20	○	weighted DTW	○

In this study, inertial sensors and bending sensors were deployed simultaneously in the hand space to collect hand shape and motion features. This method is a practical and promising solution to combine these two parts of features to recognize sign language. Thus, in this research, the Sign-Glove system was implemented, as shown in Figure 1. The development of such systems will give us a future where we wear sensors such as accessories that make it easier to communicate between a hearing person and a hearing-impaired person. When we developed the system to recognize sign language on portable devices with recent technology, for the recognition algorithm of sign language, we extended DTW to use on time series of multiple sensors. DTW is a general method for measuring the similarity between two temporal sequences. However, for data from multiple sensors, different sensors provide different recognition contributions. Thus, we proposed weighted DTW, an algorithm that improves the recognition rate by setting weighted values to raise the effect of key sensors. The contribution of the paper is as follows:

We developed a low-cost Sign-Glove system combining bending sensors and IMU for supporting communication in the context of hearing impairment. To determine hand shape and hand motion, the device supports the simultaneous collection of bending sensors and inertial sensors. We built the weighted DTW algorithm to implement multi-sensor fusion for sign language recognition. The algorithm does not limit the input data length. In addition, weights were assigned based on the contribution of sign language recognition. The contribution was analyzed based on the differences in the features of the sensors and the measurement locations of the hands. Assigning weights enhanced the influence of key sensors and reduced the errors caused by noisy data, effectively improving the recognition rate.

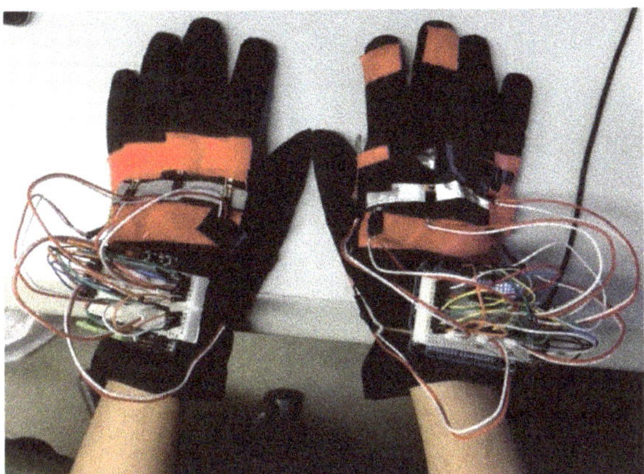

Figure 1. The Sign-Glove on hands.

The organization of this paper is as follows. The application model and sign language dataset are presented in Section 2. Our system design, implementation, and algorithms in this study are given in Section 3. The experimental setup and results evaluation are in Section 4. The discussion of the system is presented in Section 5. Finally, conclusions are given in Section 6.

2. Application Model and Sign Languages Datasets

2.1. Application Model

The system presented in this research shows the meaning of a sign language word on a PC for supporting communication between a healthy person and a hearing-impaired person. This system supposed that a user uses a pair of Sign-Gloves and shows the meaning of a sign language word on the PC.

A user wears Sign-Gloves on his/her hands. Furthermore, the user moves the motion of a sign language word. Then, the PC shows the mean of the sign language. A Sign-Glove is a glove-shaped device with a WonderSense device and a bending sensor. WonderSense is the device developed in this laboratory. This model supposed that the user wants to communicate a sign language word motion to another person. We explain the process of this system in Figure 2. A user moves the motion of the sign language word that he/she wants to communicate to another person. Sign-Glove measures the acceleration of hand motion and hand shape at this time. the WonderSense device of the Sign-Glove transmits the measured hand acceleration to WonderBox with Bluetooth low energy. WonderBox is the receiver device of WonderSense. WonderBox sends measured hand acceleration data to a PC with a serial connection. At the same time, the bending sensors of Sign-Glove measure the hand shape. Furthermore, an Arduino board sends measured data with a serial connection. An Arduino is an AVR Micon board and is used for taking data from the bending sensors and sending data to the PC. After the finished sign language gesture, the data values sent by sensors are computed to recognize a sign language word motion. This sign language word motion is converted into a message associated with the sign language word motion in the PC. Finally, the PC displays the message requested by the user. At this time, if the message is a serious one, the PC makes a sound.

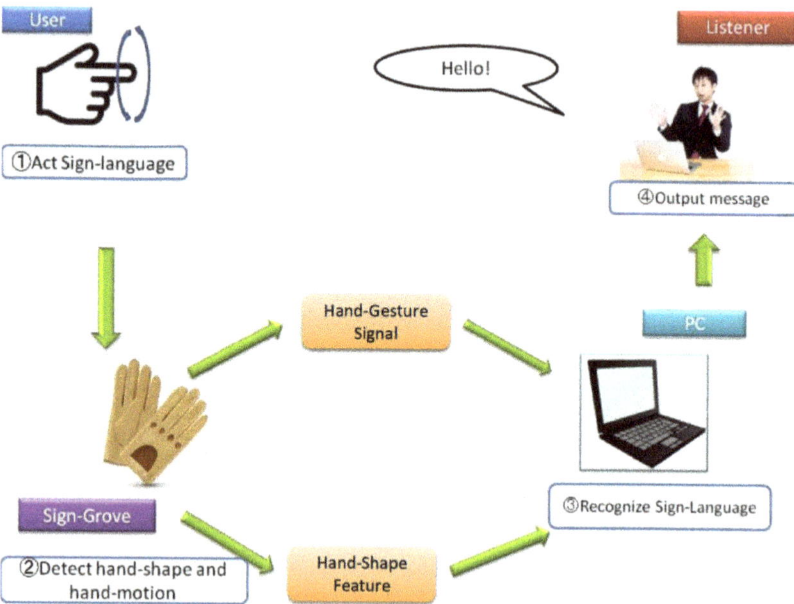

Figure 2. Usage of Sign-Glove for sign language recognition.

2.2. Sign Languages Datasets

2.2.1. Characteristics of Sign Language Data

Sign language consists of two main components in the hand part, namely, the shape of the hand and the overall movement of the hand. Static sign language is defined as a special case of dynamic sign language, which specifically means that the shape of the hand and the hand motion remain unchanged for a period of time. Figure 3 shows the hand shape parts and the hand motion parts of sign language.

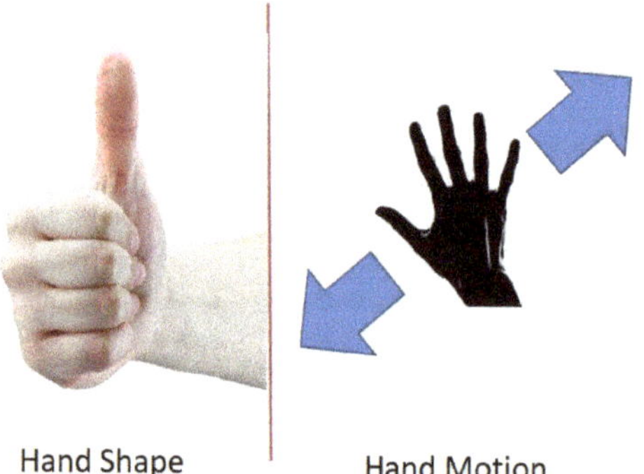

Figure 3. Hand shape and hand motion factors for sign languages.

2.2.2. Sign Language Dataset Definition

The key point to recognizing sign language is to recognize the hand shape and the hand motion at the same time. Missing one of them will significantly reduce the recognition rate, such as "please" and "good", "sick" and "obstacle", "down" and "I see", as shown in Figure 4, because the hand motion of these sign languages is the same but the shape of the hands is different. If we detect only the hand motion of these sign language words, the result is that this sign language is completely the same. In contrast, Sign-Glove used in this research can detect hand shape. Thus, we can increase the recognition rate of sign language words. Furthermore, for the same reason, we can also achieve the correct result of recognizing sign language words, which is the same hand shape and different hand motions.

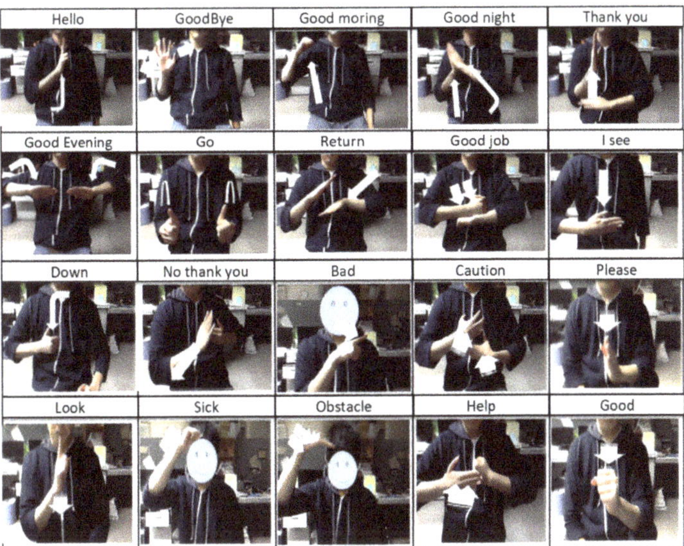

Figure 4. Selection of sign language vocabularies with both hand shape and hand motion factors.

3. Methods

The system architecture is shown in Figure 5. The data glove collecting the physical features, and the communication structure are shown in Figure 6. We explain the design of the system in Section 4.1. We explain the recognition algorithm in Section 4.2.

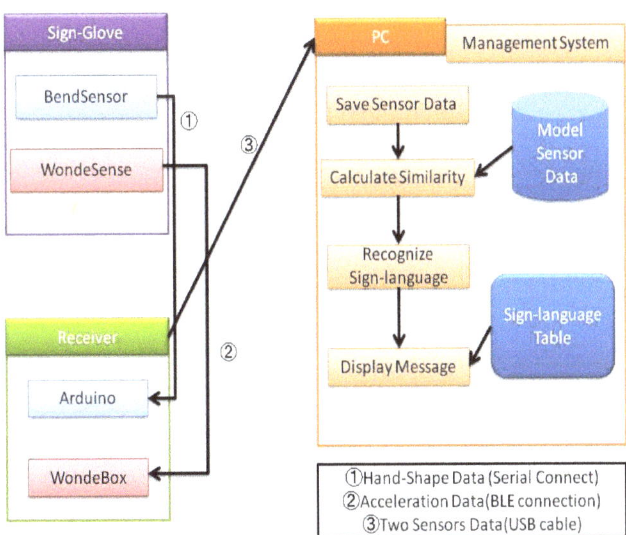

Figure 5. Architecture of the system.

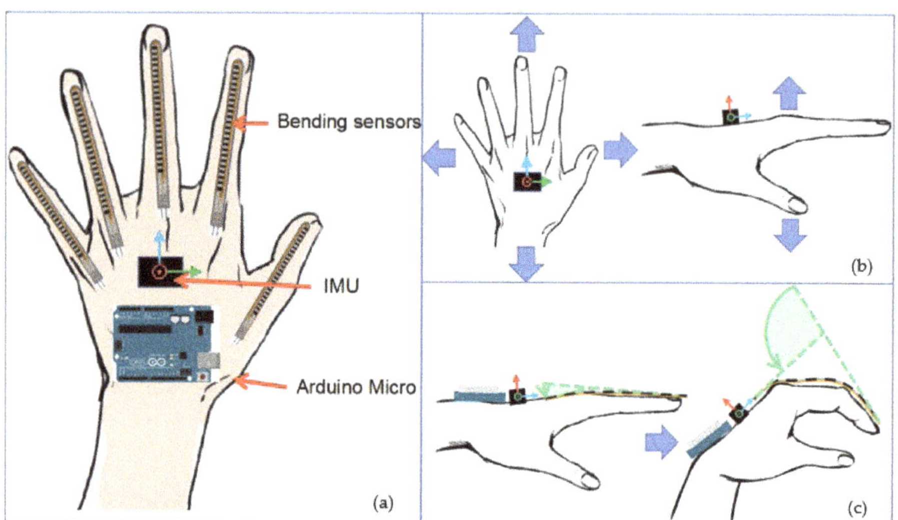

Figure 6. (**a**) System structure diagram; (**b**) IMU collecting the hand motion data; (**c**) bending sensor collecting the hand shape data.

3.1. System Design

In this section, we explain how a sign language word is recognized by the system constructed in this research and the way to detect it. First of all, we explain the operation of this system. This research system recognizes a sign language word based on bending condition data of fingers detected by a bending sensor and acceleration data of hands detected by WonderSense.

First, the Sign-Glove worn on a user's hands takes bending fingers condition data and acceleration data of hands from the bending sensor and the 3-axis acceleration sensor. The bending fingers condition data are sent from the Arduino to a PC with a serial connection via the USB cable. At the same time, the acceleration data of hands detected by WonderSense are sent to WonderBox, which is a data receiver with a BLE connection. Furthermore, WonderBox sends the data to a PC with a serial connection. Until the end of a sign language word motion, Sign-Glove continues to receive the values. Acceleration and bending finger state data are stored in the computer. After that, the model data are read.

The model data are defined with the 3-axis acceleration data and bending data of the fingers for all sign language words. For the recognition process, we use the weighted DTW algorithm to calculate the similarity between the sign language data to be recognized and the model data. The highest similarity of the model data is selected. Finally, the meaning of the sign language word is extracted from the sign language table and displayed on the screen.

As shown in Figure 6, the data glove collects the physical features of the hand. The IMU collects the motion features of the hand, as shown in Figure 6b. The bending sensor collects the shape features of the hand, as shown in Figure 6c. Both sensors are stitched to the cloth glove at the corresponding locations for fixation. The bending sensor is fixed in a special way. When the finger is bending, the skin is stretched. However, the length of the bend sensor is fixed, so we fix the top of the bend sensor to the fingertip position of the glove, and the middle of the sensor is restricted by the wire without shifting from left to right. Furthermore, the back end of the sensor is not fixed but is free to stretch, only allowing the sensor and the back of the hand to move as far as possible to fit.

3.2. Implementation

3.2.1. Hardware

We explain the construction of the hardware in this research. Sign-Glove is the device that takes the acceleration of the hand gesture and hand shape. Sign-Glove has two kinds of sensors. One of the two sensors is the bending sensor. Figure 7a shows the bending sensor. The bending sensor changes its resistance by the bending condition. The bending sensor has a polymer ink on one side if the sensor, which has about 30 KΩ of resistance when straight and increases when it is bent. The Arduino board detects the change as a voltage change value and sends it to the PC with a USB cable. The Arduino is an AVR Micon board. We use it as a banding sensor receiver. The other sensor is WonderSense. Figure 7b shows WonderSense.

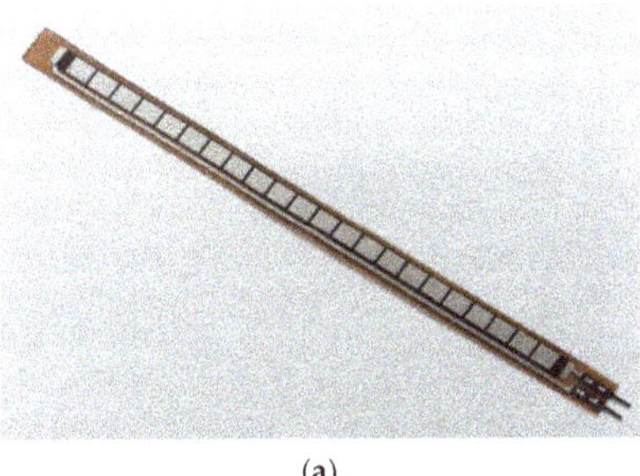

(a)

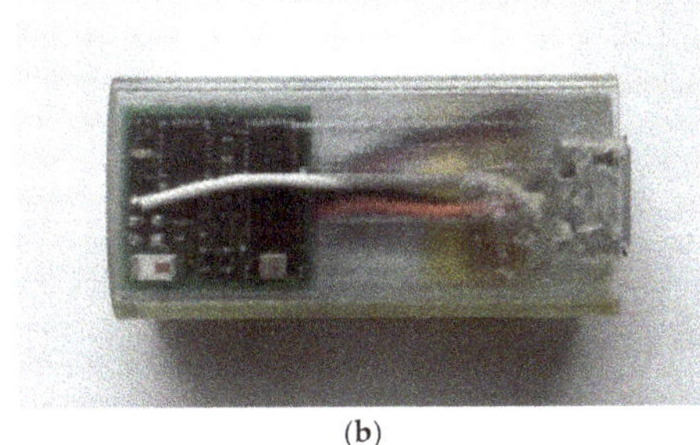

(b)

Figure 7. Sensors and data collection devices: (**a**) bending sensor; (**b**) WonderSense.

WonderSense collects acceleration data using a 9-axis inertial sensor module MPU9250. WonderBox is the data receiver of WonderSense. The core chip of the WonderBox is the PCA10040 for Bluetooth data reception. WonderBox sends data to a PC with a USB cable. A Sign-Glove device is a pair of gloves. Sign-Glove is constructed by ten bending sensors, two Arduino boards, two WonderSense sensors, and one WonderBox device. To facilitate synchronization, we sampled both the inertial and bending sensor data sampling rate to 50 Hz.

3.2.2. Software

In this research, the environment of the PC for recognizing sign language was structured by the Java language written by the integrated development environment of Eclipse in the OS of the Mac OS X 64-bit system. The DB system is MySQL of the XAMMP application. Furthermore, we used WonderTerminal to manage the data of WonderBox taken by WonderSense. WonderTerminal is software designed for controlling WonderBox and WonderSense.

3.2.3. User Interface

Figure 8 shows the user interface of our system. We explain here every function of our user interface. ① is the start button. When we click this button, recognition of sign language starts. Furthermore, when we click again, recognition of sign language stops, and ⑨ shows a recognized sign language word. ② is the save button. When the save button is clicked, sign language data that the system now contains is saved to the database. The destination database in which to save is decided by database selectors. Furthermore, what kinds of sign language words to save is decided by a sign language selector. ⑦ comprises the extract buttons. The extract AccData button is the button that reads the acceleration data of a database. The extract BendData button is the button that reads the bending data of a database. At this time, the user interface decides which database to read from and what kind of sign language word by ③. In addition, the user interface decides to read how many times data are in the data base by ⑤, ID selector. ⑥ is the Insert ModelData button. The Insert Model Data button has a function that inserts the sign language word ModelData to DataBase. Finally, ⑧ is the research button.

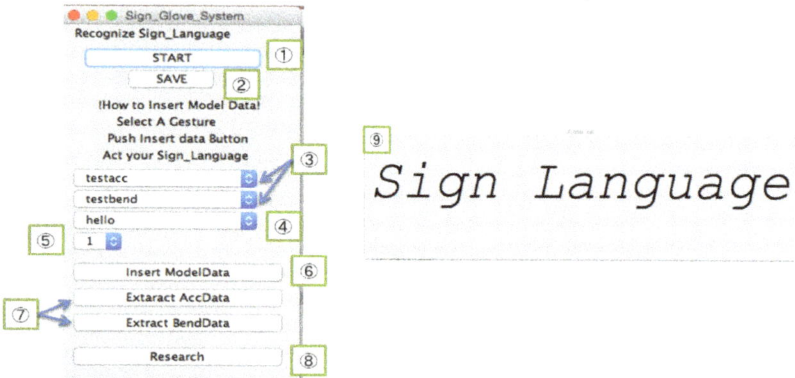

Figure 8. System user interface.

The Research button is the button that calculates the similarity between sign language data that contain current and model data.

3.2.4. Data Format

We explain here data taken from WonderSense. As above, the data taken from WonderSense are acceleration data. The acceleration data taken from WonderSense are sent to WonderTerminal on a PC through WonderBox. WonderTeminal has a function that builds a server. The server of WonderTerminal sends the acceleration data to our Sign system. The acceleration data format is String and the frequency is 50 Hz. We use the acceleration data for recognition. Our research system can save the acceleration data from WonderSense into DataBase.

We explain data here taken from bending sensors. As above, the data taken from bending sensors are the resistance values of finger bending. The resistance data taken from every bending sensor are sent to our research system in a PC through an Arduino. In this system, we can save the data from bending sensors into a database.

3.3. Recognition Method

3.3.1. Dynamic Time Warping

The dynamic time warping (DTW) algorithm is used for measuring waveform similarity. The DTW algorithm calculates the similarity of time-series data using Euclidean distance. The feature of the DTW algorithm is that the length of sample data does not become a problem for the calculation. The duration of sign language varies while expressing the same word according to habit, proficiency, and other factors. Even in this situation, the DTW algorithm can calculate similarity. Next, we explain how to calculate the DTW algorithm for a single sensor.

1. To calculate the similarity of a sequence $X = x_1, x_2, \ldots x_M \ M \in \mathbf{N}$ and sequence $Y = y_1, y_2, \ldots y_M \ N \in \mathbf{N}$, make similarity arrays $D(i, j)$ of size $M \times N$.

$$D(i,j)(i = 1, 2 \ldots M)(j = 1, 2 \ldots N) \tag{1}$$

2. Assign 0 to $D(0,0)$ and ∞ to the others.

$$D(0,0) = 0$$
$$D(i,j) = \infty(i = 1, 2 \ldots M)(j = 1, 2 \ldots N) \tag{2}$$

3. Calculate the similarity of two time-series data, $D(M, N)$, with calculation (3) for $i = 1, 2 \ldots M$ and $j = 1, 2 \ldots N$. $f\left(x_i, y_j\right)$ is the cost function.

$$D(i, j) = f(x_i, y_j) + \min(D_{i,j-1}, D_{i-1,j}, D_{i-1,j-1})$$
$$f(x_i, y_j) = \sqrt{(x_i^2 - y_j^2)} \tag{3}$$

4. The DTW distance we need is the result of calculation in all combinations $D(M, N)$.

$$D(M, N) \tag{4}$$

3.3.2. Weighted DTW

DTW can calculate model data and sensor data similarity for a single sensor. Furthermore, by assigning weights, the weighted DTW can effectively fuse data from multiple sensors. The model data are the ideal data generated by analyzing the average value of the standard action and the waveform trend of each sensor for multiple executions.

The contribution of each sensor to sign language recognition is different. In this research, we used both bending sensors and inertial sensors. Furthermore, two inertial sensors measure the movement of two hands, and ten bending sensors measure the bending of ten fingers. On the one hand, the types of sensors are different, so the effectiveness of information is different. On the other hand, even with the same sensor, for sign language recognition, the thumb, index finger, and middle finger of the right hand provide more critical information in many cases, while the other fingers most of the time make little contribution to distinguishing sign language. Due to a large number of static states, the waveform has less effective information and is more affected by noise. Thus, setting the same weight is unreasonable. We set different weights between 10 bending sensors, different weights between 2 inertial sensors, and different weights between 2 types of sensors. The weights calculation process is as follows:

1. Combine 10 bending sensors added weight β_i of 10 fingers.

$$DTW(B) = \sum_{i=1}^{10} \beta_i B_i \left(\sum_{i=1}^{10} \beta_i = 1\right) \tag{5}$$

2. Combine 2 WonderSense added weight γ_j of both hands.

$$DTW(WS) = \sum_{j=1}^{2} \gamma_j WS_j \left(\sum_{j=1}^{2} \gamma_j = 1\right) \tag{6}$$

3. Combine (1) data and (2) data added weight α.

$$DTW = \alpha DTW(WS) + (1 - \alpha) DTW(B) \tag{7}$$

4. Experiment and Evaluation

In the experiment, we evaluated the performance of the sign language data glove. In this section, we first describe the experimental setup. Next, we show how the experiments compared the recognition performance of hand shape, hand motion, and combined data of both. After that, we verify the recognition performance of our weighted DTW.

4.1. Experimental Setting

We recruited 8 volunteers and collected data on 20 sign language words. The average age of subjects was 22. Each person repeated each sign language three times, and we collected a total of $8 \times 20 \times 3$ data points. Model data was the average value of multiple executions of the standard action. Table 2 shows the weight parameters in the experiment. Next, we introduced the usage of Sign-Glove.

Table 2. Setting of the weight parameters in the experiment.

Weight	Value	Weight	Value
α	0.05	β_6	0.0002
β_1	0.2448	β_7	0.0002
β_2	0.3772	β_8	0.0002
β_3	0.3776	β_9	0.0002
β_4	0.0002	β_{10}	0.0002
β_5	0.0002	γ	0.5

Usage of the Sign-Glove

In this section, we explain how to wear the Sign-Glove and the starting position of recognizing a sign language word, as shown in Figure 4, with this system. First, the Sign-Glove is worn on the hand. The fingers should be inserted into the Sign-Glove because the Sign-Glove has a bending sensor in each finger part. Figure 6a shows the correct wearing of Sign-Gloves.

Figure 9a,b show a pose and hand position when we start to recognize a sign language word. The basic position is sitting in a chair and putting the hands on the knees. Recognizing a sign language word must begin from this basic position. In this research, we started the system during recognition. Furthermore, when a sign language word motion was finished, we stopped the system. Then, the hands were returned to the basic position.

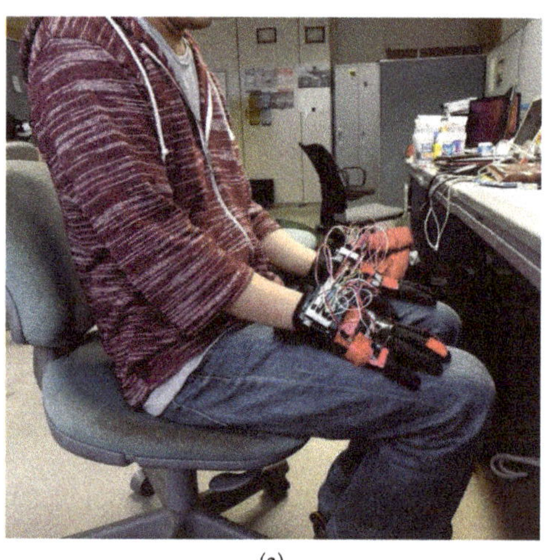

(a)

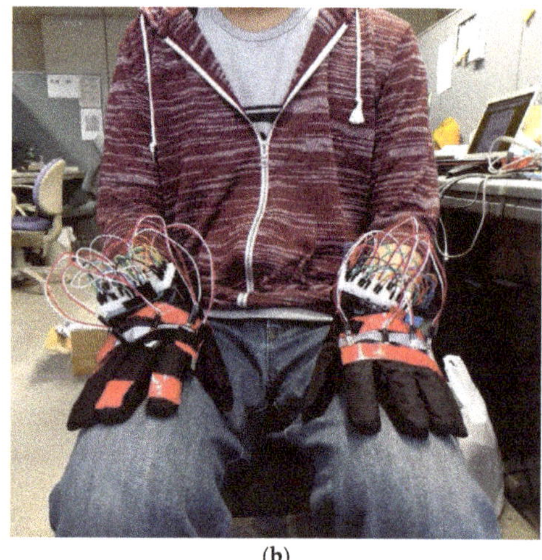

(b)

Figure 9. The basic pose and hand position during the usage of the system: (**a**) side view; (**b**) front view.

4.2. Experiment Results

4.2.1. Comparison between the Hand Shape, Hand Motion, and Combination Methods

As shown in Figure 10, the recognition ratio of twenty kinds of sign languages in this experiment was obtained. Experiments were performed to calculate the recognition rate of sign language for three kinds of feature data: combined hand motion data and hand shape data, data only based on hand motion, and data only based on hand shape. The motion data of the hand originates from the inertial sensor, which is shown as a red rectangle on the graph. The shape of the hand originates from the bending sensor, which is shown as a blue rectangle on the graph.

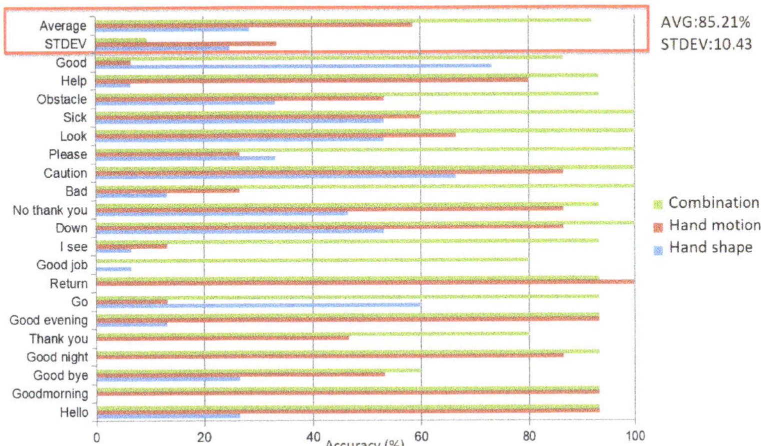

Figure 10. Comparison between the hand shape, hand motion, and combination methods. (The red boxes on the right give the values of AVG and STDEV for combination method.).

The combined sign language recognition rate was the highest, with an average recognition rate of 85.21% and a standard deviation of 10.43. The next highest recognition rate was using only the hand motion features, and the lowest recognition rate was using only the hand shape features for sign language recognition. Depending on the features of different sign languages, the contribution of hand motion data and hand shape data were different. In this dataset, hand motion features contributed more to the recognition rate.

The hand motion and hand shape parts of the data are complementary most of the time. Except for the word "return", where the combined features are not as good as just the hand motion part of the data, the hand shape part of the data plays a reverse role.

4.2.2. Comparison between Using Our Proposed Weighted DTW or Not

We obtained result of cases in which we used different weights or the same weight as data fusion, as shown in Figure 11.

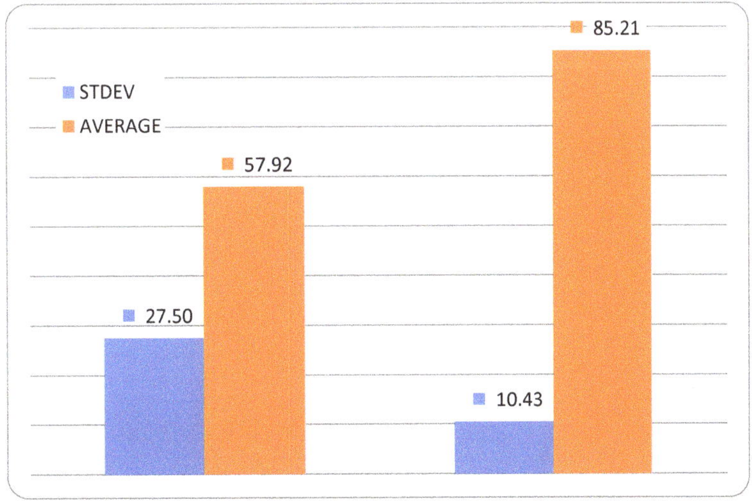

Figure 11. Comparison between use our proposed weighted DTW and the original DTW.

When we used our weight by data fusion algorithm, we could take the average recognition rate of 85.21%, and the standard deviation was about 10.43. When we did not use our weight by data fusion algorithm, the average recognition ratio was about 57.92%, and the standard deviation was 27.50. Thus, we understand that this data fusion algorithm was increasing the recognition ratio and decreasing the standard deviation. Thus, we can say that this algorithm is useful for recognizing sign language.

5. Discussion

We built a data glove based on bending sensors and inertial sensors to capture hand shape and motion features, and then it uses weighted DTW fusion features to recognize sign language. We experimentally verified that both hand shape and hand motion contribute to sign language recognition. Moreover, the two features are complementary, and a higher recognition rate can be obtained by fusing the two features to recognize sign language. Adjusting the weight values to fuse the features, we found that the quality of information provided by sensors with different placements is different. By adjusting the weights to focus on the sensors with large value changes during the execution of sign language, the recognition accuracy can be improved. We collected data for 20 dynamic sign language words from eight volunteers, and the recognition accuracy was 85.21%. The feasibility of the system was verified.

In comparison with similar systems, although there has been a large number of studies on sign language recognition, the defined sign language countries are different, and the numbers of participants in the experiments are different. The number of sign language word data contained in the dataset is different. We chose Chu's system, which is similar in structure to our system, and both use bending sensors and IMUs, and we also used Japanese sign language, for comparison. The results are shown in Table 3, which shows that the weighted DTW has a better recognition rate when the number of participants and the number of recognized sign language words are both greater.

Table 3. Our system compared with Chu's system.

Research (Years)	Subject	Number of Signs	Algorithm	Sensor	Cross-Recognition
This study	8	20	Weighted DTW	Bend and IMU	85.21%
Chu et al. [26] (2021)	3	7	DTW	Bend and IMU	82.5%

There are still many limitations of our system. The data glove prototype system uses a breadboard, so the system is rather bulky. For some palm-related sign language words, the system sometimes causes inaccurate movements. However, the semantic impact on sign language expression is minimal. It is still able to recognize sign language words in sign language communication. Regarding the impact of data collection, there will be data loss or disconnection problems during long time periods of data collection.

In addition to hand shape features and hand motion features, collecting other features in sign language has the potential to further improve recognition rates in the future, for example, the relationship between head and hand position, body posture, facial expressions, etc. In addition, the data features of some locations on the hand do not contribute much to recognition, offering the possibility of simplifying the device in the future.

6. Conclusions

In this research, we built a Sign-Glove system to recognize sign language. By analyzing the process of sign language, we noticed that sign language is composed of both hand motion and the hand shape in time. Therefore, we decided to use IMU to detect the hand motion part and the bend sensor to detect the hand shape part. Then, we combined this information and used the weighted DTW algorithm to fuse the features and recognize sign language words. In the experiments, we verified the performance of the Sign-Glove system and obtained high recognition rates of sign language. Such a wearable glove system has the potential to greatly reduce the cost of communication for people with hearing impairment.

In the future, with further improvements, we will exchange the cables for wireless connections such as BLE and Xbee. In addition, word-by-word sign language recognition was achieved, but sign language is often used to construct meaning through continuous use. We will replace the breadboard connection with a printed circuit board (PCB) and flexible flat cable (FFC) connections to achieve more stable data collection over a long period of time in daily use. We hope to build a system capable of continuous sign language recognition in the future. A more concise system will provide more convenient and complete sign language expression.

Author Contributions: Conceptualization, C.L. and S.A.; software, S.A.; validation, S.A. and C.L.; formal analysis, C.L.; investigation, C.L. and S.A.; resources, L.J.; data curation, S.A.; writing—original draft preparation, S.A. and C.L.; writing—review and editing, C.L.; visualization, C.L. and S.A.; supervision, L.J.; project administration, L.J.; funding acquisition, L.J. All authors have read and agreed to the published version of the manuscript.

Funding: This work was supported by JSPS KAKENHI Grant Number 22K12114, the JKA Foundation, and NEDO Intensive Support for Young Promising Researchers Number JPNP20004.

Informed Consent Statement: Informed consent was obtained from all subjects involved in the study.

Data Availability Statement: Dataset used in this paper is available and can be accessed in the Google Drive repository: https://drive.google.com/drive/folders/1dpd2QMel8tlRI_uMXA6l9WGyb7 hwHQyr?usp=sharing.

Conflicts of Interest: The authors declare no conflict of interest.

References

1. Ministry of Health. Labour and Welfare Home Page, "2016 Survey on Difficulty in Life (Nationwide Fact-Finding Survey on Children with Disabilities at Home) Results". Available online: https://www.mhlw.go.jp/toukei/list/dl/seikatsu_chousa_c_h28.pdf (accessed on 20 January 2023).
2. Rastgoo, R.; Kiani, K.; Escalera, S. Sign Language Recognition: A Deep Survey. *Expert Syst. Appl.* **2021**, *164*, 113794. [CrossRef]
3. Amin, M.S.; Rizvi, S.T.; Hossain, M.M. A Comparative Review on Applications of Different Sensors for Sign Language Recognition. *J. Imaging* **2022**, *8*, 98. [CrossRef] [PubMed]
4. Jiang, S.; Kang, P.; Song, X.; Lo, B.P.; Shull, P.B. Emerging Wearable Interfaces and Algorithms for Hand Gesture Recognition: A Survey. *IEEE Rev. Biomed. Eng.* **2021**, *15*, 85–102. [CrossRef] [PubMed]
5. Seçkin, A.Ç. Multi-Sensor Glove Design and Bio-Signal Data Collection. *Nat. Appl. Sci. J. Full Pap. 2nd Int. Congr. Updates Biomed. Eng.* **2021**, *3*, 87–93.
6. Seçkin, M.; Seçkin, A.Ç.; Gençer, Ç. Biomedical Sensors and Applications of Wearable Technologies on Arm and Hand. *Biomed. Mater. Devices* **2022**, *1*, 1–13. [CrossRef]
7. Aloysius, N.; Geetha, M.K. Understanding vision-based continuous sign language recognition. *Multimed. Tools Appl.* **2020**, *79*, 22177–22209. [CrossRef]
8. Sharma, S.; Singh, S. Vision-Based Sign Language Recognition System: A Comprehensive Review. In Proceedings of the 2020 International Conference on Inventive Computation Technologies (ICICT), Coimbatore, India, 26–28 February 2020; pp. 140–144.
9. Ma, Y.; Zhou, G.; Wang, S.; Zhao, H.; Jung, W. SignFi: Sign Language Recognition Using WiFi. *Proc. ACM Interact. Mob. Wearable Ubiquitous Technol.* **2018**, *2*, 23:1–23:21. [CrossRef]
10. He, W.; Wu, K.; Zou, Y.; Ming, Z. WiG: WiFi-Based Gesture Recognition System. In Proceedings of the 2015 24th International Conference on Computer Communication and Networks (ICCCN), Las Vegas, NV, USA, 3–6 August 2015; pp. 1–7.
11. Kudrinko, K.; Flavin, E.; Zhu, X.; Li, Q. Wearable Sensor-Based Sign Language Recognition: A Comprehensive Review. *IEEE Rev. Biomed. Eng.* **2020**, *14*, 82–97. [CrossRef] [PubMed]
12. Lokhande, P.M.; Prajapati, R.; Pansare, S. Data Gloves for Sign Language Recognition System. *Int. J. Comput. Appl.* **2015**, *975*, 8887.
13. Samaan, G.H.; Wadie, A.R.; Attia, A.K.; Asaad, A.M.; Kamel, A.E.; Slim, S.O.; Abdallah, M.S.; Cho, Y. MediaPipe's Landmarks with RNN for Dynamic Sign Language Recognition. *Electronics* **2022**, *11*, 3228. [CrossRef]
14. Kohei, M.; Youngha, C.; Nobuhiko, M. *Recognition of Fingerspelling in Japanese Sign Language based on Nail Detection and Wrist Position*; ITE Technical Report; ITE: Singapore, 2013; pp. 199–202.
15. Shibata, H.; Hiromitsu, N.; Hiroshi, T.; Daisuke, K. Similarity Analysis of Motion Difference for Sign Language Recognition using Colored Gloves. *Forum Inf. Technol.* **2015**, *14*, 551–554.

16. Salagar, M.; Kulkarni, P.; Gondane, S. Implementation of Dynamic Time Warping for Gesture Recognition in Sign Language Using High Performance Computing. In Proceedings of the 2013 International Conference on Human Computer Interactions (ICHCI), Chennai, India, 23–24 August 2013; pp. 1–6.
17. Korzeniewska, E.; Kania, M.; Zawislak, R. Textronic Glove Translating Polish Sign Language. *Sensors* **2022**, *22*, 6788. [CrossRef] [PubMed]
18. Na, Y.; Yang, H.; Woo, J. Classification of the Korean Sign Language Alphabet Using an Accelerometer with a Support Vector Machine. *J. Sensors* **2021**, *2021*, 9304925:1–9304925:10. [CrossRef]
19. Tateno, S.; Liu, H.; Ou, J. Development of Sign Language Motion Recognition System for Hearing-Impaired People Using Electromyography Signal. *Sensors* **2020**, *20*, 5807. [CrossRef] [PubMed]
20. Khomami, S.A.; Shamekhi, S. Persian sign language recognition using IMU and surface EMG sensors. *Measurement* **2021**, *168*, 108471. [CrossRef]
21. Abhishek, K.S.; Qubeley, L.C.F.; Ho, D. Gloved-Based Hand Gesture Recognition Sign Language Translator Using Capacitive touch sensor. In Proceedings of the IEEE International Conference on Elrctron Devices and Solid-State Circuits (EDSSC), Hong Kong, China, 3–5 August 2016; pp. 334–337.
22. Gałka, J.; Masior, M.; Zaborski, M.; Barczewska, K. Inertial Motion Sensing Glove for Sign Language Gesture Acquisition and Recognition. *IEEE Sens. J.* **2016**, *16*, 6310–6316. [CrossRef]
23. Faisal, M.A.; Abir, F.F.; Ahmed, M.U. Sensor Dataglove for Real-Time Static and Dynamic Hand Gesture Recognition. In Proceedings of the 2021 Joint 10th International Conference on Informatics, Electronics & Vision (ICIEV) and 2021 5th International Conference on Imaging, Vision & Pattern Recognition (icIVPR), Kitakyushu Virtual, Japan, 16–20 August 2021; pp. 1–7.
24. Faisal, M.A.; Abir, F.F.; Ahmed, M.U.; Ahad, M. Exploiting domain transformation and deep learning for hand gesture recognition using a low-cost dataglove. *Sci. Rep.* **2022**, *12*, 21446. [CrossRef] [PubMed]
25. Lee, B.; Lee, S.M. Smart Wearable Hand Device for Sign Language Interpretation System with Sensors Fusion. *IEEE Sens. J.* **2018**, *18*, 1224–1232. [CrossRef]
26. Chu, X.; Liu, J.; Shimamoto, S. A Sensor-Based Hand Gesture Recognition System for Japanese Sign Language. In Proceedings of the 2021 IEEE 3rd Global Conference on Life Sciences and Technologies (LifeTech), Nara, Japan, 9–11 March 2021; pp. 311–312.

electronics

MDPI

Article

A Novel Anchor-Free Localization Method Using Cross-Technology Communication for Wireless Sensor Network [†]

Nan Jing [1,2,*], Bowen Zhang [1,2] and Lin Wang [1,3]

[1] School of Information Science and Engineering, Yanshan University, Qinhuangdao 066004, China
[2] Hebei Key Laboratory of Information Transmission and Signal Processing, Qinhuangdao 066004, China
[3] Hebei Key Laboratory of Software Engineering, Qinhuangdao 066004, China
* Correspondence: jingnan@ysu.edu.cn
[†] This paper is an extended version of our paper published in IEEE ICPADS 2021, Beijing, China, 14–16 December 2021: International Conference on Parallel and Distributed Systems under the title "Anchor-Free Self-Positioning in Wireless Sensor Networks via Cross-Technology Communication".

Abstract: In recent years, wireless sensor networks have been used in a wide range of indoor localization-based applications. Although promising, the existing works are dependent on a large number of anchor nodes to achieve localization, which brings the issues of increasing cost and additional maintenance. Inspired by the cross-technology communication, an emerging technique that enables direct communication among heterogeneous wireless devices, we propose an anchor-free distributed method, which leverages the installed Wi-Fi APs to range instead of traditional anchor nodes. More specifically, for the asymmetric coverage of Wi-Fi and ZigBee nodes, we first design a progressive method, where the first unknown node estimates its location based on two Wi-Fi APs and a sink node, then once achieving its position, it acts as the alternative sink node of the next hop node. This process is repeated until the new members can obtain their positions. Second, as a low-power technology, ZigBee signal may be submerged in strong signal such as Wi-Fi. To overcome this problem, a maximum prime number is deployed to be the Wi-Fi broadcasting period based on the numerical analysis theory. Among many of prime numbers, we have the opportunity to select an appropriate one to achieve full coverage with the relatively small packet collisions. Last, simulations and experiments are performed to evaluate the proposal. The evaluation results show that the proposal can achieve decimeter level accuracy without deploying any anchor node. Moreover, the proposal demonstrates the anti-interference ability in the crowded open spectrum environment.

Keywords: Cross-Technology Communication (CTC); Wireless Sensor Network (WSN); localization; Wi-Fi; ZigBee

Citation: Jing, N.; Zhang, B.; Wang, L. A Novel Anchor-Free Localization Method Using Cross-Technology Communication for Wireless Sensor Network. *Electronics* **2022**, *11*, 4025. https://doi.org/10.3390/electronics11234025

Academic Editor: Dimitris Kanellopoulos

Received: 20 October 2022
Accepted: 23 November 2022
Published: 4 December 2022

Publisher's Note: MDPI stays neutral with regard to jurisdictional claims in published maps and institutional affiliations.

1. Introduction

In the past decade years, with the deep integration of Internet of Things (IoT), 5G (fifth-generation), Artificial Intelligence (AI), Edge Computing (EC), Block Chain (BC), etc., [1] Wireless Sensor Network (WSN) has again attracted the attention of industry and academia, playing an increasingly vital role in the fields of automated agriculture, national defense security, environmental monitoring, forest carbon sink, disaster rescue, and smart home [2], etc. In most WSN applications, the information collected by sensors is based on the location of nodes [3,4]. For instance, the location of the incident needs to be confirmed urgently when pollution or fire occurs. The corresponding strategy cannot be deployed without the specific coordinates of the nodes [5]. Therefore, it is crucial to know the exact location of the sensor node where events occurred or located for most WSN applications [6].

The most common solution to position is the anchor-based approach [7]. Localization is carried out utilizing the available position of the Anchor Nodes (AN) and the information interaction between AN and unknown nodes. According to whether the distance or angle between nodes needs to be calculated, there are two approaches toward AN-based localization: range-based method [8] and range-free method [9].

Common range-based schemes include Time of Arrival (ToA) [10] and Time Difference of Arrival (TDoA) [11], Differential Time Difference of Arrival (DTDoA) [12] and Equivalent Differential Time Difference of Arrival (E-DTDoA) [13]. These are generally preferred time-based ranging schemes. For the abundant hardware of the sensor node, the antenna array can be deployed to measure the Angle of Arrival (AoA) [14], with which the orientation of the target can be calculated. On the contrary, for the resource-constrained commodity ZigBee devices, Received Signal Strength Indicator (RSSI) is the best way to measure the distance between the transceivers [15]. Owing to easy acquisition, RSSI based methods are widely used in many indoor localization systems. The classic RSSI-based localization scheme is trilateration localization [16], which requires the locations of at least three known beacons to position the target. In order to further improve ranging accuracy, RSSI is combined with the Channel State Information (CSI) parameters to achieve multi-dimension hybrid measurements, because CSI has higher granularity than RSSI to reveal more detail about wireless channel, such as RSSI-AoA [17], ToA-RSSI [18], multiple parameters [19,20], etc. Additionally, both RSSI-based and CSI-based considers the fingerprint algorithm [21], which requires an offline training process and the dataset of fingerprints collected by anchor nodes, as known as a "Radio Map". Fingerprint algorithm estimates the likely location of the target by comparing collected data with the reference data.

The alternate and cost-effective solution is the range-free approach, which utilizes proximity measurements without any additional hardware and knowledge about distance or angle. The popular range-free schemes include the Centroid algorithm [22], DV–Hop [23], amorphous algorithm [24], approximate point-in-triangulation (APIT) [25] and Multidimensional Scaling (MDS) [26], PHL-based approach for anisotropic WSN [27]. Compare to the range-based schemes, range-free methods do not need the prior knowledge of ANs. However, the localization accuracy of range-free techniques deteriorate correspondingly. Therefore, tradeoff between the two localization techniques has to be considered.

The aanchor-based location algorithm is mature and has been widely deployed. However, the research on anchorless positioning started relatively late with fewer results, and most of them only stay at the level of algorithm design and have not been applied in practice. Although anchor-based schemes have higher accuracy, they are highly dependent on the number and the position of ANs. Moreover, regular maintenance of the ANs is also required, which leads to cost, maintainability and scalability issues. Anchor-free positioning schemes can reduce the cost and complexity without the need to pre-deploy the anchor nodes, especially in some special applications. For example, the locations of all nodes in a WSN are random and cannot be obtained in advance, or it could be used in dangerous areas where humans are unable to reach. Therefore, the anchorless location algorithm is of great importance to the field of WSN location.

The earliest anchor-free location algorithm is the ABC (assumption based coordinates) algorithm [28], which establishes the positioning order based on the communication connection between nodes and estimates the position in turn. ABC is easy to implement, but it has a large positioning error, so it is often used as an auxiliary positioning method. The AFL (anchor-free localization) algorithm [29] obtains a fold-free network model to obtains approximate estimated coordinates. However, the AFL algorithm is not fully distributed. According to the principle of multidimensional scaling, anchor-free localization can be achieved [30], but it is centralized. The GFF (GPS-free-free) algorithm [31] uses the hops from nodes to anchor nodes as the distance. GFF is a coarse-grained location algorithm with simple calculation and low hardware requirements, but large positioning error. It is generally used for initial location estimation of network deployment. In [32], the individual sensors use knowledge of received beacons and information from nearest neighbors to identify the sub-sectors in which they reside. KPS (deployment knowledge based positioning system) [33] is based on a probabilistic model of group placement. KPS has low communication overhead but is only suitable for static WSN. The possibility of using image registration techniques to achieve localization is investigated in WSN [34], but it is only applicable in case of nodes equipped with video sensors. The above are several classical anchor-free WSN localization algorithms, and their common problem is that the poor localization accuracy.

Although the existing positioning algorithms have been well developed in the past two decades, they are far from the application requirements:

1. Improving coverage. Limited by the low power, a sensor device can only achieve about 20 m transmission. In order to enlarge the coverage, a large quantity of ANs has to be deployed. Except for the unacceptable cost, deployment of a large number of ANs will increase the complexity and maintenance of the network.
2. Anti-interference. In indoor scenarios with complex radios, as the lower-power device, sensor is vulnerable to interference from the signals with strong-power, such as Wi-Fi and Bluetooth.
3. Avoiding packet collisions. ANs have to broadcast their locations and identities in WSN to enable the ZigBee Node (ZN) to be aware of their positions. When a large number of ANs is connecting, WSN suffers from packet collisions.

Fortunately, Cross-Technology Communication (CTC) has emerged and flourished to solve the coexistence and spectrum efficiency problem among the heterogeneous devices, which has attracted many attentions both in industry and academia [35,36]. CTC is significant for the scalability, availability and reliability of the diverse IoT.

The key challenge for CTC is to deal with the heterogeneity among devices caused by the diversity of IoT, including incompatibility of communication standards and asymmetry of communication and computing power. The existing works are divided into packet-level CTCs and physical-level CTCs. Early works on CTC are packet-level CTCs. It manipulates packet and encodes certain channel parameters as the carrier to convey messages for transmissions, such as packet energy, packet length and timing [37]. These packet-level methods are relatively easy to be accessed. However, their communication efficiency, quality, and throughput are extremely low. In addition, the asymmetry of the communication protocols results in a large amount of wasted bandwidth. Physical-level CTCs can effectively alleviate the limitations of packet-level CTCs, physical-level CTCs emerge as the times require. WEBee [36] is the first Wi-Fi-to-ZigBee physical-level CTC technology. Figure 1 illustrates the architecture of WEBee, where the ZigBee time domain waveform is emulated by redesigning the payload of the Wi-Fi packet on the sending side. When the ZigBee node receives these choreographed signals, the header, preamble, and trailer of Wi-Fi packets are ignored as noise, then the Wi-Fi payload is recognized as a legitimate ZigBee packet. The ZigBee nodes can demodulate the heterogeneous transmitters' signals without any hardware modifications. Following WEBee, quantity of physical-level CTC technologies is emerging to improve communication performance and address the emulation errors [37–39].

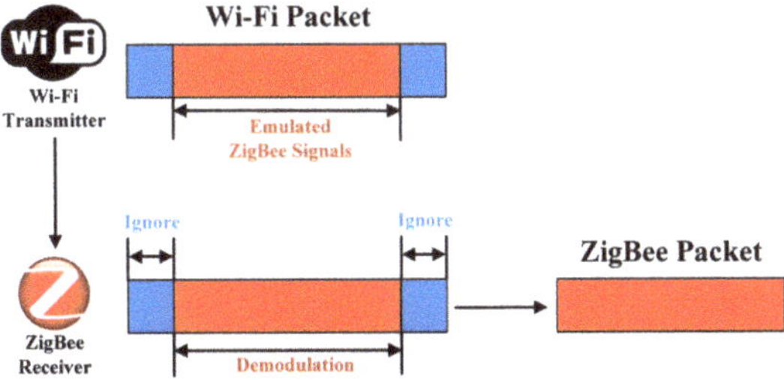

Figure 1. The architecture of WEBee.

Inspired by CTC between Wi-Fi and ZigBee, we propose a novel WSN self-positioning method by leveraging the installed Wi-Fi infrastructure to achieve distance calculation instead of traditional ANs in WSN, named as Wi-Fi based ZigBee nodes Localization (WiZiLoc [40]). Specifically, first, two Wi-Fi APs and a Sink Node (SN) is deployed to locate the one-hop node. Any ZigBee node can be considered as the one-hop node once it accesses the intersection area of the two Wi-Fi APs and SN. When one-hop node gets positioned, it becomes the reference of the next Unknown ZigBee Node (UZN). Second, two Wi-Fi APs and the positioned one-hop node are used to locate the next hop UZN, which is not located in the mentioned intersection area. The first and second step is repeated until the new members can achieve their positions. Finally, in order to avoid the signal conflicts caused by the strong Wi-Fi signal due to the crowded unlicensed spectrum (e.g., ISM bands), we design a proper Wi-Fi broadcasting period to send CTC packet based on a numerical calculation, where the broadcasting period is set to a maximum prime number. Through WiZiLoc, ZigBee nodes can overcome the hardware limitations and access a larger coverage, meanwhile, the number of the ANs decreases dramatically with only a few Wi-Fi APs.

In summary, our contributions is outlined as follows:

- For the first time, we apply the advanced CTC technique to WSN localization. Instead of anchor-based localization, we propose an anchor-free mechanism by leveraging the installed Wi-Fi infrastructure to calculate the distance.
- We design WiZiLoc, a distributed self-positioning method for localization in WSN. To achieve this, we face two challenges, including the asymmetric coverage and the strong interference introduced by the environmental Wi-Fi. A distributed progressive method and a reasonable Wi-Fi broadcasting period are investigated to address the challenges.
- We implement WiZiLoc on the SDR platform USRP N210 and the commodity ZigBee device, CC2530. The experimental results demonstrate that WiZiLoc can achieve decimeter-level accuracy within two Wi-Fi APs coverage area, without deploying any ANs.

2. Preliminaries

2.1. Overview

As shown in Figure 2, WiZiLoc are composed of two Wi-Fi APs, an SN and several UZNs, where W_1 and W_2 denote the Wi-Fi APs, UZN_1 to UZN_k denote k UZNs accessing to WSN. Specifically, W_1 and W_2 deploy WEBee and send CTC packets periodically. Therefore, all UZNs in the coverage area of W_1 and W_2 can successfully receive CTC packets. During positioning with two Wi-Fi APs, an artifact is introduced, we eliminate the artifact by utilizing an SN.

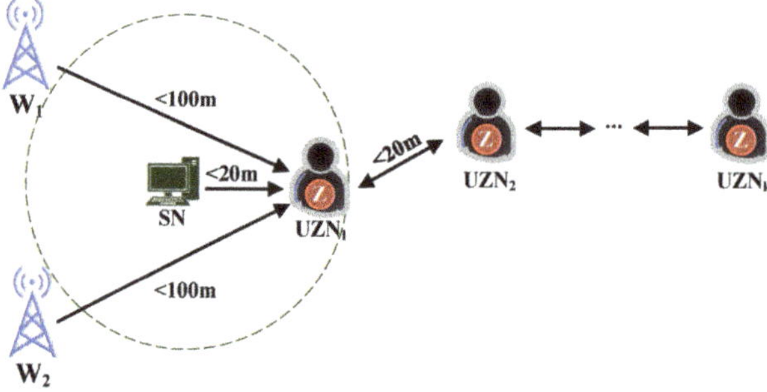

Figure 2. WiZiLoc overview.

SN is a special ZN. It is responsible for data aggregation and fusion. Additionally, it saves and uploads data to the server. However, as for the normal ZN, its communication distance is below 20 m. Limited by the SN coverage ability, UZNs are divided into two groups. One group is located within the intersection area of W_1, W_2 and SN, shown as the green dotted circle in Figure 2. In this intersection area, the UZNs can receive packets not only from W_1 and W_2, but also from the SN. We call such UZNs one-hop nodes. The other group lies in the coverage area of W_1 and W_2, but locates outside the coverage radius of the SN (outside the green dotted circle). These UZNs cannot receive the packets sent by the SN, but they are still able to receive from other UZNs. We name them as multi-hop nodes.

According to the above discussions, obviously, the premise of WiZiLoc is that there must be an intersection of the coverage areas of W_1, W_2 and SN, and there at least is a one-hop node (denoted by UZN_1) in this intersection.

2.2. Challenges

To achieve WiZiLoc, we have to face the following challenges, as shown in Figure 3. Challenge 1, the asymmetric coverage caused by the discrepant transmission power between Wi-Fi and ZigBee. In terms of the standard 802.11 and 802.15.4, the coverage of Wi-Fi signal is up to 200 m, while that of ZigBee is only around 20 m. That means ZigBee node UZN_1 can hear Wi-Fi, however the Wi-Fi is deaf to UZN_1. As a result, the conventional centralized location methods fail to work. Although the distributed methods may be employed to solve the above problem, these methods only achieve that by use of a large amount of ANs. Therefore, communication from UZN to Wi-Fi without any help of ANs is still challenging.

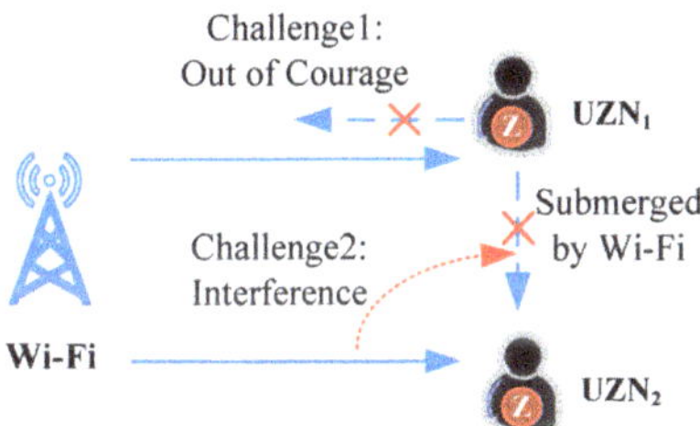

Figure 3. Technical challenges of WiZiLoc.

Challenge 2, in this paper, Wi-Fi APs replace the ANs in ZigBee WSN positioning system and therefore form a heterogeneous network. The strong and weak devices in the heterogeneous network will inevitably affect each other. Compared with the strong Wi-Fi signal, ZigBee signal is weak. The communications among UZNs in WSN are easily submerged and interfered with by the environmental installed Wi-Fi infrastructures. In order to demonstrate the impact from the Wi-Fi interference on the ZNs communications, we conduct several experiments to evaluate the communication performance of UZN receiving CTC signals in heterogeneous networks.

We carried out the following experiments in an indoor barrier free environment: when one Wi-Fi AP sends CTC signals to communicate with UZN, another Wi-Fi AP sends Wi-Fi signals as interference. BER (bit error rate) and PRR (packet reception rate) of UZN are recorded to measure the performance. Both CTC technology used (WEBee) and the traditional Wi-Fi AP adopt IEEE 802.11 g. We tested BER and PRR with the same transmission power at different distances of the transmitter and receiver. As shown in Figure 4, the communication performance of ZigBee WSN deteriorates rapidly, where the BER goes up sharply and the PRR declines to 0.3 when the distance is beyond 7 m. In addition, we also present the physical-level CTC performance in the presence of the Wi-Fi interference for comparison. Compared with WSN communication, CTC can obtain lower BER and more stable PRR at the same distance, physical-level CTC outperforms the conventional ZigBee WSN, which also demonstrates the superiority of substituting the ANs in traditional WSN with Wi-Fi. Thus, the interference introduced by the environmental installed Wi-Fi is another challenging task.

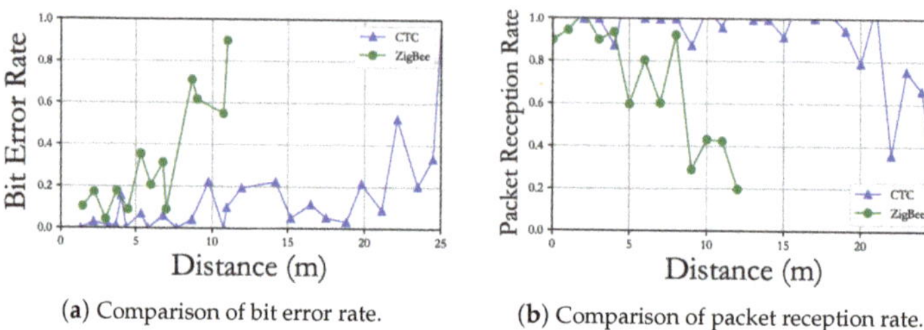

(**a**) Comparison of bit error rate.   (**b**) Comparison of packet reception rate.

Figure 4. Performance of ZigBee and CTC in presence of interference.

3. WiZiLoc Design

3.1. Workflow

Workflow of WiZiLoc is depicted in Figure 5. The system is divided into three units: Wi-Fi unit, SN unit and UZN unit. The signal transmission direction between the three units is indicated by the dotted arrow line. The key works in this paper are marked with a red dotted line.

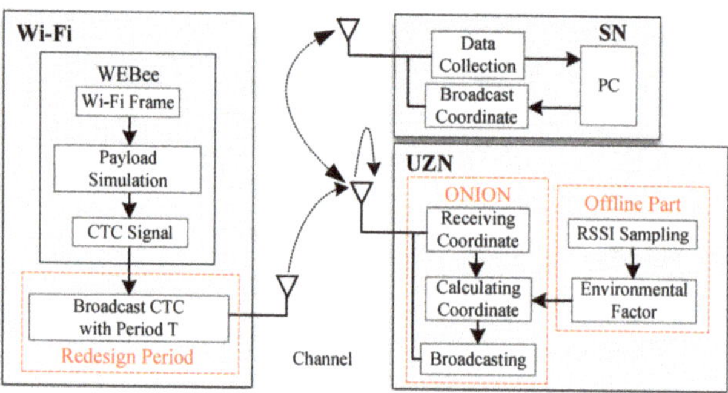

Figure 5. Workflow of WiZiLoc.

In the Wi-Fi unit, at first the physical-level CTC signal is constructed through WE-Bee [36]. The working principle of WEBee is briefly outlined here: WEBee does not modify the header and tail of the Wi-Fi frame, but simulates a complete ZigBee frame through the payload part of the Wi-Fi frame. It can be correctly received and decoded by ZigBee node. The coordinates and identity information of the Wi-Fi AP are loaded into the CTC signal.

After receiving the CTC signal, UZN unit performs RSSI ranging and calculates its own coordinates according to the ONION algorithm, which we discuss next. Additionally, the corresponding environmental factors is calculated offline. In addition to the auxiliary positioning function, SN unit can also collect and summarize the system data, update the location information of nodes in the network in real time.

Finally, WSN usually adopts the sleep and wake-up mode alternatively to save energy consumption. Inspired by that, we redesign a broadcasting period for Wi-Fi APs, which is explored next. It can not only reduce interference to ensure the internal communication quality of WSN, but also guarantee that all the UZNs are covered, none of which is missed.

3.2. ONION Algorithm

WiZiLoc is composed of two Wi-Fi APs, one SN and the UZNs. According to the one-hop node and multi-hop node, the procedure is conducted in two steps: one-hop node positioning and multi-hop node positioning, as shown in Figure 6.

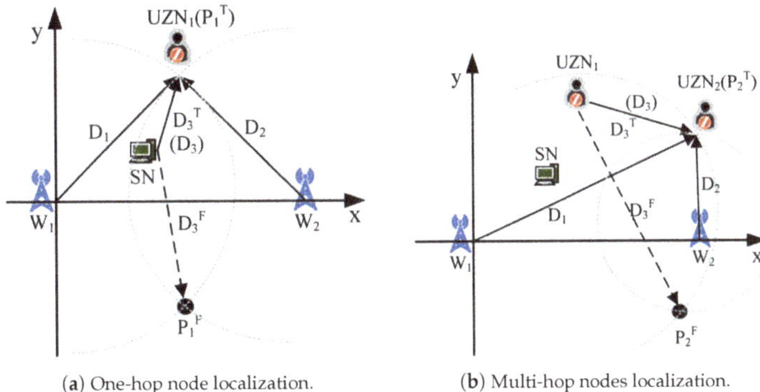

(**a**) One-hop node localization. (**b**) Multi-hop nodes localization.

Figure 6. Basic idea of ONION.

First, we use two Wi-Fi APs and a SN to position the one-hop node. To facilitate analysis, we assume that the coverage of the SN is intersected with that of W_1 and W_2, and there is only one one-hop node accessing to this intersection area. If there are multiple one-hop nodes, they are equal in localization. For the one-hop node, due to the CTC packets is also the legitimate ZigBee frame, thus one-hop node can receive and demodulate the CTC packets and calculate its coordinate with Wi-Fi APs locations and identities contained by these CTC packets. However, two Wi-Fi APs will give two coordinates for one target, one of which is the pseudo coordinate, and we call it the artifact. In order to eliminate the artifact, we use a SN to detect the true one-hop node coordinate. UZN_1 can also receive ordinary ZigBee signals from SN, so the trilateration method can be implemented to obtain the coordinates of UZN_1.

Second, since the two-hop node is outside the coverage of the SN, once one-hop node gets its coordinate, it acts as the SN, and then floods its coordinate into the network. Meanwhile, the second-hop node UZN_2, that can access to the SN only via forwarding by the one-hop node, calculates its coordinate in terms of the coordinates of W_1 and W_2, and removes the artifact based on one-hop node. In the same manner, all multi-hop nodes, that cannot directly access the SN, eliminate their artifacts by utilizing the previous hop node. We define such progressive computing procession as ONION algorithm, which is investigated in this subsection.

3.2.1. Basic Idea of ONION

ONION is proposed to address the asymmetry problem of the communication range between Wi-Fi and ZigBee. Intrinsically, ONION is a distributed self-positioning method, and its basic idea is presented in Figure 6.

It can be seen from Figure 6 that the Cartesian coordinate system is established to model the WiZiLoc, where W_1 lies in the origin and W_2 is in the positive x-axis, the SN is located in the upper half. The ONION algorithm is implemented in two steps: the first step is to calculate the coordinates of one-hop nodes, the second step is to calculate the coordinates of multi-hop nodes. To make ONION easy to understand, we assume that there is only one one-hop node accessing to the SN, denoted by UZN_1, and only one second-hop node connecting to UZN_1, represented by UZN_2.

The trilateration algorithm based on RSSI is to reach the coordinates of the unknown point by the coordinates of three known single points. The RSSI value of the signal from the unknown point to these three points can be used to calculate distance. After obtaining the distance from the unknown point to the three known points, the rest is to solve the coordinates of the unknown point. We know that two circles will intersect at one or two points (if they intersect), then if three circles intersect, they must intersect at one point (when the centers of three circles are on a straight line, it is possible to intersect at two points, which is not considered here). Therefore, the unknown point we need to calculate is exactly the intersection of the three circles drawn with the three known points as the center and the distance between them and the unknown point as the radius.

However, our proposed approach is slightly different. We use two known points (i.e., W_1 and W_2) to get two intersecting circles to obtain two possible coordinates, and then use the distance from the third known point (SN) to remove the artifact. According to the Pythagorean theorem, the formula for calculating the position of the unknown point is obtained, and the problem of positioning is transformed into the problem of finding the coordinates of the intersection of the circles. Figure 6a demonstrates the first step. UZN_1 calculates the distances from UZN_1 to W_1, W_2 and SN via the RSSI of the received signal, denoted as D_1, D_2, D_3 respectively. Then UZN_1 calculates two possible coordinates P_1^T and P_1^F according to (1), where P_1^T is the correct coordinate of UZN_1, while P_1^F is the artifact, illustrated by black solid circles in the Figure 6.

$$P_i(x_i, y_i) = \begin{cases} x_i{}^2 + y_i{}^2 = D_1{}^2 \\ (x_i - x_{W_2})^2 + y_i{}^2 = D_2{}^2 \end{cases} \tag{1}$$

where i represents the i-th UZN, x_i and y_i denote the horizontal and vertical coordinates of UZN_i. Particularly, for UZN_1, $i = 1$. x_{W_2} is the x-coordinate of W_2.

Then the correct coordinate of UZN_1 is achieved according to (2).

$$P_1^T = (x_i, y_i) = \underset{(x_i, y_i)}{\arg\min} \left| D_3 - \sqrt{(x_s - x_i)^2 + (y_s - y_i)^2} \right| \tag{2}$$

where x_s and y_s are the horizontal and vertical coordinates of SN. Let

$$D_3^{T/F} = \sqrt{(x_s - x_i)^2 + (y_s - y_i)^2} \tag{3}$$

where D_3^T is the distance between P_1^T and SN, D_3^F is the distance between P_1^F and SN. Obviously, $|D_3 - D_3^T|$ is definitely smaller than $|D_3 - D_3^F|$. Hence the true coordinate P_1^T can be distinguished from the artifact P_1^F. Similarly, when UZN is in the negative half of the y-axis, the artifact can still be removed by this method because $|D_3 - D_3^T|$ is still smaller than $|D_3 - D_3^F|$.

Figure 6b illustrates the second step. Whether there are one or multiple one-hop nodes in the coverage area, once UZN_1 achieves its coordinate, it floods its coordinate immediately. Although UZN_2 is located outside the coverage area of SN, it can receive the coordinate of UZN_1, W_1 and W_2. Same as the first step, UZN_2 can calculate D_1, D_2, D_3, P_1^T and P_1^F, D_1^T and D_3^F respectively. According to (1) and (2), UZN_2 can obtain its correct coordinate. It can be seen from the UZN_2 positioning process, UZN_1 acts as the SN of UZN_2, and so on. Each multi-hop node can only hear the adjacent previous hop node that has been located, multi-hop nodes can eliminate their artifact by treating the previous hop node as the SN.

3.2.2. Inherent Error of ONION

However, we find that the positioning error increases suddenly when UZN_1 is around the x-axis in terms of implementing a large number of simulation experiments. We deploy 100 UZNs in an area of 10,000 square meters for evaluations, and the simulation results are shown in Figure 7. Figure 7a is the simulation map with an area of 100×100 square meters. The x-axis is length and the y-axis is width, while the origin is Wi-Fi AP W_1.

Only when the black dot (the actual coordinate of the node) and the star dot (the estimated node coordinate) are completely coincident, the localization is accurate. It can be seen from the two red boxes marked in Figure 7a that the positioning accuracy in these two areas decreases dramatically, which concentrate around the x-axis and the lower right corner. It also can be found from Figure 7b that the maximum error is up to 60 m. The two areas occurring errors are referred to as inherent errors of ONION.

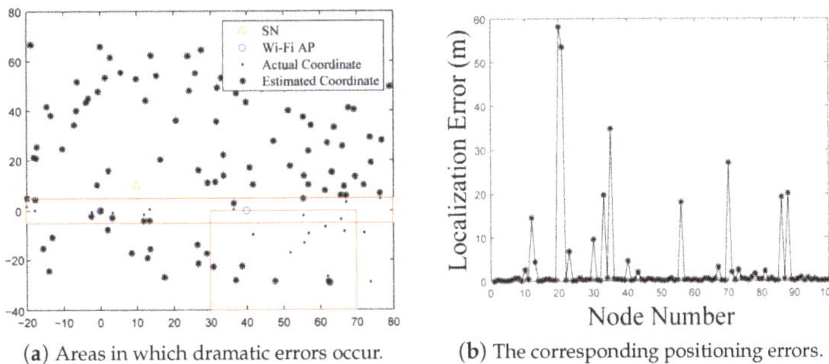

(**a**) Areas in which dramatic errors occur.　　(**b**) The corresponding positioning errors.

Figure 7. Inherent errors of ONION.

Intrinsically, the first area, around x-axis, is the stable error region, which means serious positioning errors must be occurring as long as UZN_1 is located in this area. The second area varies randomly around the first area because the next hop node positioning is dependent on the previous hop node coordinate. If the previous hop node is inaccurate, the next hop node is wrong inevitably. Unfortunately, such errors are accumulative. To address the above problem, we first analyze the causes of the inherent errors, then the improvement schemes are explored as follows.

Around the x-Axis

As shown in Figure 8. The values of D_3^T and D_3^F are very close when UZN_1 is located around the x-axis. WiZiLoc adopts RSSI to calculate the distance, as we know, RSSI is easily disturbed by the environmental noise. Hence, it is difficult to distinguish the P_1^T and P_1^F for UZN_1. To solve this problem, we stipulate that as long as the positioning coordinates estimated by UZN_1 are within εm around the x-axis, UZN_1 is considered to be on the x-axis.

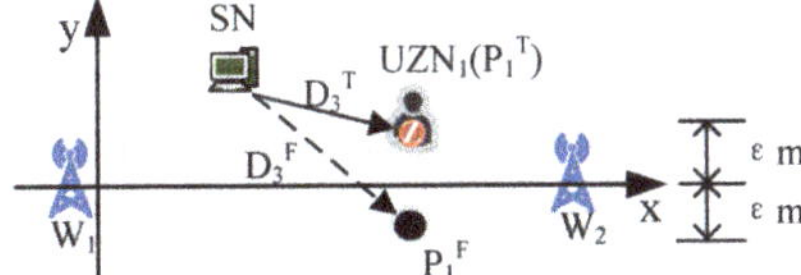

Figure 8. Cause of the error near x-axis.

In the actual indoor environment, when calculating the environmental factor, the interference and multipath effect at different distance d are complicated, and thus their impact on the RSSI measurement is also different. Therefore, under different environments and calculation trajectories, ε is also different. In this paper, the CDF (cumulative distribution function) of localization error under different environmental factor and calculation trajectories is obtained by analyzing the actual experimental results, as shown in Figure 9. The average value of path 1 is 0.8 m, so $\varepsilon = 0.8$ is also taken in the simulation experiment to achieve the minimum positioning error. It is worth noting that after switching scenarios and environments, ε should be determined according to the actual statistical data.

Random Areas near the First Area

If P_1^F is considered to be the current correct coordinate of UZN_1 and flooded to the network, it will enable UZN_2 mistakenly use P_1^F as the true coordinate of the previous hop node, then P_2^F is considered as the current correct coordinate of UZN_2. As shown in Figure 10, P_1^F is a wrong coordinate, which causes P_2^F of UZN_2 is also wrong. By parity of reasoning, errors keep accumulating and eventually cause the algorithm to fail to work. To address this problem, in the rectangular coordinate system shown in Figure 10, WiZiLoc chooses the previous hop node with the largest vertical coordinate to be the SN for the next-hop node. This has the advantage of extending the coverage, meanwhile, ensuring to reduce the inherent error due to that the next-hop node is as far away from the x-axis as possible.

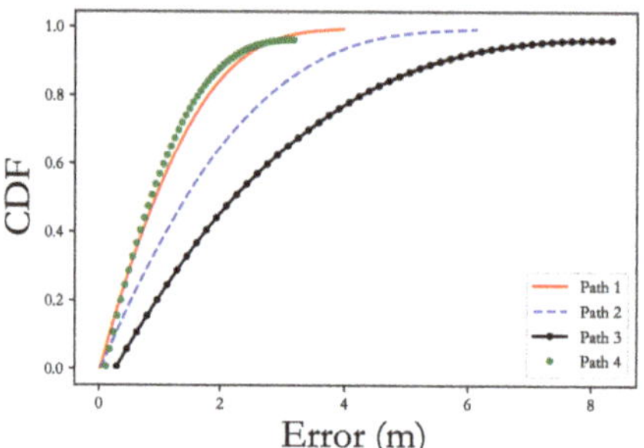

Figure 9. CDF of actual scene localization error.

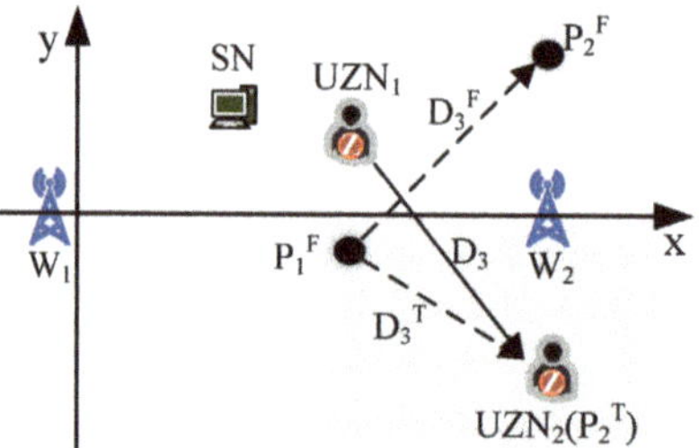

Figure 10. Cause of the error in the random area around the x-axis.

We modify the ONION according to the above analyses, and the corresponding experimental simulation results are shown in Figure 11. After averaging the positioning errors of all 100 nodes, the average error is only 0.62 m. The enhanced ONION algorithm greatly suppresses the inherent errors and improves the average positioning accuracy of WiZiLoc. In summary, the process of the ONION algorithm is shown in Algorithm 1.

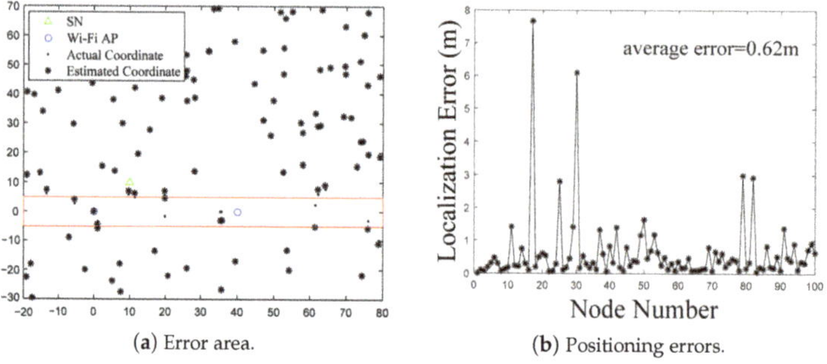

(**a**) Error area.

(**b**) Positioning errors.

Figure 11. Enhanced ONION after modifications.

Algorithm 1: ONION

Input: RSSI and coordinates of W_1, W_2, SN (or the previous hop coordinates)
received by UZN_i

Output: Coordinate P_i of UZN_i

1 calculate D_1, D_2 between UZN_i and W_1, W_2
2 calculate P_i^T and P_i^F according to (1)
3 **if** UZN_i *is the one-hop node* **then**
4   calculate D_3 between UZN_i and SN
5   select P_i in (2), let $P_i = P_i^T$
6   **if** UZN_i *is the multi-hop node* **then**
7   choose the previous hop with the maximum y_i to substitute the SN in (2)
8   **end**
9   calculate D_3 between UZN_i and the previous hop node
10   select P_i in (2), let $P_i = P_i^T$
11 **end**
12 **if** $P_i < 0.8\text{m}$ **then**
13   let $y_i = 0$
14 **end**
15 flooding P_i to the network

3.3. Broadcasting Period

In this section, we comprehensively discuss the relationship between the broadcasting period of Wi-Fi AP and the wake-up period of ZigBee. Then we design the Wi-Fi broadcasting period according to the discussion to address the second challenge: the weak ZigBee is vulnerable to strong Wi-Fi interference.

As wearable equipment, ZigBee devices usually use batteries to provide energy, so its power management is a crucial issue. To save energy consumption, ZigBee devices adopt a sleep-and-wake-up mechanism when waiting for transmission. The main goal of the sleep-and-wake-up mechanism is to keep the connection between the ZigBee devices and the network while reducing the energy consumption of the waiting ZigBee nodes.

Due to the sleep-and-wake-up mechanism, the Wi-Fi broadcasting period must be carefully designed. WiZiLoc requires ZNs to receive both CTC packets from Wi-Fi APs and ZigBee packets separately at their wake-up times, which needs to schedule the broadcasting period of Wi-Fi APs to keep synchronizing with the wake-up period of ZNs. Otherwise, the whole network communication performance may be degraded.

If the broadcasting period is consistent with the wake-up period, ZNs cannot receive any ZigBee signal except the CTC packets. If the broadcasting period is selected randomly, the ZNs may not receive any CTC packets. Furthermore, if the Wi-Fi broadcasting period is too short, the 2.4 GHz frequency band will be continually occupied, causing interference to the communication within the network.

In order to solve the above problem, a reasonable broadcasting period of Wi-Fi APs is requisite. The broadcasting period needs to be synchronized with the specific wake-up time to avoid occupying all the wake-up times of the ZNs and jamming communications among ZNs, in the meantime, to ensure all ZNs in the network can receive the CTC packets.

Therefore, in order to reduce packet collisions and ensure the communication quality between UZNs, the principle for designing the broadcasting period is that the broadcasting period is as large as possible within a certain range. Therefore, it is assumed that the broadcasting period is not less than the wake-up period. Furthermore, since when the larger of two numbers is prime, the two numbers must be coprime. This is also the reason for our assumption, which will be discussed later.

We now comprehensively discuss the relationship between the broadcasting period and the wake-up period in several cases, and design how to select the broadcasting period according to their relationship. As shown in Figure 12, we introduce a time axis model, set the start time of a certain broadcasting period as the origin of the coordinate axis, the time unit as Δt, the moment when UZN enters the network as t_{in}, the broadcasting period as T, and the wake-up period of UZN as D_c, there are $T \geq D_c$. The downward black solid arrow indicates the times when Wi-Fi broadcast CTC packets. The upward gray dotted arrow is the moment when ZNs is awakened. Only when they coincide can UZN receive CTC packets.

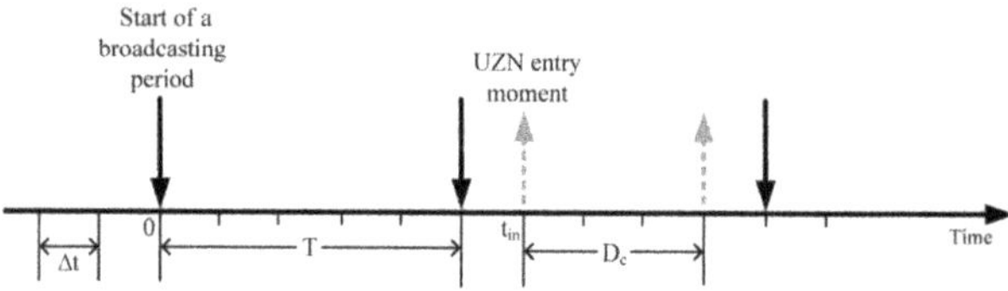

Figure 12. Time coordinate axis model.

Case 1

T is a multiple of D_c, i.e.,

$$T = k \cdot D_c, k = 1, 2, 3 \cdots \tag{4}$$

if $k = 1$, that is, $T = D_c$, then obviously only when the wake-up time and the broadcast period coincide, i.e.

$$t_{in} = p \cdot T, p = 0, 1, 2, 3 \cdots \tag{5}$$

can ZigBee receive the CTC signal, and only the CTC signal can be received, which means no extra space left for ZigBee signals. If waking up at other times, ZigBee cannot receive any CTC signal.

If $k \neq 1$, we will use an example to illustrate. As shown in Figure 13, the broadcasting period of Wi-Fi AP broadcasting is $T = 6 \cdot \Delta t$, the wake-up period of ZNs is $D_c = 3 \cdot \Delta t$. Similarly, the signal can be successfully received only when the wake-up period and the broadcast period coincide. Therefore, there is no guarantee that every UZN can receive the CTC signal in this case.

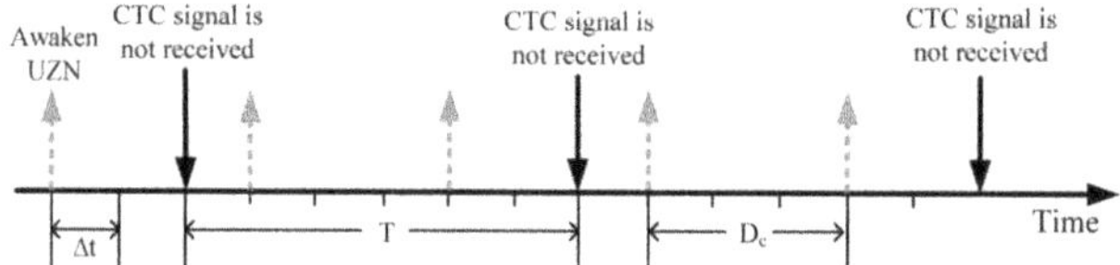

Figure 13. Example that ZNs cannot receive CTC packets.

Case 2

T is not a multiple of D_c and T and D_c are not coprime. We set

$$T = q \cdot D_c + r \tag{6}$$

where q is the quotient and r is the remainder. Since T and D_c are not coprime, we set the greatest common divisor of T and D_c as $\gcd(T, D_c)$

In this case, we analyze only when t_{in} is at some specific moment, i.e.,

$$t_{in} = \begin{cases} e \cdot D_c \\ f \cdot r \\ g \cdot \gcd(T, D_c) \end{cases} \tag{7}$$

where e, f, g are all natural numbers.

Under this circumstance, the signal can be received successfully only when the network access time t_{in} is an integer multiple of the wake-up period D_c, the remainder r or the greatest common divisor $\gcd(T, D_c)$. Obviously, when t_{in} is equal to an integer multiple of D_c or r, the wake-up period D_c and the broadcast period T will always coincide after several periods, i.e., the signal is received. When $t_{in} = g \cdot \gcd(T, D_c)$, since when two numbers are divided by their greatest common divisor, the obtained quotients must be relatively prime which causes CTC signal must be able to be received at some moment in the future. The reason for coprime is also the third case we will discuss next.

Case 3

T and D_c are relatively prime. Now we will prove that when T and D_c are relatively prime, no matter what the network access time t_{in} is, the wake-up period and the broadcast period can always coincide, i.e., to prove that there are m and n to make

$$m \cdot D_c + t_{in} = n \cdot T \tag{8}$$

established. Among them m and n are both integers. In other words, after ZigBee enters the network at t_{in}, after m wake-up periods D_c, it may coincide with n broadcasting periods T.

Since T and D_c are relatively prime, it can be known from the Euclidean Algorithm that there must be two integers i and j such that

$$i \cdot T + j \cdot D_c = 1 \tag{9}$$

We let

$$\begin{cases} i = n/t_{in} \\ j = -(m/t_{in}) \end{cases} \tag{10}$$

then the formula $m \cdot D_c + t_{in} = n \cdot T$ can be workable. In conclusion, when T and D_c are relatively prime, ZigBee must be able to receive the CTC signal after m wake-up periods.

Obviously, since T and D_c are relatively prime, there is no common factor other than 1 for T and D_c, so the greatest common divisor is 1, namely

$$\gcd(T, D_c) = 1 \tag{11}$$

According to the theorem: The product of the greatest common divisor and the least common multiple of two numbers is equal to the product of these two numbers, which is proved next.

Let there exist two numbers called x and y, and the greatest common divisor of them $\gcd(x, y) = a$.

Then their least common multiple $\mathrm{lcm}(x, y)$ is given as

$$\mathrm{lcm}(x, y) = (x/a) \times (y/a) \times a = (xy)/a \tag{12}$$

hence the greatest common divisor and the least common multiple of these two numbers

$$\gcd(x, y) \cdot \mathrm{lcm}(x, y) = (x \times y)/a \times a = xy \tag{13}$$

The proof is done. In combination with this theorem,

$$\gcd(T, D_c) \cdot \mathrm{lcm}(T, D_c) = T \cdot D_c \tag{14}$$

we have

$$\mathrm{lcm}(T, D_c) = T \cdot D_c \tag{15}$$

which means ZigBee received CTC signals every $T \cdot D_c$ time units.

When the larger of two numbers is prime, these two numbers must be coprime. Therefore, we only need to set T as a prime number, and it is greater than the wake-up periods of all ZigBee nodes in the network (different ZigBee nodes may have different wake-up periods), then no matter when the UZN accesses the network and no matter what D_c it has, UZN can receive CTC data broadcast by the Wi-Fi AP at a specific wake-up time. Moreover, to prevent too many signal collisions between CTC signal and UZN signal, T is designed as the largest prime number among the wake-up periods of ZNs, we set T to be

$$T > \max\{D_{c,i}, i = 1, 2, 3 \cdots\}, T \text{ is a prime number} \tag{16}$$

where $D_{c,i}$ is the wake-up period of the i-th ZN in the network. The reason why we don't set $T = D_{c,i}$ is to avoid that in some extreme cases (Case 1), Wi-Fi may happen to synchronize with some ZigBee nodes completely, thus affecting the communication of the ZigBee system.

With the help of the prime number design, we ensure that all of the UZN can receive CTC signals while allowing enough time for ZigBee network communication without the Wi-Fi interference at other wake-up times.

An example of a prime number-based Wi-Fi broadcasting period is shown in Figure 14, where $T = 7 \cdot \Delta t$, $D_c = 3 \cdot \Delta t$. The downward black dotted arrow is the moment when the CTC packets are received by UZNs. It can be seen from the Figure 14 that whenever UZN enters the network, it can receive the CTC packets at some instants, while many spare wakeup times remain to enable ZNs communicate with each other without being interfered by CTC packets.

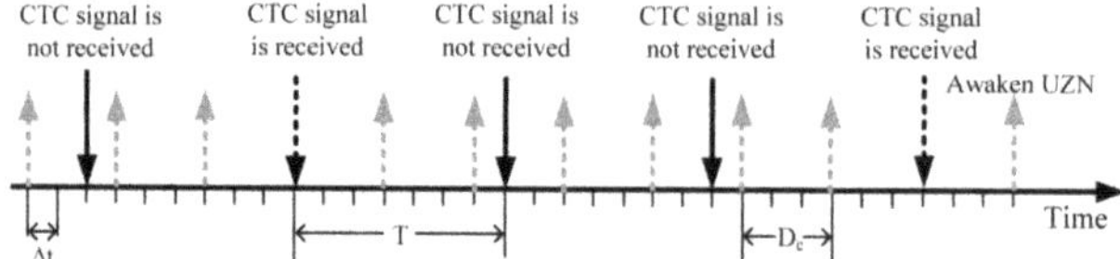

Figure 14. Example with prime number design.

3.4. Offline Part

The RSSI ranging in the ZigBee WSN localization system uses a standard logarithmic attenuation model as shown in (17)

$$RSSI = RSSI_0 - 10\mathrm{n}\lg\frac{d}{d_0} \tag{17}$$

where $RSSI$ is the RSSI value of the signal received when the distance between the transceivers is d, $RSSI_0$ is the RSSI when the distance is d_0, usually $d_0 = 1$ m, and n is the environmental factor. First, we use the given d_0, d and their corresponding RSSI to calculate n. n is a constant when the environment does not change. Subsequently, we measure d with n in terms of the (17).

As the MAC layer information, RSSI describes the overall characteristics of signals received from multiple paths. As we know, the amplitude and phase of each path will change with the environment, and the RSSI will change accordingly. The complex and variable multipath environment enables RSSI extremely unstable, which is reflected in n in the RSSI measurement model (17). We also find this through a lot of experiments that as long as the environment changes, n need to be recalculated. Therefore, this paper focuses on the invariable multipath environment and assumes that n is a constant due to static multipath.

4. Experiment

We implement the prototype of WiZiLoc on USRP N210 and the commercial USB CC2530. Two USRP N210 are deployed to send CTC packets with 802.11 b/g PHY, another USRP N210 is the SN with 802.15.4 PHY, and the receiver uses the commercial USB CC2530 with 802.15.4 PHY. The experimental settings are shown in Figure 15, where three USRP N210 are static and CC2530 are mobile. WEBee is adopted to send CTC packets, thus, we set the ZigBee channel at 21 to 24 and set the central frequency of the Wi-Fi channel at 2462 MHz.

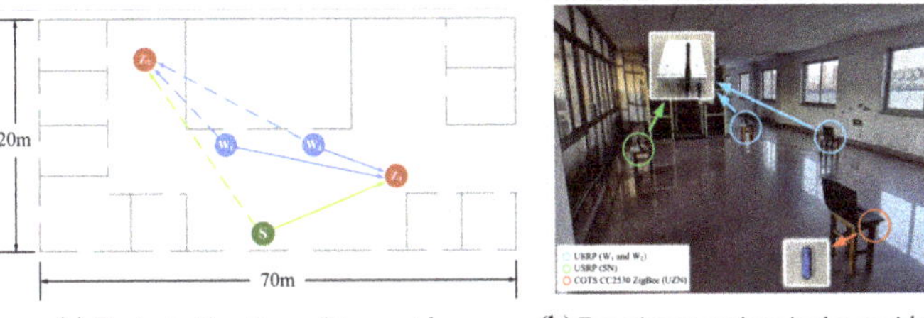

(a) The tested locations of the corridor.   (b) Experiment settings in the corridor.

Figure 15. Schematic diagram of the environment.

Our evaluations include the robustness, anti-interference, coverage and the broadcasting period. To ensure statistical validity, five experiments are performed to calculate the average results, each of which send 500 CTC packets in different scenarios, such as the corridor/canteen with short and long distance, Line-of-Sight (LOS)/Non-Line-of-Sight (NLOS), dense/sparse deployment.

4.1. Anti-Interference

First, we present the performance comparison of WiZiLoc and conventional WSN localization in the presence of Wi-Fi interference. We can see from the Figure 16 that the average positioning errors of WiZiLoc and WSN are almost same if there is no Wi-Fi interference. However, the accuracy of WSN decreases sharply from 0.5m to 1.6m. By comparison, the accuracy of WiZiLoc degrades to 0.8m. Obviously, WiZiLoc shows the better anti-interference ability.

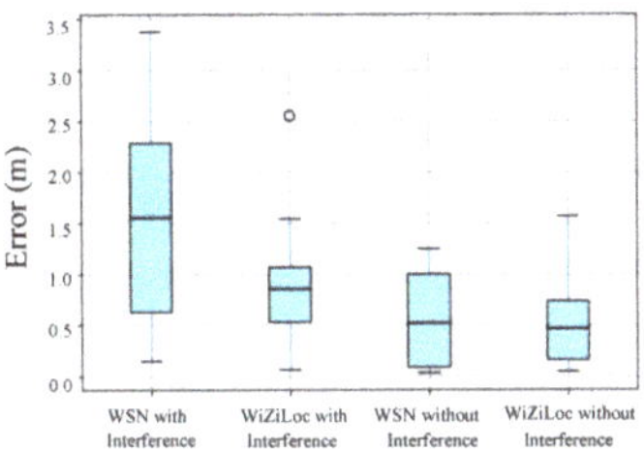

Figure 16. Performance with interference.

In the conventional WSN, a large amount of ANs has to be deployed to ensure the WSN connections, which limits the network capacity. In addition, collision and interference cannot be avoided, accordingly, the network performance deteriorates. We evaluate the anti-interference of WiZiLoc under high density ZNs by conducting numerical simulations. Three kinds of ZN density are considered to verify the anti-interference performance of WiZiLoc. In addition, Wi-Fi interference, LOS and NLOS environments are also performed for comparison. The experimental results are shown in Figure 17. In the same experimental settings, the positioning accuracy of the three densities is very similar, which indicates that WiZiLoc is capable of resisting the various interference in high-density ZNs environment.

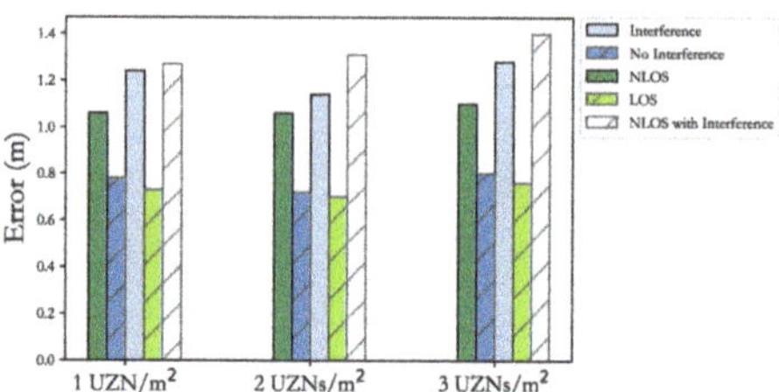

Figure 17. Density vs. accuracy.

4.2. Robustness

Next, we evaluate the robustness of WiZiLoc under LOS and NLOS environment, which is carried out in the corridor. For comparison, we also conduct the experiments for WSN upon the same testbed. The floor plan of the corridor is demonstrated in Figure 15a, and the corresponding experimental setting is presented in Figure 15b. The distance between two Wi-Fi APs is 5 m, and the distance between Wi-Fi APs and the SN is about 3 m.

As shown in Figure 18, in the LOS environment, the average positioning error of WSN and WiZiLoc is comparable, 0.7 m and 0.6 m respectively. However, in the NLOS environment, the performance of WSN degrades to 1.6 m, in contrast, WiZiLoc decreases only 0.4m. From this aspect, WiZiLoc shows the better robustness to LOS/NLOS environment than the conventional WSN.

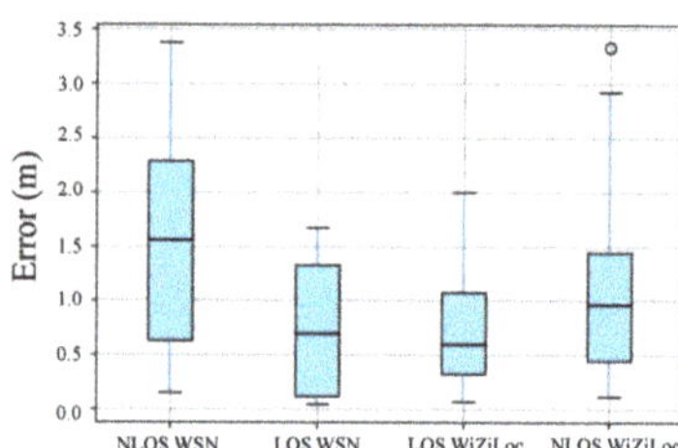

Figure 18. LOS vs. NLOS.

Furthermore, in order to further evaluate the robustness of WiZiLoc to the environment, the positioning experiments are investigated under different ambient environment, the canteen and laboratory corridor, the canteen is divided into crowded meal time and idle afternoon time. The experimental results are shown in Figure 19, where the average positioning errors in the empty/crowded canteen and corridor are 1.31 m, 1.77 m and 1.15 m respectively. We find the accuracy in corridor fluctuate slightly, and that in canteen shows greater fluctuation. The reason is that more moving targets carrying wireless device in canteen lead to the interference as the physical obstruction and signal interference. In addition, in the same canteen environment, the error does not increase significantly even when it becomes crowded during meal time. It shows that WiZiLoc has strong robustness.

4.3. Coverage

Limited by the coverage of ZNs, the conventional WSN needs a large number of ANs to enlarge the network coverage. In practical, the number of ANs cannot be unlimited, and evenly distributed in the monitoring area, thus a blind area will be appeared when positioning. The main contributions of WiZiLoc is to leverage the high-power Wi-Fi instead of the ANs to improve the network coverage. Hence, we investigate the coverage of WiZiLoc based on the simulations, as the simulation map with an area of 100 × 100 square meters illustrated in Figure 20. The x-axis is length and the y-axis is width. For comparison, we also depict the coverage of the traditional WSN. There are 30 ANs and 100 UZNs. It can be seen from the Figure 20a that even if arranged AN accounts for 30% of UZN, there are still a large number of dead corners causing node uncovered.

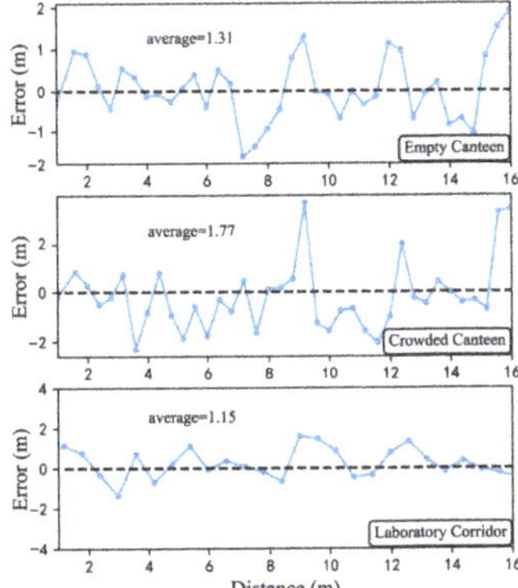

Figure 19. Location errors in canteen and laboratory corridor.

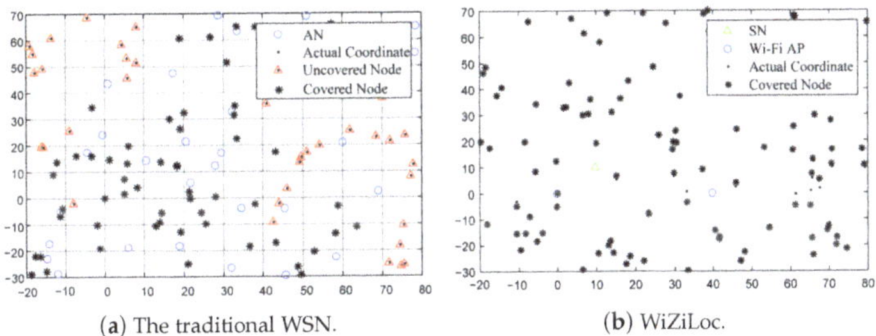

(**a**) The traditional WSN.

(**b**) WiZiLoc.

Figure 20. Coverage comparison of WSN and WiZiLoc.

Compared with the WSN localization system, which requires a large number of ANs to achieve a certain coverage, WiZiLoc can achieve full coverage without any AN. As can be seen from Figure 20b where there are also 100 UZNs, WiZiLoc can achieve almost 100% coverage without deploying any AN in advance.

4.4. The Broadcasting Period

In order to reduce the impact of Wi-Fi interference on the communication among ZNs, we design a broadcasting period of Wi-Fi APs by choosing a prime number which is larger than the maximum wake-up period of ZNs. To verify the effectiveness of the design, we conduct experiments in the corridor under NLOS ambient environment. One USRP is deployed as the ZN sender with wake-up period 12 ms, and another USRP is placed beside to broadcast CTC packets with period 30 ms and 53 ms, respectively. A commodity CC2530 is used to be a ZigBee receiver.

Since there are multiple common divisors between 12 and 30, such as 2, 3, 4 and 6, according to our broadcast period analysis, this multiple common divisor situation will lead to relatively serious packet collisions, which will lead to poor communication performance (BER and PRR). For example, 53 is a prime number significantly larger than 12, so it is reasonable to get better communication performance. However, the broadcast period should not be too large, otherwise the throughput and system efficiency will be reduced. The experimental results are shown in Figure 21.

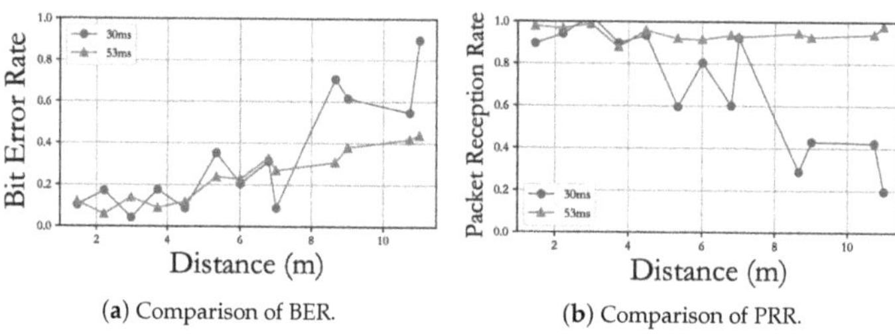

(**a**) Comparison of BER.   (**b**) Comparison of PRR.

Figure 21. Performance with different broadcasting period.

It can be seen from the Figure 21 that when the broadcasting period of CTC is the prime number 53 ms, the BER of ZNs is stable at 0.4 and the PPR is no less than 0.85. However, when the broadcasting period of CTC is 30 ms, the BER of ZNs degrades greatly when the distance beyond 7 m. By comparison, we find that the proposed prime number broadcasting period can improve the communication performance of ZNs in the network without affecting CTC packets reception.

4.5. Comparison

This paper compares WiZiLoc and the existing WSN localization system from four aspects: system structure, localization precision, system cost and algorithm complexity. The comparison results are shown in Table 1. Obviously, WiZiLoc outperforms the other methods in aspect of precision, cost and complexity.

Table 1. Comparison between WiZiLoc and the existing WSN localization system.

System	Structure	Precision	Cost	Complexity
[41]	Centralized	High	High	High
[13,42]	Distributed	Low	Low	High
[15,43]	Distributed	High	Low	High
WiZiLoc	Distributed	High	Low	Low

5. Conclusions

In this paper, we propose WiZiLoc, an anchor-free, distributed localization system for WSN with decimeter accuracy. The main contribution of WiZiLoc is that no ANs are required. WiZiLoc leverage physical-level CTC to achieve localization in WSN system for the first time. To address the asymmetric coverage and Wi-Fi interference problems, a new progressive self-positioning method, ONION, is proposed. The CTC broadcasting period is designed to be the maximum prime number of the wake-up cycle of ZNs in the network. Extensive experiments and simulations have been performed on the prototype of WiZiLoc, the localization accuracy can be up to 0.6m, 0.8m in the presence of Wi-Fi interference and 1m in NLOS scenario. Experimental results show that the proposed WiZiLoc can achieve better performance than the conventional anchor-based WSN in terms of accuracy, robustness, coverage and cost, etc.

Because of the anchor-free characteristic, the cost of arranging anchor nodes in all kinds of traditional WSN application scenarios can be reduced (such as factory warehouse management, terminal container monitoring, etc.). CTC technology for localization is only a modest display of a master. If CTC is applied beyond positioning, such as real-time interaction, giving instructions and supervisory control, a greater vision can be achieved. However, the security of CTC communication needs attention. To resist illegal invasion, the information encryption and identity authentication of CTC need to be deliberated.

Author Contributions: Conceptualization, N.J. and B.Z.; methodology, B.Z.; software, B.Z.; validation, N.J. and L.W.; formal analysis, N.J. and B.Z.; investigation, B.Z.; resources, L.W.; data curation, B.Z.; writing—original draft preparation, B.Z.; writing—review and editing, N.J.; visualization, L.W.; supervision, N.J.; project administration, L.W.; funding acquisition, N.J. All authors have read and agreed to the published version of the manuscript.

Funding: This work was funded in part by the National Natural Science Foundation of China [Grant No. 61772453], the Natural Science Foundation of Hebei Province [Grant No. F2020203074, F2022203045].

Acknowledgments: Our grateful thanks go to the anonymous referees for detailed reviews and insightful comments that have helped to improve this article substantially. The funding's support for our work is greatly appreciated.

Conflicts of Interest: The authors declare no conflict of interest.

References

1. Shahraki, A.; Taherkordi, A.; Haugen, Ø.; Eliassen, F. A survey and future directions on clustering: From wsns to iot and modern networking paradigms. *IEEE Trans. Netw. Serv. Manag.* **2021**, *18*, 2242–2274. [CrossRef]
2. Akyildiz, I.F.; Su, W.; Sankarasubramaniam, Y. Wireless sensor networks: A survey. *Comput. Netw.* **2002**, *38*, 393–422. [CrossRef]
3. Souza, F.L.; Nakamura, E.F.; Pazzi, R.W. Target tracking for sensor networks: A survey. *ACM Comput. Surv.* **2016**, *49*, 1–31. [CrossRef]
4. Annepu, V.; Rajesh, A.; Bagadi, B. Radial basis function-based node localization for unmanned aerial vehicle-assisted 5g wireless sensor networks. *Neural Comput. Appl.* **2021**, *33*, 12333–12346. [CrossRef]
5. Yassin, A.; Nasser, Y.; Awad, M.; Al-Dubai, A.; Liu, R.; Yuen, C.; Raulefs, R.; Aboutanios, E. Recent advances in indoor localization: A survey on theoretical approaches and applications. *IEEE Commun. Surv. Tutor.* **2016**, *19*, 1327–1346. [CrossRef]
6. He, W.; Cheng, R.; Mao, K.; Yan, K.; Wei, J.; Xu, Y. A novel anchorless node positioning method for wireless sensor network. *J. Sens.* **2022**, *2022*, 5385393. [CrossRef]
7. Liu, W.; Luo, X.; Wei, G.; Liu, H. Node Localization Algorithm for Wireless Sensor Networks Based on Static Anchor Node Location Selection Strategy, Computer Communications. Available online: https://www.sciencedirect.com/science/article/pii/S0140366422002122 (accessed on 1 August 2022). [CrossRef]
8. Huang, H.; Miao, W.; Min, G.; Huang, C.; Wang, C. Resilient range-based d-dimensional localization for mobile sensor networks. *IEEE/ACM Trans. Netw.* **2020**, *28*, 2037–2050. [CrossRef]
9. Singh, S.P.; Sharma, S. Range free localization techniques in wireless sensor networks: A review. *Procedia Comput. Sci.* **2015**, *57*, 7–16. [CrossRef]
10. Zhao, S.; Zhang, X.-P.; Cui, X.; Lu, M. A new toa localization and synchronization system with virtually synchronized periodic asymmetric ranging network. *IEEE Internet Things J.* **2021**, *8*, 9030–9044. [CrossRef]
11. Sun, Y.; Zhang, F.; Wan, Q. Wireless sensor network-based localization method using tdoa measurements in mpr. *IEEE Sens. J.* **2019**, *19*, 3741–3750. [CrossRef]
12. Liu, J.; Hu, M.; Zhou, M.; Ding, Z. Performance analysis and comparison for tdoa and dtdoa. In Proceedings of the 2021 3rd International Conference on Information Technology and Computer Communications, Shanghai, China, 23–25 April 2021; pp. 103–108.
13. Vashistha, A.; Law, C.L. E-dtdoa based localization for wireless sensor networks with clock drift compensation. *IEEE Sens. J.* **2019**, *20*, 2648–2658. [CrossRef]
14. Zheng, Y.; Sheng, M.; Liu, J.; Li, J. Exploiting aoa estimation accuracy for indoor localization: A weighted aoa-based approach. *IEEE Wirel. Commun. Lett.* **2019**, *8*, 65–68. [CrossRef]
15. Sahota, H.; Kumar, R. Maximum-likelihood sensor node localization using received signal strength in multimedia with multipath characteristics. *IEEE Syst. J.* **2016**, *6*, 1–10. [CrossRef]
16. Xie, Y.; Xiong, J.; Li, M.; Jamieson, K. md-track: Leveraging multi-dimensionality in passive indoor wi-fi tracking. In Proceedings of the 25th Annual International Conference, Athens, Greece, 15–18 May 2019; pp. 1–16.
17. Costa, M.S.; Tomic, S.; Beko, M. An socp estimator for hybrid rss and aoa target localization in sensor networks. *Sensors* **2021**, *21*, 1731. [CrossRef]
18. Panwar, K.; Katwe, M.; Babu, P.; Ghare, P.; Singh, K. A majorization-minimization algorithm for hybrid toa-rss based localization in nlos environment. *IEEE Commun. Lett.* **2022**, *26*, 1017–1021. [CrossRef]
19. Panwar, K.; Fatima, G.; Babu, P. Optimal sensor placement for hybrid source localization using fused toa-rss-aoa measurements. *arXiv* **2022**, arXiv:2204.06198.
20. Panwar, K.; Babu, P. Majorization-minimization based hybrid localization method for high precision localization in wireless sensor networks. *arXiv* **2022**, arXiv:2205.03881.
21. Hosseinianfar, H.; Brandt-Pearce, M. Performance limits for fingerprinting-based indoor optical communication positioning systems exploiting multipath reflections. *IEEE Photonics J.* **2020**, *12*, 1–16. [CrossRef]

22. Kaur, A.; Kumar, P.; Gupta, G.P. A weighted centroid localization algorithm for randomly deployed wireless sensor networks. *J. King Saud Univ. Comput. Inf. Sci.* **2019**, *31*, 82–91. [CrossRef]
23. Chen, T.; Hou, S.; Sun, L. An enhanced dv-hop positioning scheme based on spring model and reliable beacon node set. *Comput. Netw.* **2022**, *209*, 108926. [CrossRef]
24. Shen, S.; Yang, B.; Qian, K.; Wang, W.; Jiang, X.; She, Y.; Wang, Y. An improved amorphous localization algorithm for wireless sensor networks. In Proceedings of the 2016 International Conference on Networking and Network Applications (NaNA), Hokkaido, Japan, 23–25 July 2016; pp. 69–72. [CrossRef]
25. Liu, J.; Wang, Z.; Yao, M.; Qiu, Z. Vn-apit: Virtual nodes-based range-free apit localization scheme for wsn. *Wirel. Netw.* **2016**, *22*, 867–878. [CrossRef]
26. Saeed, N.; Nam, H.; Al-Naffouri, T.Y.; Alouini, M.S. A state-of-the-art survey on multidimensional scaling-based localization techniques. *IEEE Commun. Surv. Tutor.* **2020**, *21*, 3565–3583. [CrossRef]
27. Zhang, S.; Liu, X.; Wang, J.; Cao, J.; Min, G. Accurate range-free localization for anisotropic wireless sensor networks. *ACM Trans. Sens. Netw.* **2015**, *11*, 1–28. [CrossRef]
28. Savarese, C.; Rabaey, J.M.; Beutel, J. Location in distributed ad-hoc wireless sensor networks. In Proceedings of the 2001 IEEE International Conference on Acoustics, Speech, and Signal Processing (Cat. No. 01CH37221), Salt Lake City, UT, USA, 7–11 May 2001; Volume 4, pp. 2037–2040.
29. Priyantha, N.B.; Balakrishnan, H.; Demaine, E.; Teller, S. Anchor-free distributed localization in sensor networks. In Proceedings of the 1st International Conference on Embedded Networked Sensor Systems, Los Angeles, CA, USA, 5–7 November 2003; pp. 340–341.
30. Shang, Y.; Ruml, W.; Zhang, Y.; Fromherz, M.P.J. Localization from mere connectivity. In Proceedings of the 4th ACM International Symposium on Mobile Ad Hoc Networking and Computing, Annapolis, MD, USA, 1–3 June 2003; pp. 201–212.
31. Benbadis, F.; Friedman, T.; De Amorim, M.D.; Fdida, S. Gps-free-free positioning system for wireless sensor networks. In Proceedings of the Second IFIP International Conference on Wireless and Optical Communications Networks, Dubai, United Arab Emirates, 6–8 March 2005; pp. 541–545.
32. Qu, H.; Wicker, S.B. Anchor-free localization in rapidly-deployed wireless sensor networks. In Proceedings of the 2006 IEEE International Conference on Mobile Ad Hoc and Sensor Systems, Vancouver, BC, Canada, 9–12 October 2006; pp. 627–632.
33. Fang, L.; Du, W.; Ning, P. A beacon-less location discovery scheme for wireless sensor networks. In Proceedings of the IEEE 24th Annual Joint Conference of the IEEE Computer and Communications Societies, Miami, FL, USA, 13–17 March 2005; pp. 161–171.
34. Fuiorea, D.; Gui, V.; Pescaru, D.; Toma, C. Using registration algorithms for wireless sensor network node localization. In Proceedings of the 2007 4th International Symposium on Applied Computational Intelligence and Informatics, Timisoara, Romania, 17–18 May 2007; pp. 209–214.
35. He, Y.; Guo, X.; Zheng, X.; Yu, Z.; Zhang, J.; Jiang, H.; Na, X.; Zhang, J. Cross-technology communication for the internet of things: A survey. *ACM Comput. Surv.* **2023**, *55*, 1–29. [CrossRef]
36. Li, Z.; Tian, H. Webee: Physical-layer cross-technology communication via emulation. In Proceedings of the 23rd Annual International Conference, Staten Island, NY, USA, 7–9 June 2017; pp. 2–14.
37. He, Y.; Guo, X.; Zhang, J.; Jiang, H. Wide: Physical-level ctc via digital emulation. *IEEE/ACM Trans. Netw.* **2021**, *29*, 1567–1579. [CrossRef]
38. Xia, D.; Zheng, X.; Liu, L.; Wang, C.; Ma, H. c-chirp: Towards symmetric cross-technology communication over asymmetric channels. *IEEE/ACM Trans. Netw.* **2021**, *29*, 1169–1182. [CrossRef]
39. Zheng, X.; Xia, D.; Guo, X.; Liu, L.; Ma, H. Portal: Transparent cross-technology opportunistic forwarding for low-power wireless networks. In Proceedings of the Mobihoc'20: The Twenty-first ACM International Symposium on Theory, Algorithmic Foundations, and Protocol Design for Mobile Networks and Mobile Computing, Boston, MA, USA, 11–14 October 2020; pp. 241–250.
40. Jing, N.; Zhang, B.; Liu, G.; Yang, L.; Wang, L. Anchor-Free Self-Positioning in Wireless Sensor Networks via Cross-Technology Communication. In Proceedings of the 2021 IEEE 27th International Conference on Parallel and Distributed Systems (ICPADS), Beijing, China, 14–16 December 2021; pp. 915–922.
41. Kristalina, P.; Pratiarso, A.; Badriyah, T.; Putro, E.D. A wireless sensor networks localization using geometric triangulation scheme for object tracking in urban search and rescue application. In Proceedings of the 2016 2nd International Conference on Science in Information Technology (ICSITech), Yogyakarta, Indonesia, 27–28 October 2016; pp. 254–259.
42. Dutta, S.; Obaidat, M.S.; Dahal, K.; Giri, D.; Neogy, S. M-memhs: Modified minimization of error in multihop system for localization of unknown sensor nodes. *IEEE Syst. J.* **2019**, *13*, 215–225. [CrossRef]
43. Kumar, A. Karampal, Optimized range-free 3d node localization in wireless sensor networks using firefly algorithm. In Proceedings of the 2015 International Conference on Signal Processing and Communication (ICSC), Singapore, 25–27 February 2015; pp. 14–19. [CrossRef]

MDPI

Article

MetaEar: Imperceptible Acoustic Side Channel Continuous Authentication Based on ERTF

Zhuo Chang [1,2], Lin Wang [1], Binbin Li [3] and Wenyuan Liu [1,*]

1 School of Information Science and Engineering, Yanshan University, Qinhuangdao 066004, China
2 School of Cyber Security and Computer, Hebei University, Baoding 071000, China
3 School of Economics and Management, Yanshan University, Qinhuangdao 066004, China
* Correspondence: wyliu@ysu.edu.cn

Abstract: With the development of ubiquitous mobile devices, biometrics authentication has received much attention from researchers. For immersive experiences in AR (augmented reality), convenient continuous biometric authentication technologies are required to provide security for electronic assets and transactions through head-mounted devices. Existing fingerprint or face authentication methods are vulnerable to spoof attacks and replay attacks. In this paper, we propose *MetaEar*, which harnesses head-mounted devices to send FMCW (Frequency-Modulated Continuous Wave) ultrasonic signals for continuous biometric authentication of the human ear. CIR (channel impulse response) leveraged the channel estimation theory to model the physiological structure of the human ear, called the Ear Related Transfer Function (ERTF). It extracts unique representations of the human ear's intrinsic and extrinsic biometric features. To overcome the data dependency of Deep Learning and improve its deployability in mobile devices, we use the lightweight learning approach for classification and authentication. Our implementation and evaluation show that the average accuracy can reach about 96% in different scenarios with small amounts of data. *MetaEar* enables one to handle immersive deployable authentication and be more sensitive to replay and impersonation attacks.

Keywords: acoustic sense; continuous authentication; Ear Related Transfer Function; FMCW; CIR

Citation: Chang, Z.; Wang, L.; Li, B.; Liu, W. MetaEar: Imperceptible Acoustic Side Channel Continuous Authentication Based on ERTF. *Electronics* **2022**, *11*, 3401. https://doi.org/10.3390/electronics11203401

Academic Editor: Akshya Swain

Received: 27 September 2022
Accepted: 19 October 2022
Published: 20 October 2022

Publisher's Note: MDPI stays neutral with regard to jurisdictional claims in published maps and institutional affiliations.

1. Introduction

Ubiquitous smart devices, such as smart phones, smart headphones, and smart watches, with their rich built-in sensors, have become an important access carrier for the Cyber–Physical Human System (CPHS) [1]. On the one hand, the requirements of work and entertainment during the mobile process make wearing headphones a daily behavior, which spawns a new computing model, earable computing [2]. On the other hand, taking mobility, ubiquity, and human-centricity into account, CPHS and even the Metaverse have higher requirements for continuous identity authentication [3,4]. The global biometric authentication market size is projected to reach 30.5 billion U.S. dollars by 2026 [5].

Digital assets and commercial transactions require highly secure and reliable continuous authentication to protect user property. These transactions require ongoing multi-factor authentication to ensure their security and validity. In addition, continuous, safe and efficient biometric authentication can also be applied to scenarios such as telemedicine, disabled services, and virtual production.

Biometric authentication has the advantages of stochastic variation within the different people and no dependence on shared secrets. The existing face, fingerprint, iris, and voiceprint authentication methods are based on extrinsic biometrics. Authentication based on face, fingerprint, and iris is vulnerable to presentation attacks and also raises privacy concerns [6], using a photo [7], video [8], mask [9], or silicone fingertip [10] to impersonate a victim. In addition, the voiceprint-based authentication method is vulnerable to replay attacks [11,12] of voice recording. So, can we design a continuous and transparent biometric authentication method to ensure the system's security?

At present, ultrasonic authentication through the human ear is a novel solution. The head-mounted device sends ultrasonic sound to the human ear and performs continuous identity authentication through the feedback of the human ear. Collecting the FMCW (Frequency-Modulated Continuous Wave) ultrasonic signals reflected from the human ear uses the signal channel characteristics to describe the human body's anatomy and extract the unique biological features [13] for authentication. The inaudible ultrasound can perform continuous passive perceptual authentication without affecting human hearing and without interfering with the user's immersive experience, releasing the user from the authentication frequent interaction process. Furthermore, unlike the existing face identification and fingerprint authentication, ultrasonic authentication can prevent counterfeiting and replay attacks, thus ensuring the security of transactions in Metaverse. On this basis, we aim to build an explainable acoustic authentication model that can be applied in daily scenarios with zero effort and high accuracy.

However, we face three major technical challenges to achieving a one-fits-all model. First of all, the features originally used in acoustic action recognition, such as doppler [14] and phase [15], are aimed at dynamic behaviors and cannot be used to describe static object features. So for the uniqueness of the biological structures of auricles and ear canals, how do we design a model that extracts the unique biological geometric features of static auricles and ear canals? Second, the FMCW acoustic signal received by two microphones in real time is not synchronized. How can the multi-microphone signal be effectively synchronized? Third, ultrasound is a vibration wave, and for a monostatic sensing device with co-located transceivers, vibration will cause self-interference of the received signal. How do we remove the self-interference and extract high-resolution characteristic signals?

To overcome these challenges, we propose *MetaEar*, a continuous authentication system based on imperceptible acoustic fingerprints, which uses FMCW ultrasound to model the unique biometrics of the human ear and conduct authentication, as shown in Figure 1. The key component of *MetaEar* is ERTF (Ear Related Transfer Function), which uses ERTF to identify and extract the unique biometrics of the human ear. Our observation is that each person's auricle and the ear structure have different responses to the delay and magnitude of FMCW ultrasound. The auricle and ear canal modulate the sound signal to produce different delays. The eardrum and the cochlea convert the different magnitude mechanical sound waves to neural electrical signals.

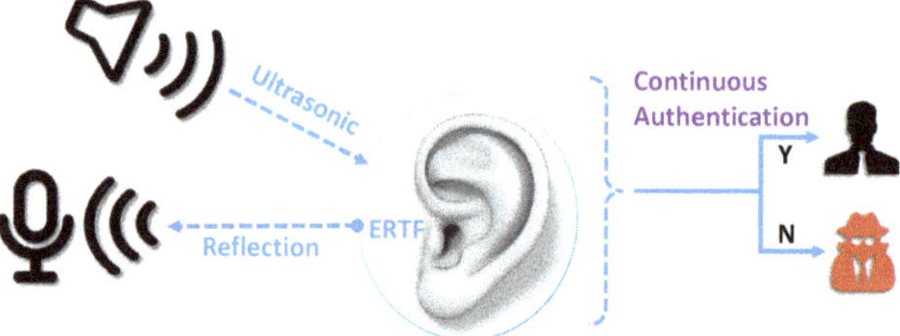

Figure 1. MetaEar, an acoustic side channel continuous authentication system.

Furthermore, the channel impulse response of the human ear can be modeled to extract the ear's unique features and authenticate it. The speaker on the headset sends inaudible FMCW ultrasonic waves, and the co-located microphone receives the reflected sound signal. After passing the band-pass filter, the phase slope change is harnessed to align the signals due to the problem of fuzzy peak judgment of cross-correlation. MetaEar employs a dual-microphone differential denoising to eliminate the self-interference caused by the co-located physical transmission of the vibration wave. Finally, the channel impulse response of the signal is calculated by modeling the ERTF so that the unique biometric vector of the human ear can be achieved. This process is equivalent to model-based human ear anatomy feature embedding. For better application deployment, instead of using a Deep Learning model that requires numerous data, we utilize a traditional SVM (Support Vector

Machine) to perform one-class authentication. Due to *MetaEar*'s modeling of the complex structure of the human ear and the relatively slow transmission speed and low power of ultrasonic signals, it prevents co-located signal collection and resists counterfeiting attacks and replay attacks.

In a nutshell, our core contributions are three-fold.

- We propose the *MetaEar* system, which uses the ultrasonic reflection signals of the auricle and ear canal to continuously authenticate users with ERTF, which effectively prevents replay attacks and counterfeiting attacks.
- We design a characteristic function to represent the auricle and ear canal biometric features through the principle of the impulse response of channel estimation, which effectively expresses the unique biological characteristics of the human auricle and ear canal.
- We build a prototype of *MetaEar* with commercial off-the-shelf smartphones and evaluate its effectiveness and security in different settings. Extensive experiments show that it authentication accuracy over 96.8%.

The rest of this paper is organized as follows. In Section 2, we review the related works. Section 3 introduces the adversary model. Section 4 gives the mathematical derivation on how ERTF could model the unique feature of the human ear, and demonstrates the feasibility study. Section 5 presents the MetaEar architecture design. Section 6 is the implementation setup. Section 7 show the evaluation result, followed by a conclusion in Section 8.

2. Related Works

2.1. Earable Computing

The era of earable computing is coming, and "earable" [2,16] refers to wearable devices around the ears, such as mobile phones, headsets, headphones, and smart glasses. These devices can utilize acoustic signals to interact with people, such as listening to music, sensing oral activities [14,17,18], and even using the ear canal for authentication [19–21].

2.2. Acoustic Authentication

Acoustic waves can measure temperature [22], tracking [23–25], gesture sensing [26,27], and activity recognition [28] to breathe monitoring [15]. The authentication based on the sound signal mainly uses the acoustic signal to extract corresponding unique biometric features to achieve authentication, including using audible sounds, such as voiceprints [29], for authentication. Some use the FMCW ultrasonic signal to extract the feature of teeth actions for authentication [20,30], and some use the sound signal to extract the characteristics of the throat movement [31] for authentication. The FMCW ultrasonic signal is also used for lip motion [32] authentication or face liveness [33] detection.

2.3. Continue Authentication

As augmented reality evolves, continuous authentication ensures system security without interfering with the user's immersive experience. Some use WiFi signals for continuous authentication [34,35], and the sensitivity of RF signals to location and orientation seriously reduces generalization. Some existing works using heart rate [36–38] and respiration [39,40] require the user to remain motionless, prone to severe interference from multipath environments, and cannot be applied to multi-person scenarios. Behaviors [41] can also be used for continuous authentication but are vulnerable to impersonation attacks. Continuous authentication uses eye movement [42,43] but is sensitive to ambient light and requires high equipment costs. There is also work using the ear canal [19,21], but the types of equipment are all earbuds, which will not only cause irreversible damage to the user's hearing but also have higher needs on the depth and position of the earplugs in the ear canal.

3. Threat Model

We assume that the attackers are not resourceful in the headset hardware. The attack succeeds if the attacker can implant malware in the headset and obtain any private data he wants. In other words, augmented reality's software and hardware resources are secure. Based on the above assumptions, two types of adversarial behaviors are considered below.

Replay attack: The replay attack scenario in which the adversary is physically near the victim and his/her enrolled handset, such as in a crowded campus cafe or public vehicle. The collected signals can be disguised as the victim entering the system to invade the victim's Metaverse property and privacy [44].

Impersonate attack: The adversary sends and receives sound wave signals from its ears in different modes, impersonating the victim, intending to deceive the security authentication into attacking the system. Alternatively, various silicone bionic materials are used to imitate the biological structure of the human body, for example, imitating faces and imitating fingerprints with silicone. However, both faces and fingerprints are explicit biometrics that can be easily forged. The counterfeiting attack uses forged 3D printed artificial dummy ears of particular anthropomorphic materials to complete the invasion. It deceives the system to destroy and take the virtual property of the owner [45,46].

4. ERTF Model

4.1. Ear Structure

The human ear is an essential auditory organ of the human body and has unique biological characteristics that are different for each person [47]. The human ear structure is divided into three parts, as shown in Figure 2, in which the auricle and the ear canal collect and transmit sound. The human ear canal is about 25~35 mm long and about 8 mm diameter. The eardrum and cochlea are the perceptions of sound. Different people have different geometric shapes of the external auricle and ear canal, and the response of the eardrum and cochlea to sound is unique for different people. Therefore, continuous authentication can be realized by modeling the characteristics of the ear's intrinsic physiological structure and extrinsic geometric shape.

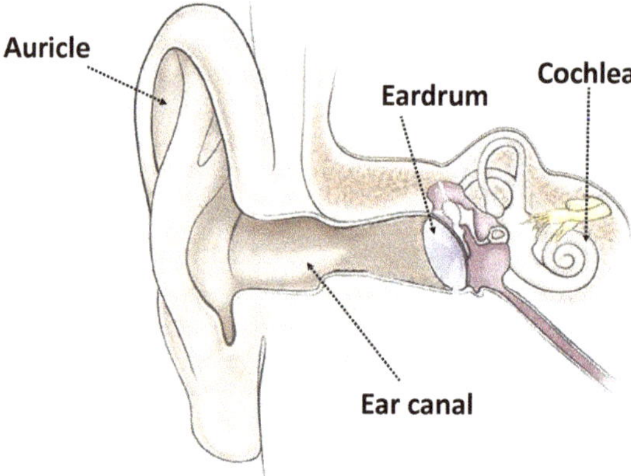

Figure 2. The physiological structure of human ear.

4.2. ERTF

In Metaverse, in addition to vision, sound immersion is also essential, allowing users to maintain the immersion of sound and perceive spatial positioning. Individualized or generic HRTFs (Head Related Transfer Functions) are usually employed to render spatialized sounds within an AR headset.

HRTF is a phenomenon that describes how an ear and head receive sound from a sound source. ERTF is very similar but describes the ear profile to the FMCW acoustic signal. Most notably, the shape of the pinna, the length and the diameter of the ear canal, and the subtle differences in biological properties influence the incoming ultrasonic signal by boosting some delays and phases. When the reflected sound signal reaches the microphone, the intrinsic unique biometric features of the human ear can be collected. So ERTF is the change of the sound's response profile to the unique characteristics of the user's ear. It is mainly produced by the pinna, ear canal, and cochlea, and we define it as ERTF:

$$h_{\mathrm{ERTF}}(t) = IFFT(H(f)). \tag{1}$$

where $H(f)$ is the channel frequency response produced by the disparate anatomy of the ear, and IFFT is Inverse Fast Fourier Transform.

We imagine the whole human acoustic sensing process as communication through an RF signal. The acoustic sensing channel can be modeled as a linear time-invariant system, which effectively models propagation delay and signal attenuation along multiple propagation paths. So, all the channel parameter changes by the user's organic textures are modeled for the channel state estimation. In this way, we could easily extract the unique features of the subject's ear.

To achieve that, we borrow the idea from channel estimation to determine the fading and path loss of the wireless channel. The received signal can be mathematically represented as $r(t) = h(t) \times s(t)$, where $h(t)$ represents the CIR (Channel Impulse Response) of the acoustic channel; $r(t)$ and $s(t)$ represent the received signal and transmitted signal, respectively. The linear frequency of the FMCW acoustic signal is expressed as:

$$f(t) = f_c + \frac{Bt}{T}, \tag{2}$$

where T is FMCW chirp duration time, B is the chirp bandwidth, and f_c is initial frequency. Phase $u(t)$ is derived as:

$$u(t) = \int_0^t f(t')dt' = 2\pi(f_c t + B\frac{t^2}{2T}), \tag{3}$$

So the send signal is:

$$s_{FMCW}(t) = e^{-j2\pi\left(f_c t + \frac{B}{2T}t^2\right)}, \tag{4}$$

Then the frequency domain expression is:

$$S_k = \sum_{i=0}^{N-1} e^{-j2\pi\left(f_c t + \frac{B}{2T}t^2\right)} e^{-j\frac{2\pi}{N}kn}. \tag{5}$$

Ultrasonic sound signals are transmitted in the ear canal and inner ear in a multipath environment and reflected in the microphone for the reception. Assuming there are P paths, the delay of each path is $\tau_i, i \in [1, P]$, and the multipath delay of the received signal is:

$$r_{FMCW}(t) = \sum_{i=1}^{P} \theta_i e^{-j2\pi\left(f_c(t-\tau_i) + \frac{B}{2T}(t-\tau_i)^2\right)}, \tag{6}$$

where θ_i is attenuation coefficient. The frequency-domain formula is:

$$R_k = \sum_{n=0}^{N-1} \sum_{i=1}^{P} \theta_i e^{-j2\pi\left(f_c(t-\tau_i) + \frac{B}{2T}(t-\tau_i)^2\right)} e^{-j\frac{2\pi}{N}kn}$$

$$= \sum_{i=1}^{P} \theta_i S_k e^{-j\frac{2\pi}{N}k\tau_i}, \tag{7}$$

S_k represents the transmitted acoustic signal. CIR-based ERTF is:

$$h_{\text{ERTF}}(t) = IFFT\left(\frac{R_k}{S_k}\right) = IFFT\left(\sum_{i=1}^{P}\theta_i e^{-j\frac{2\pi}{N}k\tau_i}\right)$$

$$= \sum_{n=0}^{N-1}\sum_{i=1}^{P}\theta_i e^{-j\frac{2\pi}{N}k\tau_i}e^{j\frac{2\pi}{N}kn} \tag{8}$$

$$= \sum_{n=0}^{N-1}\sum_{i=1}^{P}\theta_i e^{-j\frac{2\pi}{N}k(n-\tau_i)}.$$

In practice, CIR measurement is represented with a set of complex values. Each complex value measures the channel information of a specific propagation delay range and the corresponding magnitudes and phases of the CIR. It can be seen from Equation (8) that ERTF is related to attenuation θ_i and time delay τ_i. θ_i expresses the magnitude attenuation caused by the conversion of sound mechanical waves into electronic signals by the intrinsic biological structure, and τ_i expresses the multipath delay caused by the extrinsic geometric structure. That is why ERTF can express the human ear's physiological and physical unique characteristics.

4.3. Feasibility Analysis

In a quiet hall, we used the Samsung NEXUS 6 mobile phone to collect the data on the FMCW sound signals. The frequency band of the ultrasonic wave is 18~22 kHz. By calculating the received signal's PSD (Power Spectral Density) $P_r(\omega)$ of different subjects, as shown in Figure 3, it can be seen that the features can be distinguished.

$$P_r(\omega) = \left|H_{ERTF}\left(e^{j\omega}\right)\right|^2 P_s(\omega). \tag{9}$$

Because in the channel estimation, for the autoregressive time series model, the relationship between the PSD and the ERTF amplitude is as in Equation (9), we can show the unique features of different subjects by the PSD.

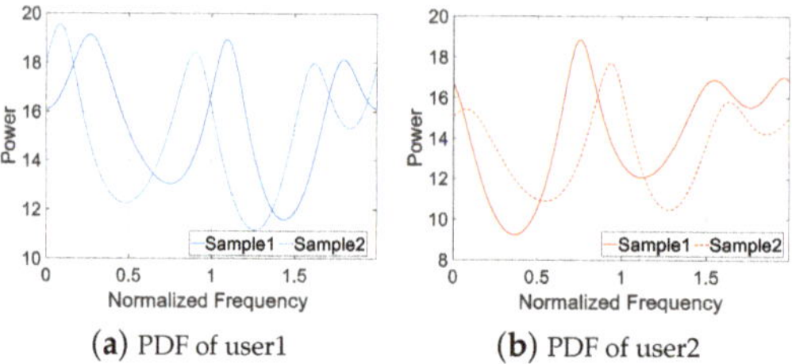

(**a**) PDF of user1　　　(**b**) PDF of user2

Figure 3. Feasibility analysis by power spectral density.

5. System Design

5.1. Overview

MetaEar is a system for continuous biometric-based authentication using FMCW ultrasound. As shown in Figure 4, the system consists of four modules, data collection, signal processing, feature extraction, and authentication. The registration process is the same as the authentication process. First, the speaker sends out FMCW ultrasonic waves with a time slot facing the human ear, and the two microphones receive the reflected chirp signal.

MetaEar contains two primary parts: signal processing and feature extraction modules. Since the device's microphones have a hardware startup time, the front empty window period caused by the hardware delay should be removed first. Then, for the co-located transmitting and receiving devices, the self-interference cancellation is performed using the differential technique of the dual microphones. After denoising, it is critical to align each chirp to perform feature extraction. Here we use a method based on the phase slope so that the two received signals can be satisfactorily aligned. We only sample the signal within the corresponding frequency band and carry out a Hanning filter to remove the noise signal outside the frequency band. There is a particular time slot between every two chirps, and the actual signal segment of each chirp needs to be extracted. The segmentation is implemented through a power spectrum envelope. Subsequently, the segments of these chirps are directly input into the feature extraction module.

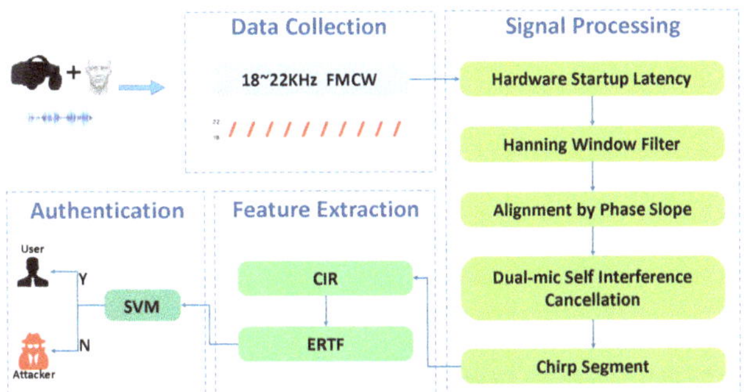

Figure 4. The authentication framework of *MetaEar* consists of four components: data collection, signal processing, feature extraction, and authentication.

The feature extraction module mainly uses the proposed ultrasonic chirp of a specific frequency band to perform feature extraction. First, the calculation of the channel frequency response is fulfilled. Secondly, the channel impulse response is converted into the channel impulse response through IFFT. The uniqueness of the human ear's geometric structure and endoplasmic structure can be represented as the channel impulse parameters of the sound transmission channel. Then PCA (Principle Component Analysis) extracts principal components to form feature vectors. These feature vectors are fed to the SVM of the authentication module for one-classification to achieve efficient continuous authentication.

5.2. Acoustic Design

On the one hand, the ultrasonic signal cannot produce audible noise that interferes with the user. Due to the limitation of mobile phone hardware, the highest sound frequency of prevalent mobile phones is 24 kHz, and the maximum sample rate is 48 kHz. In order to increase the bandwidth of the sound signal, we employ the ultrasonic signal above 18 kHz. The FMCW ultrasonic signal frequency is 18~22 kHz, $FC = 18$ kHz, $B = 4$ kHz, and the wavelength is 16~21 mm. The ear canal length is about 25~35 mm. Furthermore, combined with the reflection distance, the overall length can reach the range of 1~2 sound wavelengths, which naturally satisfies the near-field requirements.

As shown in Figure 5, the chirp duration of FMCW ultrasound is 10 ms because the smaller the FMCW period, the higher the range resolution. To avoid ISI (Inter Symbol Interference) between each chirp, we added two 10 ms idle time slots between every two chirps. The time delay between the transmitted signal and the received signal is τ.

5.3. Denoise

Hardware delay elimination. Because transceivers of mobile phones have a hardware run-up time, the 10 ms period FMCW sound signal is sensitive to the short startup time. Therefore, we need to remove the empty sampling points at the beginning of the acquisition signal. According to experience, this part of the time is 500 samples, which is $500/48 = 10.4167$ ms.

Ambient noise. Typically, common noise in the daily circumstances declines sharply in the frequency band above 8 kHz and lower than 18 kHz [48]. Therefore, we use the Hanning window filter for the corresponding filtering and only maintain the signal within

18~22 kHz, as shown in Figure 6, to remove the high- and low-frequency noise in the environment and reduce the frequency leakage.

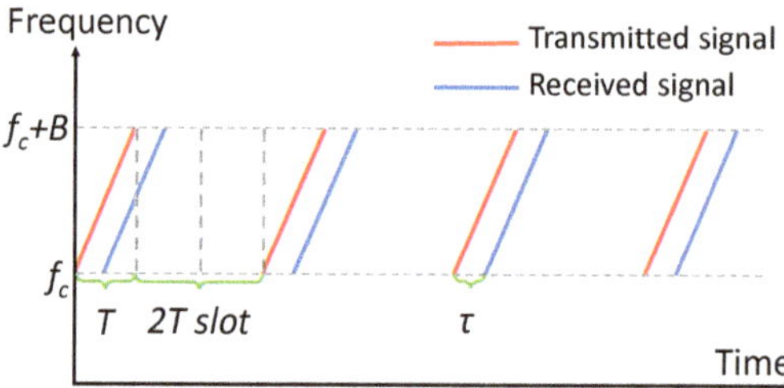

Figure 5. The FMCW chirp signal design.

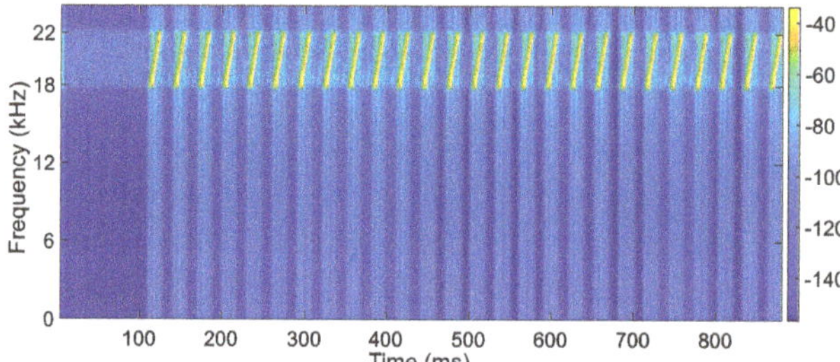

Figure 6. The frequency domain denoised FMCW chirp signal.

Dynamic interference. Because the sound wave propagates slowly, the transmission power is weak. It attenuates quickly, the propagation distance is exceptionally compact, and the transmission and reception area is close to the ear. There is almost no need to remove the interference of moving objects in the long-distance environment to the signal. For very close moving objects, such as other people doing random behaviors close to the user's ears, since the action frequency is generally lower than 18 kHz, the above Hanning bandpass filter can filter out this part of the noise.

5.4. Synchronization

The two microphone channels of sound signals have a time difference due to the distance between the two microphones, so time alignment and synchronization are required. Time alignment is obtained by transforming the signals into the frequency domain and applying the linear phase shifts corresponding to the time delay.

Existing [49] work in RF sensing uses two transmit antennas to cancel the direct path signal at the receiver. However, this approach does not work for acoustic-based ERTF. The reasons are two-fold. First, a smartphone typically has two pairs of co-located speakers and microphones. Playing sounds with any speaker may saturate the corresponding microphone by directly transferring the acoustic signal, making it infeasible to sense reflections. Second, the speakers on the smartphone are designed for different usages (e.g., communication, playing sound) and thus are highly heterogeneous. It is hard to perform equalization on a commercial audio system for FMCW chirp signals.

Instead, we leverage two microphones available on smartphones to achieve self-interference cancellation. Specifically, suppose one speaker plays the FMCW signal, and two microphones receive $r_1(t)$ and $r_2(t)$, respectively. Then ERTF estimates the subsample delay ε_t with the phase slope changing in the frequency domain and further calculates the correlation between two aligned signals [37].

$$\varepsilon_t = \min_{\varepsilon} |\angle(\mathcal{F}[r_1(t)]\mathcal{F}^*[r_2(t)]) + 2\pi f\varepsilon|$$

$$r_2^{\text{shift}}(t) = \mathcal{F}^{-1}\left[\mathcal{F}[r_2(t)] \cdot e^{-j2\pi f\varepsilon_t}\right], \tag{10}$$

where $\mathcal{F}[\cdot]$ denotes Fourier transform. Since the direct signal from the speaker to the microphones is the most potent component, we can approximate the estimation above as the delay and amplitude ratio of the direct acoustic of the two microphones.

5.5. Dual-Mic Subtraction

After alignment, we want to remove the self-interfering signal received by the microphone due to direct physical propagation. Since this part of the signal propagates through solid hardware, the propagation speed is 15 times the speed in the air, so the first received signal component is the self-interference signal. The most commonly used method utilizes autocorrelation [39,50] to eliminate the self-interference through the corresponding tap peak search. However, due to autocorrelation, there will be the disadvantage that the tap peak is ambiguous. The distance is too close to the regular reflection signal peak, so the blur is too high to distinguish the peak of the direct path. We use the dual microphone cancellation method. Differentiating the signals of the two microphones after alignment can eliminate the self-interference signal and retain the relevant dynamic and adequate information about the ERTF. For the noise in the environment, both microphones can receive it, so the dual-microphone differential method can also remove the ambient noise. From the above, we know that:

$$\rho = \frac{r_1(t) \cdot r_2^{\text{shift}}(t)}{|r_1(t)||r_2^{\text{shift}}(t)|}, \tag{11}$$

Thus, ERTF scales $r_2^{\text{shift}}(t)$ with ρ, and subtracts it from $r_1(t)$:

$$r(t)^{\text{cancel}} = r_1(t) - \rho r_2^{\text{shift}}(t), \tag{12}$$

Then the signals from each microphone are time synchronization. The synthesized signal in the time domain is formulated below.

$$r(t)^{\text{cancel}} = \frac{1}{N}\sum_{i=1}^{N}\theta_i\omega_i x_i(t - \Delta t_{\varepsilon_t}). \tag{13}$$

where Δt_{ε_t} indicates the time delay.

5.6. Chirp Segment

After completing the above procedures, each cycle transmitted FMCW sound signal has been segmented and extracted. Because the guard gap slot in the middle is futile, it is only necessary to calculate the ERTF for the transmit and receive chirps in each cycle. After trying several segment methods such as VAD (Voice Activity Detection), variance, and power accumulation, we eventually chose signal envelope, which is more efficient in computation. First, find all peaks in the signal, and obtain local maxima that are separated by N points at least, then use spline interpolation to return the peak envelope of the signal, as formulated in Equation (14).

$$Envelope_{r(t)} = \left|r(t)_{\text{cancel}}\right|. \tag{14}$$

After the power spectrum envelope is obtained, we set the threshold to be the overall expectation and detect the start and end points of the chirp signal of each cycle. The algorithm is shown in Algorithm 1.

Algorithm 1: Chirp Segmentation.

Input: Denoised & Synchronized FMCW Sequence: $r[n]$
Output: Segmented Chirp Sequence: $Chirp\_Matrix[n]$

```
// Get the envelope of r[n].
```
1 $Peak[i] = FindTop(r[i])$;
2 $Envelope[i] = Interpolation(Peak[i])$;

```
// Find peaks of envelope[i].
```
3 $up\_peaks[j] = FindPeaks\_Envelope(Envelope[i])$;
4 $threshold = Average(r[n])$;

```
// Detect start/end points of chirp segment .
```
5 $m = n = up\_peaks[i]$;
6 **while** $r[m] >= threshold$ **do**
7 $\quad m--$;
8 **end**
9 $start\_point[k] = m$;
10 **while** $r[n] >= threshold$ **do**
11 $\quad n++$;
12 **end**
13 $end\_point[k] = n$; $Segments = Fusion(start\_point, end\_point)$;
14 $Chirp\_Matrix[n] = Segments[0:19]$;

The result is shown in Figure 7, which can segment each chirp ideally.

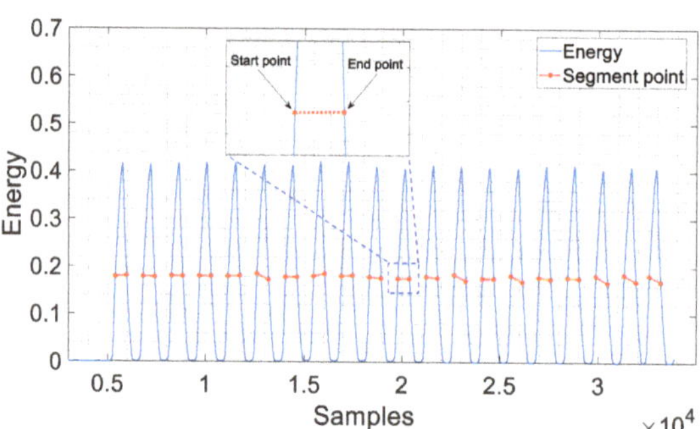

Figure 7. The energy envelope and boundary points of segmentation.

5.7. ERTF

Previous works [51,52] have demonstrated that the microphone's frequency response (especially in the ultrasonic band) is a stable and feasible feature over time, even sensitive enough to distinguish among millions of smartphones of the same brand and model. Considering it from another view, attenuation and time delay still express the geometric structure and intrinsic biological features of the auricle and ear canal besides the frequency response. The overall biologically unique characteristics of the expression occupy a principal place.

However, the time-domain equalization calculation complexity is too high, and the result is not accurate enough. In addition, the CIR calculation in the time domain utilizes a complex cross-correlation calculation. Moreover, most acoustic signal processing is implemented in the frequency domain. In order to avoid multiple conversions between the frequency and time domain and obtain the corresponding CIR using FFT (Fast Fourier Transform) for high efficiency, we first calculate the frequency-domain response CFR (Channel Frequency Response) and then use the computationally efficient IFFT to obtain $CIR = IFFT(CFR) = IFFT(H(f))$. $H(f)$ represents the acoustic frequency-domain signal.

The CIRs obtained from a set of chirp signals are stacked to form a matrix, which is the echo matrix of the signal generated by a sound cycle sequence.

To this end, the system applies a PCA (Principal Component Analysis) to compress the twenty-channel matrix signals into one channel. Each of these channels is input as an observation of the PCA algorithm. We use the covariance analysis of PCA to decorrelate the data, projecting the data to the direction with the most significant variance as the main feature, to achieve the purpose of dimensionality reduction. The component with the highest eigenvalue carries the most critical information: the user's unique biometric vector embedding.

5.8. Authentication

Finally, the feature vector obtained by PCA is saved to assemble the unique feature embedding of the human ear. Since the authentication issue is typically a classification problem, SVM is the most suitable method. We do not choose a data-driven deep neural network for three reasons. First, it can ensure the efficiency and applicability of the *MetaEar*. Second, it avoids much heavy work of collecting user data. Third, users can quickly sign up without retraining the network or fine-tuning. Transfer learning and these Deep Learning algorithms are almost unattainable to apply. We input the feature vector into the SVM for training and eventually output the authentication result of 1 (legitimate user) or 0 (illegal user). MetaEar uses LibSVM [53] and sets up an RBF (Radial Basis Function) kernel with the one-classification function. The dataset is divided into two parts, the training set and the test set, which account for 75% and 25%, respectively, and then utilize 10-fold cross-validation to train the model, and finally obtain the certification accuracy of the test set.

6. Implementation Setup

6.1. Mobile APP

We developed an Android APP to send and collect FMCW acoustic signals within 18~22 kHz, as shown in Figure 8. The Android APP could config the sample rate to 48 kHz and the highest frequency to 20 kHz. The lowest frequency is 18 kHz, the chirp period is 0.01 s, and the upper speaker is employed to send the FMCW acoustic signal. Before starting to collect data, the preparation time is 1000 ms, and the total number of samples to be collected is 1.

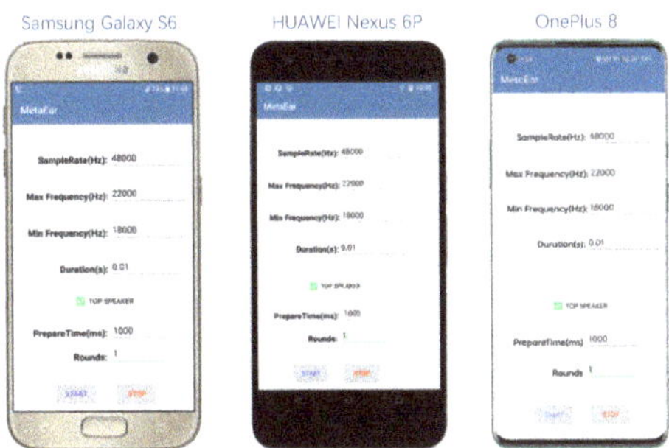

Figure 8. The Android APP and three smartphones.

As shown in Figure 8, we chose three smartphones, Huawei Nexus 6P, Samsung Galaxy S6, and OnePlus 8. They have more device diversity than headsets and can collect data at different angles, distances, and postures so that the ERTF model can be effectively verified in multiple dimensions. On the one hand, there will not be a significant deviation due to the depth or angle of the inserted earbuds. Furthermore, it also avoids the harmful effect of long-term in-ear headphones on the physiological function of the ear. In the experiment, to verify our point of view, we used a desktop computer in the background, with AMD Ryzen 3 2200G, 16G RAM, which can ensure that our data is thoroughly and efficiently processed.

6.2. Environment

We collected FMCW sound signals with mobile phones in three scenarios, hall, laboratory, and street, as shown in Figure 9. The laboratory is relatively clean, with less noise and few people walking around. There is human noise in the hall and interference of the sound signal by moving objects. The street is the scene with the most complicated environment. Not only is there a large number of moving object interference, but also the most significant noise interference.

(**a**) The hall scenario   (**b**) The laboratory scenario   (**c**) The street scenario

Figure 9. Three scenarios for data collection.

7. Evaluation

7.1. Dataset

We recruited 17 people aged 22 to 25, eleven males and six females. Each person uses three mobile phones to collect data at different distances, angles, and behaviors in each scene. The collection was performed 780 times by each person, and a total of $780 \times 17 = 13260$ samples were collected. The dataset is sufficient to train the SVM for classification. During SVM training, the data of the current legitimate user are labeled as positive, while the data of other users are labeled as negative. These negative samples, randomly selected from all labeled negative users, have the same number of positive ones. We divide the current dataset into three parts, 70% for training, 15% for validation, and 15% for testing.

7.2. Overall Accuracy

Because the system aims to authenticate users, it can be attributed to a one-classification problem with traditional OC-SVM (One-Class SVM). The overall confusion matrix is shown in Figure 10. It is worth noting that this confusion matrix is different from the traditional one. The x-axis represents different users, while the ordinate represents different individual models. Each row in the figure denotes the average authentication accuracy produced after inputting different user data into models. Therefore, for each system legitimate user, a corresponding individual model is trained to constitute a model library. It can be seen that the minimum authentication accuracy is 92.8%, the highest accuracy is 100%, and the average accuracy is 96.48%.

7.3. Impact of Environment

We collected data in three different scenarios, as shown in Figure 9: the hall, the laboratory, and the noisy street. In the environment, there are different noises and disturbances, respectively. The indoor environment is relatively less noisy, but there will be reflection interference from close-range moving objects. The outdoor environment selected the noisiest and most complex street environment, which aims to test whether the system can work typically in the most complex and noisy environment. These three environments basically cover daily production and life scenarios. As shown in Figure 11, no matter what kind of environment, our system can denoise sufficiently and perform safety authentication. In most cases, the authentication accuracy in the hall and lab is better than in the street environment. In the hall and the laboratory, people often walk, and there are human voices and keyboard and mouse tapping noises in the laboratory. The accuracy in the hall is the highest, achieving on average 96.19%. Noisy street authentication accuracy is slightly lower, and average accuracy is also achieved at 92.02%. The loud street is not only contained by high-decibel motor vehicle noise but also the voice of people. For user7 and user9, the authentication accuracy of the street environment is also relatively good due to some uncontrollable factors in the environment noise. For example, there are few vehicles and low noise during data collection.

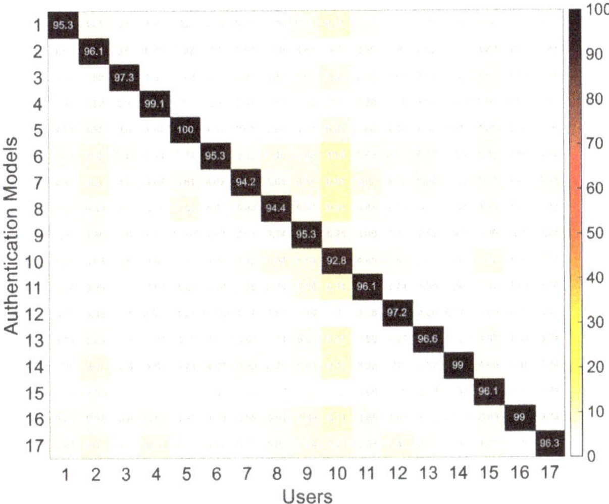

Figure 10. The fusion matrix of overall accuracy.

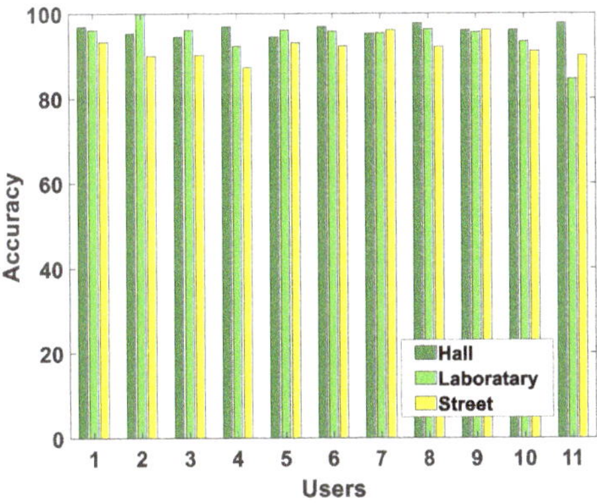

Figure 11. Accuracy at different circumstances.

7.4. Impact of Angel

Different angles of the mobile phone have different effects on the authentication accuracy. We measured different angles with a semicircular protractor in the hall and collected data when the mobile phone was at four different angles, as shown in Figure 12. The coordinate system is set to the auricle plane, the x-axis is from back to front, and the y-axis is from top to bottom. When the mobile phone and the y-axis coincide, the angle is 0 degrees. The corresponding rotation to the x-axis can reach 30 degrees, 60 degrees, and 90 degrees, four angles for experimentation in total. As can be seen from the figure, most users hold the highest authentication accuracy at 30 degrees and 60 degrees, and the average accuracy is about 93.54%. The reason is that the FMCW ultrasonic signal is most apparent in terms of fetching the characteristic expression of the pinna and ear canal at 30 degrees and 60 degrees. In addition, the mobile phone's microphone also receives the highest signal strength, and the SNR (Signal to Noise Ratio) is optimal so that it can achieve the best authentication effect. However, some users have the highest accuracy at 90-degree angles, primarily because different users hold their mobile phones in different postures. At 90 degrees, the mobile phone is not very close to the face, so the expression of human ear biometrics is accurate.

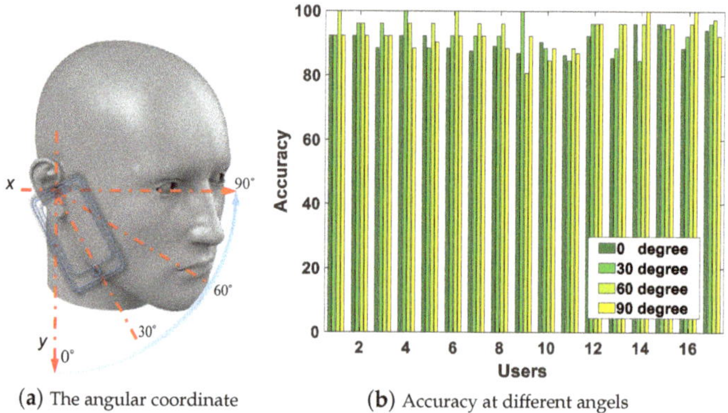

(**a**) The angular coordinate   (**b**) Accuracy at different angles

Figure 12. Impact of different angels.

7.5. Impact of Distance

We verified the authentication sensitivity to the distance between the human ear and the transceiver. The authentication accuracy was measured when the mobile phone was 1 cm, 3 cm, and 5 cm away from the human ear, as shown in Figure 13. It can be concluded from the figure that the accuracy at 1 cm is the highest, and the average accuracy is achieved at 92.65%. At a 5 cm distance, the accuracy is lower, with an average of 92.3%. This is mainly because when the device is far away from the human ear, the SNR decreases, and the characteristic signals of the ear canal and auricle feedback cannot be well received. Furthermore, hair and earrings will increase negative influence so that the accuracy will decrease. We did not test more extended distance scenarios because if the signal selective fading increases sharply with distance, the SNR decreases, and the accuracy drops severely.

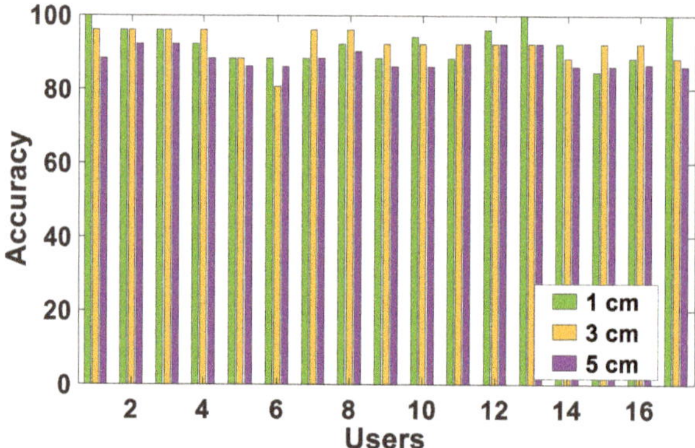

Figure 13. Accuracy at different distances.

7.6. Impact of Different Behavior

During the authentication process, users may be sitting motionless and doing some other body movements in daily life. We test the most potential activities, including static state, shaking, and walking. as shown in Figure 14. Among them, the authentication accuracy in the static state is 96.89%. It is 96.15% in the shaking head state, and while walking, the average accuracy is 91.60%. Because the vibration of the bones and muscles of the human body will affect the feedback of the signal during the walking process, the accuracy will decline. The authentication accuracy is reduced due to the change of the mobile phone position caused by non-vigorous head shaking. In the static state, the authentication effect is the best.

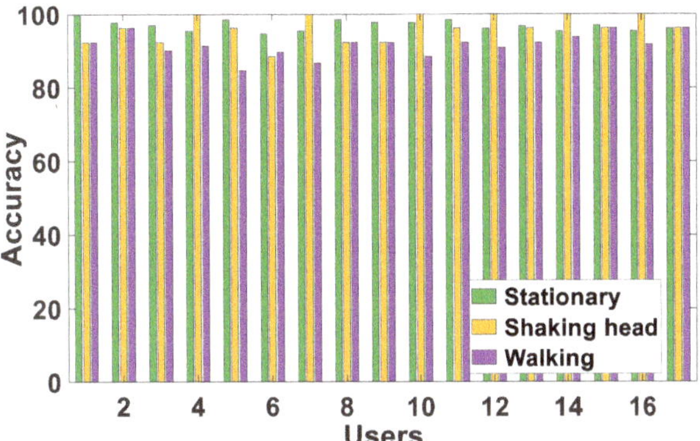

Figure 14. Accuracy of different behaviors.

7.7. Impact of Different Devices

In order to measure the impact of different hardware on the authentication, we use three brands of mobile phones for experimental verification, Oneplus 8, Samsung galaxy s6, and Huawei nexus 6p. When collecting data in the early stage, due to the difference in the microphone ADC (Analog-to-Digital Converter) circuit and nonlinear processing methods and numerical precision of different devices, the collected data will be different, so different devices will cause accuracy differences. We use AMD Ryzen PC to simulate the same technical data process. As shown in Figure 15, the average accuracy of the three devices is almost the same, showing that our system is adaptable to different hardware. Overall, the effect of Oneplus 8 is slightly better, with accuracy reaching 98.44%, principally because the smartphone was produced in 2020 with improved hardware and performance. However, the result of user10 using Oneplus 8 is low, which is mainly caused by the extensive construction noise interference in the environment when the user collects data.

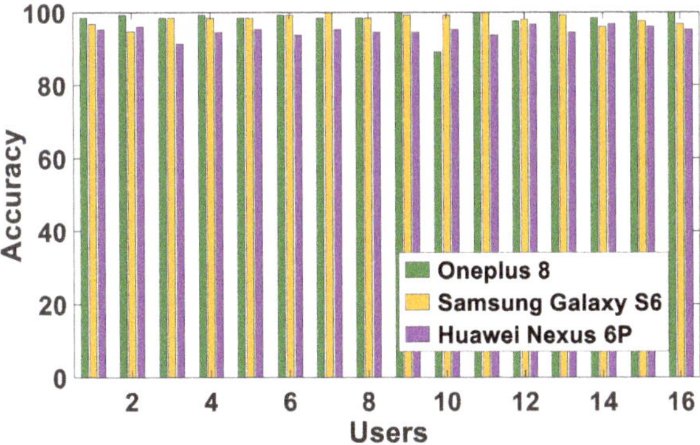

Figure 15. Authentication accuracy by different smartphones.

7.8. Impact of Data Quantity

Machine learning is a data-driven learning model. Although OC-SVM uses much less data for training than DNN, it also obtains different authentication results for different data amounts. We train the model with different amounts of data and then use the test data to test the authentication accuracy of the model, as shown in Figure 16. As the volume of data increases, the accuracy steadily increases. When using 30 data samples, the accuracy reaches 86%, and when the number of training data samples is 90, the test accuracy reaches 95.56%, which can meet the daily requirements. Ultimately, when using 590 training samples, the accuracy can reach 98.98%.

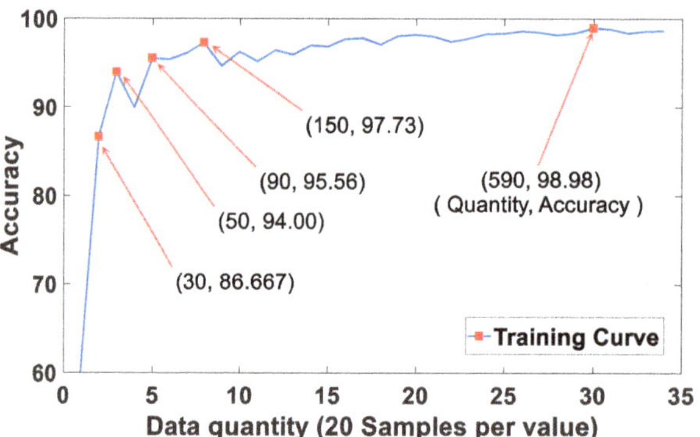

Figure 16. Training accuracy by different data quantities.

7.9. Efficiency of Attack Defense

We conducted an attack defense test on the *MetaEar*. For impersonation attacks, the adversaries utilize their biological signal to imitate the signal of the legitimate user, intending to deceive the system. We take one person as a legitimate user, and then 16 other people imitate this person's habits and behavioral characteristics to collect data. In the hall, the voice data of 17 people were collected at the same location and time. For the model trained for a legitimate user, we take the data of the remaining 16 people as the attacker, input it into the model, and the output is shown in Figure 17. It is concluded that the median FAR of other users is below 4%, except for user2, user9, user10, and user15. In particular, the FAR of user10 is 11%. The main reason for the deviation of these users is that the legitimate users did not follow the dictated actions when collecting data, which caused the distance between the device and the ear to change, and the extracted biometric characteristics of the human ear changed significantly, thereby increasing attack success probability. Nonetheless, attackers have a low probability of successfully executing impersonation attacks. Compared with face or fingerprint authentication, the registration process also requires a fixed posture and process. If our system can allow legitimate users to develop fixed habits when collecting data and adopt dictated actions, the results of resisting attacks will be better.

7.10. Analysis of Time Efficiency

We measured the time efficiency of each system module, as shown in Table 1. The table is divided into two stages, one is the registration stage, and the other is the login stage. In the registration stage, the collected data need to be trained by SVM, so the time in the table is the consumption for training 90 samples. The time statistics of each phase in the login is the time consumption of one sampling. 'pre' denotes data preprocessing, including some data access time and the time to remove the hardware delay, and its time efficiency is 0.364 s. 'Denoise' runtime is 0.414 s. The signal alignment described by 'Ali' has a running time of 0.88 s. 'SIC' is self-interference cancel, consuming 0.883 s. Furthermore, 'seg' is the chirp segment, which takes 20.056 s, and includes the total running time of segmenting samples of 90 times. This time consumption is slightly higher than the other parts. However, it is much less than the time spent on fingerprint scanning when performing fingerprint authentication registration. So it still has an overall advantage in time efficiency. 'FE' is feature extraction, which takes 0.544 s. The SVM training phase took 1.335 s, which shows that the training efficiency of SVM is very high. Finally, the total time spent in the training stage is about 24 s, which is less than the registration time consumed by some existing fingerprint or face biometric authentication methods. When logging in, the total time consumption is 0.939 s. Moreover, in the MALAB execution environment, the complexity of all programs is O(n). After program optimization, the processing time should be able to meet the actual needs. Overall, *MetaEar* is an efficient and deployable system.

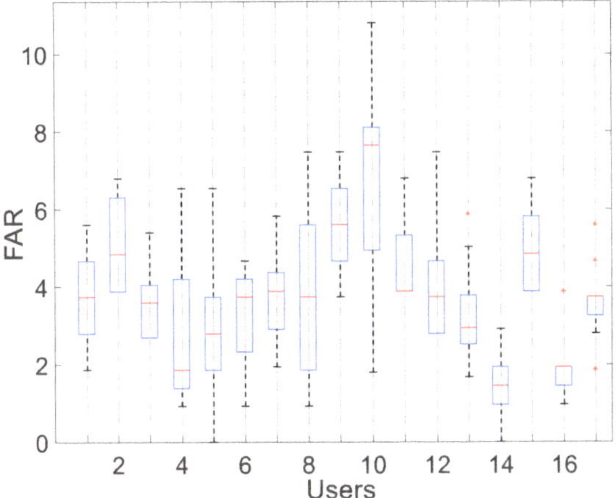

Figure 17. Resistance capability to impersonate attack (The red plus signs indicate outliers).

Table 1. Average Time Efficiency.

Phase	Pre	Denoise	Ali	SIC	Seg	FE	Train	Auth	Total
Reg(s)	0.364	0.414	0.880	0.883	20.056	0.544	1.335	None	**24.476**
Login(s)	0.015	0.018	0.008	0.009	0.273	0.022	None	0.594	**0.939**

7.11. Comparison of Different System

Furthermore, we compared the other four systems in three main aspects: device usage, biometric location, and average accuracy, as shown in Table 2. The average authentication accuracy of our system is the highest, and we use not only the intrinsic biometrics of the ear canal but also the extrinsic biometrics of the auricle, making the biometrics' uniqueness more accurate. Moreover, the smartphones we use do not need to be modified and use the machine learning algorithm SVM, which can be efficiently integrated with existing systems and deployed in practical scenarios.

Table 2. Comparison of Different Acoustic Biometric Authentication.

AcousticAuth	EarPrint [46]	VocalLock [31]	EarEcho [19]	EarDynamic [54]	MetaEar
Devices	Earphone	Smartphone	Earphone	Earphone	Smartphone
Features	Body sound	Vocal Tract	Ear Canal	Ear Canal	Ear Canal+Pinna
Accuracy	96.36%	91.1%	94.2%	93.04%	96.48%

8. Conclusions

This paper proposes *MetaEar* for modeling and authenticating human ear ERTF biometrics using FMCW ultrasonics. By sending FMCW ultrasonic waves to the ear, the dual microphones receive the feedback sound wave, extract the features through ERTF, and then feed into the SVM for one-class authentication. A large number of experiments verify that our average authentication accuracy can reach 96.48%, which can effectively strengthen biometric authentication and resist replay attacks and imitation attacks. From the overall authentication accuracy result, we do not achieve the best authentication accuracy, and the attack resistance test FAR also does not achieve the optimal level. Therefore, the next step is further improving the authentication accuracy and achieving practical deployment. First, we will use the combinatorial optimization method to perform more accurate feature extraction on ultrasound to improve the authentication accuracy; second, through multi-modal fusion biometric authentication, we will obtain a more robust performance against attacks.

Author Contributions: Conceptualization, methodology, software, validation, writing—original draft preparation, Z.C.; formal analysis, investigation, resources, funding acquisition, writing—review and editing, L.W.; data curation, visualization, B.L.; supervision, project administration, W.L. All authors have read and agreed to the published version of the manuscript.

Funding: This research was funded by the National Science Foundation of China (NSFC) grant number 61772453, and the Natural Science Foundation of Hebei Province grant number F2022203045, F2018203444.

Institutional Review Board Statement: Not applicable.

Informed Consent Statement: Not applicable.

Conflicts of Interest: The authors declare no conflict of interest.

References

1. Chen, K.; Zhang, D.; Yao, L.; Guo, B.; Yu, Z.; Liu, Y. Deep Learning for Sensor-Based Human Activity Recognition: Overview, Challenges, and Opportunities. *ACM Comput. Surv.* **2021**, *54*, 1–40. [CrossRef]
2. Choudhury, R.R. Earable Computing: A New Area to Think About. In Proceedings of the 22nd International Workshop on Mobile Computing Systems and Applications, Virtual, 24–26 February 2021; Association for Computing Machinery: New York, NY, USA, 2021; HotMobile '21, pp. 147–153.
3. Bhalla, A.; Sluganovic, I.; Krawiecka, K.; Martinovic, I. MoveAR: Continuous Biometric Authentication for Augmented Reality Headsets. In Proceedings of the 7th ACM on Cyber–Physical System Security Workshop, Hong Kong, China, 7 June 2021; pp. 41–52.
4. Lee, L.H.; Braud, T.; Zhou, P.; Wang, L.; Xu, D.; Lin, Z.; Kumar, A.; Bermejo, C.; Hui, P. All one needs to know about metaverse: A complete survey on technological singularity, virtual ecosystem, and research agenda. *arXiv* **2021**, arXiv:2110.05352.
5. *Biometric Authentication & Identification Market 2022 with Growth Opportunities, Top Countries Data, Future Trends and Business Size and Share with Revenue Forecast to 2026*; MarketWatch: New York, NY, USA, 2022.
6. Acquisti, A.; Gross, R.; Stutzman, F.D. Face recognition and privacy in the age of augmented reality. *J. Priv. Confid.* **2014**, *6*, 1. [CrossRef]
7. Raghavendra, R.; Raja, K.B.; Busch, C. Presentation attack detection for face recognition using light field camera. *IEEE Trans. Image Process.* **2015**, *24*, 1060–1075. [CrossRef] [PubMed]
8. Boulkenafet, Z.; Komulainen, J.; Li, L.; Feng, X.; Hadid, A. OULU-NPU: A mobile face presentation attack database with real-world variations. In Proceedings of the 2017 12th IEEE International Conference on Automatic Face & Gesture Recognition (FG 2017), Washington, DC, USA, 30 May–3 June 2017; IEEE: Piscataway, NJ, USA, 2017; pp. 612–618.
9. Komkov, S.; Petiushko, A. Advhat: Real-world adversarial attack on arcface face id system. In Proceedings of the 2020 25th International Conference on Pattern Recognition (ICPR), Milan, Italy, 10–15 January 2021; IEEE: Piscataway, NJ, USA, 2021; pp. 819–826.
10. ISO—ISO/IEC 30107-1:2016, I.I. Information Technology—Biometric Presentation Attack Detection—Part 1: Framework. Available online: https://www.iso.org/standard/53227.html (accessed on 26 September 2022).
11. Zhang, L.; Tan, S.; Yang, J.; Chen, Y. Voicelive: A phoneme localization based liveness detection for voice authentication on smartphones. In Proceedings of the 2016 ACM SIGSAC Conference on Computer and Communications Security, Vienna, Austria, 24–28 October 2016; pp. 1080–1091.
12. Zhang, L.; Tan, S.; Wang, Z.; Ren, Y.; Wang, Z.; Yang, J. Viblive: A continuous liveness detection for secure voice user interface in iot environment. In Proceedings of the Annual Computer Security Applications Conference, Austin, TX, USA, 7–11 December 2020; pp. 884–896.

13. Tuyls, P.T.; Verbitskiy, E.; Ignatenko, T.; Schobben, D.; Akkermans, T.H. Privacy-protected biometric templates: Acoustic ear identification. In *Biometric Technology for Human Identification*; International Society for Optics and Photonics: Bellingham, DC, USA, 2004; Volume 5404, pp. 176–182.

14. Xu, X.; Gao, H.; Yu, J.; Chen, Y.; Zhu, Y.; Xue, G.; Li, M. ER: Early recognition of inattentive driving leveraging audio devices on smartphones. In Proceedings of the IEEE INFOCOM 2017-IEEE Conference on Computer Communications, Atlanta, GA, USA, 1–4 May 2017; IEEE: Piscataway, NJ, USA, 2017; pp. 1–9.

15. Xu, X.; Yu, J.; Chen, Y.; Zhu, Y.; Kong, L.; Li, M. Breathlistener: Fine-grained breathing monitoring in driving environments utilizing acoustic signals. In Proceedings of the 17th Annual International Conference on Mobile Systems, Applications, and Services, Seoul, Korea, 17–21 June 2019; pp. 54–66.

16. Yang, Z.; Choudhury, R.R. Personalizing Head Related Transfer Functions for Earables. In Proceedings of the 2021 ACM SIGCOMM 2021 Conference, Virtual, 23–27 August 2021; Association for Computing Machinery: New York, NY, USA, 2021; SIGCOMM '21, pp. 137–150.

17. Prakash, J.; Yang, Z.; Wei, Y.L.; Hassanieh, H.; Choudhury, R.R. EarSense: Earphones as a Teeth Activity Sensor. In Proceedings of the 26th Annual International Conference on Mobile Computing and Networking, London, UK, 21–25 September 2020; Association for Computing Machinery: New York, NY, USA, 2020.

18. Yang, Z.; Wei, Y.L.; Shen, S.; Choudhury, R.R. Ear-AR: Indoor Acoustic Augmented Reality on Earphones. In Proceedings of the 26th Annual International Conference on Mobile Computing and Networking, London, UK, 21–25 September 2020; Association for Computing Machinery: New York, NY, USA, 2020.

19. Gao, Y.; Wang, W.; Phoha, V.V.; Sun, W.; Jin, Z. EarEcho: Using ear canal echo for wearable authentication. *Proc. ACM Interact. Mob. Wearable Ubiquitous Technol.* **2019**, *3*, 1–24. [CrossRef]

20. Wang, Z.; Ren, Y.; Chen, Y.; Yang, J. Earable Authentication via Acoustic Toothprint. In Proceedings of the 2021 ACM SIGSAC Conference on Computer and Communications Security, Virtual, 15–19 November 2021; Association for Computing Machinery: New York, NY, USA, 2021; CCS '21, pp. 2390–2392.

21. Wang, Z.; Tan, S.; Zhang, L.; Ren, Y.; Wang, Z.; Yang, J. An Ear Canal Deformation Based Continuous User Authentication Using Earables. In Proceedings of the 27th Annual International Conference on Mobile Computing and Networking, New Orleans, LA, USA, 25–29 October 2021; Association for Computing Machinery: New York, NY, USA, 2021; MobiCom '21, pp. 819–821.

22. Cai, C.; Pu, H.; Ye, L.; Jiang, H.; Luo, J. Active Acoustic Sensing for Hearing Temperature under Acoustic Interference. In *IEEE Transactions on Mobile Computing*; IEEE: Piscataway, NJ, USA, 2021; p. 1.

23. Yun, S.; Chen, Y.C.; Zheng, H.; Qiu, L.; Mao, W. Strata: Fine-grained acoustic-based device-free tracking. In Proceedings of the 15th Annual International Conference on Mobile Systems, Applications, and Services, New York, NY, USA, 19–23 June 2017; pp. 15–28.

24. Wang, W.; Liu, A.X.; Sun, K. Device-free gesture tracking using acoustic signals. In Proceedings of the 22nd Annual International Conference on Mobile Computing and Networking, New York, NY, USA, 3–7 October 2016; pp. 82–94.

25. Cai, C.; Pu, H.; Wang, P.; Chen, Z.; Luo, J. We Hear Your PACE: Passive Acoustic Localization of Multiple Walking Persons. *Proc. ACM Interact. Mob. Wearable Ubiquitous Technol.* **2021**, *5*, 1–24. [CrossRef]

26. Ling, K.; Dai, H.; Liu, Y.; Liu, A.X.; Wang, W.; Gu, Q. Ultragesture: Fine-grained gesture sensing and recognition. *IEEE Trans. Mob. Comput.* **2020**, *21*, 2620–2636. [CrossRef]

27. Ruan, W.; Sheng, Q.Z.; Yang, L.; Gu, T.; Xu, P.; Shangguan, L. AudioGest: Enabling fine-grained hand gesture detection by decoding echo signal. In Proceedings of the 2016 ACM International Joint Conference on Pervasive and Ubiquitous Computing, Heidelberg, Germany, 12–16 September 2016; pp. 474–485.

28. Zheng, T.; Chao, C.; Chen, Z.; Luo, J. Sound of Motion: Real-time Wrist Tracking with A Smart Watch-Phone Pair. In Proceedings of the IEEE INFOCOM 2022-IEEE Conference on Computer Communications, London, UK, 2–5 May 2022.

29. Feng, H.; Fawaz, K.; Shin, K.G. Continuous Authentication for Voice Assistants. In Proceedings of the 23rd Annual International Conference on Mobile Computing and Networking, Snowbird, UT, USA, 16–20 October 2017; Association for Computing Machinery: New York, NY, USA, 2017; MobiCom '17, pp. 343–355.

30. Xie, Y.; Li, F.; Wu, Y.; Chen, H.; Zhao, Z.; Wang, Y. TeethPass: Dental Occlusion-based User Authentication via In-ear Acoustic Sensing. In Proceedings of the IEEE INFOCOM 2022-IEEE Conference on Computer Communications, London, UK, 2–5 May 2022.

31. Lu, L.; Yu, J.; Chen, Y.; Wang, Y. Vocallock: Sensing vocal tract for passphrase-independent user authentication leveraging acoustic signals on smartphones. *Proc. ACM Interact. Mob. Wearable Ubiquitous Technol.* **2020**, *4*, 1–24. [CrossRef]

32. Lu, L.; Yu, J.; Chen, Y.; Liu, H.; Zhu, Y.; Kong, L.; Li, M. Lip Reading-Based User Authentication Through Acoustic Sensing on Smartphones. *IEEE/ACM Trans. Netw.* **2019**, *27*, 447–460. [CrossRef]

33. Zhou, M.; Wang, Q.; Li, Q.; Jiang, P.; Yang, J.; Shen, C.; Wang, C.; Ding, S. Securing Face Liveness Detection Using Unforgeable Lip Motion Patterns. *arXiv* **2021**, arXiv:2106.08013.

34. Kong, H.; Lu, L.; Yu, J.; Chen, Y.; Xu, X.; Tang, F.; Chen, Y.C. MultiAuth: Enable Multi-User Authentication with Single Commodity WiFi Device. In Proceedings of the Twenty-Second International Symposium on Theory, Algorithmic Foundations, and Protocol Design for Mobile Networks and Mobile Computing, Shanghai, China, 26–29 July 2021; Association for Computing Machinery: New York, NY, USA, 2021; MobiHoc '21, pp. 31–40.

35. Shi, C.; Liu, J.; Liu, H.; Chen, Y. WiFi-Enabled User Authentication through Deep Learning in Daily Activities. *ACM Trans. Internet Things* **2021**, *2*, 1–25. [CrossRef]

36. Zhao, T.; Wang, Y.; Liu, J.; Chen, Y. Your Heart Won't Lie: PPG-Based Continuous Authentication on Wrist-Worn Wearable Devices. In Proceedings of the 24th Annual International Conference on Mobile Computing and Networking, New Delhi, India, 29 October–2 November 2018; Association for Computing Machinery: New York, NY, USA, 2018; MobiCom '18, pp. 783–785.

37. Qian, K.; Wu, C.; Xiao, F.; Zheng, Y.; Zhang, Y.; Yang, Z.; Liu, Y. Acousticcardiogram: Monitoring heartbeats using acoustic signals on smart devices. In Proceedings of the IEEE INFOCOM 2018-IEEE Conference on Computer Communications, Honolulu, HI, USA, 16–19 April 2018; IEEE: Piscataway, NJ, USA, 2018; pp. 1574–1582.

38. Lin, F.; Song, C.; Zhuang, Y.; Xu, W.; Li, C.; Ren, K. Cardiac Scan: A Non-Contact and Continuous Heart-Based User Authentication System. In Proceedings of the 23rd Annual International Conference on Mobile Computing and Networking, Snowbird, UT, USA, 16–20 October 2017; Association for Computing Machinery: New York, NY, USA, 2017; MobiCom '17, pp. 315–328.

39. Wang, T.; Zhang, D.; Zheng, Y.; Gu, T.; Zhou, X.; Dorizzi, B. C-FMCW Based Contactless Respiration Detection Using Acoustic Signal. *Proc. ACM Interact. Mob. Wearable Ubiquitous Technol.* **2018**, *1*, 1–20. [CrossRef]

40. Chen, Y.; Xue, M.; Zhang, J.; Guan, Q.; Wang, Z.; Zhang, Q.; Wang, W. ChestLive: Fortifying Voice-based Authentication with Chest Motion Biometric on Smart Devices. *Proc. ACM Interact. Mob. Wearable Ubiquitous Technol.* **2021**, *5*, 1–25. [CrossRef]

41. Eberz, S.; Rasmussen, K.B.; Lenders, V.; Martinovic, I. Evaluating Behavioral Biometrics for Continuous Authentication: Challenges and Metrics. In Proceedings of the 2017 ACM on Asia Conference on Computer and Communications Security, Abu Dhabi, UAE, 2–6 April 2017; Association for Computing Machinery: New York, NY, USA, 2017; ASIA CCS '17, pp. 386–399.

42. Zhang, Y.; Hu, W.; Xu, W.; Chou, C.T.; Hu, J. Continuous Authentication Using Eye Movement Response of Implicit Visual Stimuli. *Proc. ACM Interact. Mob. Wearable Ubiquitous Technol.* **2018**, *1*, 1–22. [CrossRef]

43. Liu, J.; Li, D.; Wang, L.; Xiong, J. BlinkListener: "Listen" to Your Eye Blink Using Your Smartphone. *Proc. ACM Interactive Mob. Wearable Ubiquitous Technol.* **2021**, *5*, 1–27. [CrossRef]

44. Kinnunen, T.; Sahidullah, M.; Delgado, H.; Todisco, M.; Evans, N.; Yamagishi, J.; Lee, K.A. The ASVspoof 2017 Challenge: Assessing the Limits of Replay Spoofing Attack Detection. In Proceedings of the 18th Annual Conference of the International Speech Communication Association, Stockholm, Sweden, 20–24 August 2017; pp. 2–6.

45. Huang, W.; Tang, W.; Jiang, H.; Luo, J.; Zhang, Y. Stop Deceiving! An Effective Defense Scheme Against Voice Impersonation Attacks on Smart Devices. *IEEE Internet Things J.* **2022**, *9*, 5304–5314. [CrossRef]

46. Gao, Y.; Jin, Y.; Chauhan, J.; Choi, S.; Li, J.; Jin, Z. Voice In Ear: Spoofing-Resistant and Passphrase-Independent Body Sound Authentication. *Proc. ACM Interact. Mob. Wearable Ubiquitous Technol.* **2021**, *5*, 1–25. [CrossRef]

47. Kates, J.M. A computer simulation of hearing aid response and the effects of ear canal size. *J. Acoust. Soc. Am.* **1988**, *83*, 1952–1963. [CrossRef] [PubMed]

48. Rappaport, T.S. *Wireless Communications: Principles and Practice*; Prentice Hall PTR: Hoboken, NJ, USA, 1996; Volume 2.

49. Adib, F.; Katabi, D. See through walls with WiFi! In Proceedings of the ACM SIGCOMM 2013 Conference on SIGCOMM, Hong Kong, China, 12–16 August 2013; pp. 75–86.

50. Sun, K.; Zhao, T.; Wang, W.; Xie, L. Vskin: Sensing touch gestures on surfaces of mobile devices using acoustic signals. In Proceedings of the 24th Annual International Conference on Mobile Computing and Networking, New Delhi, India, 29 October–2 November 2018; pp. 591–605.

51. Zhou, Z.; Diao, W.; Liu, X.; Zhang, K. Acoustic fingerprinting revisited: Generate stable device id stealthily with inaudible sound. In Proceedings of the 2014 ACM SIGSAC Conference on Computer and Communications Security, Scottsdale, AZ, USA, 3–7 November 2014; pp. 429–440.

52. Han, D.; Chen, Y.; Li, T.; Zhang, R.; Zhang, Y.; Hedgpeth, T. Proximity-proof: Secure and usable mobile two-factor authentication. In Proceedings of the 24th Annual International Conference on Mobile Computing and Networking, New Delhi, India, 29 October–2 November 2018; pp. 401–415.

53. Chang, C.C.; Lin, C.J. LIBSVM: A library for support vector machines. *ACM Trans. Intell. Syst. Technol. (TIST)* **2011**, *2*, 1–27. [CrossRef]

54. Wang, Z.; Tan, S.; Zhang, L.; Ren, Y.; Wang, Z.; Yang, J. EarDynamic: An Ear Canal Deformation Based Continuous User Authentication Using In-Ear Wearables. *Proc. ACM Interact. Mob. Wearable Ubiquitous Technol.* **2021**, *5*, 1–27. [CrossRef]

MDPI AG
Grosspeteranlage 5
4052 Basel
Switzerland
Tel.: +41 61 683 77 34

Electronics Editorial Office
E-mail: electronics@mdpi.com
www.mdpi.com/journal/electronics